Dynamics Exam File

Charles Smith, Oregon State University, Editor; Nazeer Ahmed, University of Louisville; Leon Bahar, Drexel University; Henry Bernstein, University of Houston; Frank Bosworth, South Dakota School of Mines; Feridun Dalale, Drexel University; Clarence de Silva, Carnegie-Mellon University; Robert Fopma, University of Cincinnati; John Giger, Rose State College; Biswa Ghosh, College of the Finger Lakes; M. R. Hansen, Oregon Institute of Technology; S. Graham Kelly, The University of Akron; Erol Kilik, California State University, Long Beach; J. Dale Pounds, Western Carolina University; Weston Smedley, Idaho State University; William Stiefel, Western Michigan University; Wallace Venable, West Virginia University; Alok Verma, Old Dominion University; Richard Witt, University of Wisconsin - Eau Claire

ENGINEERING PRESS, INC. SAN JOSE, CALIFORNIA 95103-0001

Donald G. Newnan, Ph.D.
Exam File Series Editor

Printed in the United States of America

Library of Congress Cataloging in Publication Data
Main entry under title:

Dynamics exam file.

1. Dynamics--Problems, exercises, etc. I. Smith, Charles Edward, 1932-
TA352.D94 1985 620.1'04'076 84-24699
ISBN 0-910554-44-7

5 4 3 2

Engineering Press, Inc. P.O. Box 1 San Jose, California 95103-0001

Contents

Foreword

Many students find that previously-used examination problems are a useful supplement to their textbooks and lecture notes. Although copies of such examinations are often made available by professors, there is usually a need for more of this material than can be easily supplied by one teacher. Further, that material which *is* made available to a particular class reaches only a small fraction of the students in the country who would find it useful. Thus, the many, well-considered problems residing in professors' files across the country are unavailable to the people who could best use them. The purpose of this book is to make some of this material available.

Professors from around the country have agreed to open their exam files and to allow their problems and solutions to be published. These professors all teach an introductory course in dynamics and use one of the half dozen popular textbooks. The problems for publication were carefully selected to cover the fundamentals of dynamics found in these textbooks. In general, the problems are just as they appeared in an actual course examination. Similarly, each solution is that prepared by the professor who wrote the exam problem. Thus this book is a combination of authentic examination problems, together with the professors' own solutions.

Each contributor has been careful to eliminate errors, so we hope there are none and expect but a few. If you find any, a note, mailed to the Engineering Press address, would be appreciated. I hope you find this material helps improve both your understanding of dynamics and your examination scores.

Charles E. Smith
Editor

How To Raise Your Exam Score

Prof. Wallace Venable, West Virginia University

As a general rule, mechanics professors are looking for more than just an answer to a problem. They are looking for evidence that your answer was reached through an orderly and logical process. Since they can only infer your train of thought from what you have written, they are, in effect, basing your grade on your ability to *communicate* as well as your ability to calculate. The problems in this book are examples of the process of communicating a solution as well as of the solution itself.

Many students say, "I don't have time to write all that junk during an exam." Most instructors, on the other hand, find that they have more time for the serious side of problem solving if they have the details clearly written down. They find it easier to say "ah-ha, I'll just substitute this in here" when both pieces are right in front of them. Professors find they can usually work faster by writing more, not less. With practice, you may improve your speed as well as your grade by writing as you think.

No two instructors will grade in exactly the same way, but there is probably more similarity in the scores they give than you would expect. At a meeting of engineering professors, a mechanics professor passed out copies of student work without any explanation, and asked the audience to grade them. Of course some took off two points for this, and others gave more credit for that, but when the scores were compared, the mechanics teachers were in general agreement. Overall, we look for the same things—the ones on the Check List.

Check List

- ☐ **Sketch of problem (free body diagram)**
- ☐ **Basic equation(s) stated in general form**
- ☐ **Necessary assumptions stated**
- ☐ **Necessary and sufficient equations written, using problem notation**
- ☐ **Substitutions or simultaneous solutions labelled**
- ☐ **Units converted properly**
- ☐ **Final answer clearly indicated**
- ☐ **"Garbage" erased or crossed-out**

Many simple mechanics problems can be adequately presented in ten lines or so, and using only one basic equation. In some areas, it may be necessary to work the problem in several ways using different assumptions and then select the final answer from among the sets of calculations. All of the assumptions and their related calculations are part of the solution of the problem.

In short, the best way to maximize your exam score is to explain your work to the grader as you work the problem. The *explanations* which the professors have included in this book will provide you with a good model for your own work.

1 BASIC CONCEPTS

1-1

Complete the following table listing the appropriate units for mass, length, time and force:

	SI System	English System
Mass		
Length		
Time		
Force		

What are the base units for mass in the English system?__________

What are the base units for force in the SI system?__________

****

	SI System	English System
Mass	kg	slug
Length	meter	ft.
Time	sec.	sec.
Force	Newton	Lb.

The base units for mass in the English system $\frac{\text{Lb-sec}^2}{\text{Ft.}}$

The base units for force in the SI system $\frac{\text{kg M}}{\text{sec}^2}$

1-2

Differentiate between the following terms.
(a) Kinematics
(b) Kinetics
(c) Particle
(d) Rigid Body

****

Kinematics - Kinematics is the study of motion variables like displacement velocity and acceleration and the relationship between them. Forces are not included in the analysis.

Kinetics - Kinetics is the study of forces and the resulting motion. Newton's second law is used to study the effect of forces on the motion variables.

Particle - Particle has a definite mass but no shape and size. Hence rotation of a particle has no meaning.

Rigid Body - A body in which distance between two arbitrary points is always constant ie. even in presence of large external forces, the body remains undeformed.

1-3

Knowing that 1 Joule = 1 N - m, (a) find the number of ft.-lbs. in 1 Joule and (b) the number of kg. - m^2 in 1 lb.-ft.-s^2 (moment of inertia of a mass) given the following information: 1 slug = 1 $\frac{\text{lb-s}^2}{\text{ft}}$, 1 ft. = .305m, 1 lb. = 4.45 N, 1 slug = 14.6 kg.

****

(a) $1\,J = 1\,N\text{-}m = \frac{1}{4.45}\,lb \cdot \frac{1}{.305}\,ft = \frac{1}{1.357} \cong .737\ ft\text{-}lbs.$

(b) $1\ lb.\text{-}ft.\text{-}s^2 = 1\ \frac{lb\text{-}s^2}{ft} \cdot ft^2 = 1\ slug \cdot ft^2$

$= 14.6\,Kg \cdot (.305m)^2$

$= 1.36\ Kg\text{-}m^2$

2 KINEMATICS OF PARTICLES

ONE-DIMENSIONAL MOTION

2-1

The mechanism shown is called a "scotch yoke." Pin A is fixed on arm OA and is 2.4 inches from O. A slides in the slot as the arm rotates at a constant 3 radians per second, counter-clockwise. This causes shaft B to move up and down. Find the acceleration of B when θ = 30 degrees.

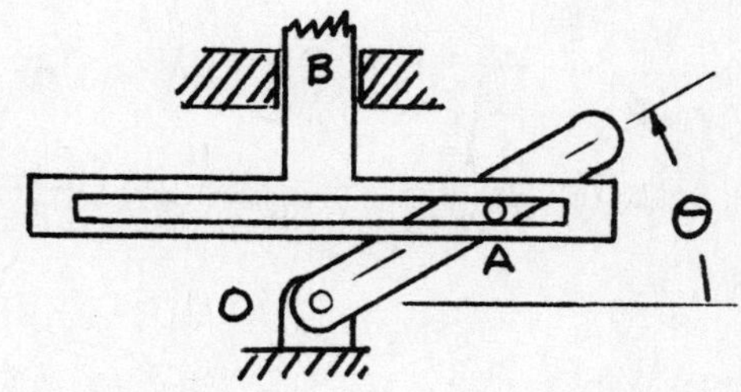

****

B moves following the "y" part of A's motion

$$y = 2.4\sin\theta \qquad \theta = 30$$

$$v = \dot{y} = 2.4\cos\theta\,\dot{\theta} \qquad \dot{\theta} = 3$$

$$a = \ddot{y} = 2.4\cos\theta\,\ddot{\theta} - 2.4\sin\theta\,\dot{\theta}^2 \qquad \ddot{\theta} = 0$$

$$a = -2.4\left(\tfrac{1}{2}\right)(3)^2$$

$a = 10.8$ down
or
$\bar{a} = -10.8\,\bar{j}$

2-2

A box is pulled in the upward direction along an inclined plane with slope 5/12, by passing a rope over a small pulley at B and pulling the end C (along the line CD) with constant horizontal velocity. Determine the velocity of the box at any time t in terms of h, v_o and t.

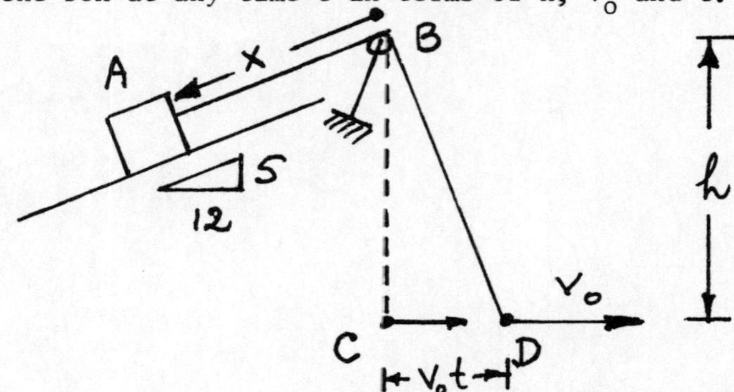

****

Length ABD is constant during the motion.

It is given by $\underbrace{x}_{AB} + \underbrace{(h^2 + v_o^2 t^2)^{1/2}}_{BD} = \text{const}$.

Differentiating w.r.t. time gives

$$\frac{dx}{dt} = -\frac{v_o^2 t}{(h^2 + v_o^2 t^2)^{1/2}} \quad \text{or} \quad v_A = \frac{v_o^2 t}{(h^2 + v_o^2 t^2)^{1/2}}$$

up the incline.

NOTES: 1) The velocity is negative, which indicates that x decreases w.r.t. time, i.e. A is moving upward.

2) The result does not depend on the slope.

3) The result can be written as $V_A = V_o \cos\alpha$ where V_A is the upward velocity of A. This shows that V_A is the projection of V_o along the inextensible rope.

2-3

A particle travels in a straight line and its position is defined by the formula $s = t^3 - t^2 - 8t + 12$, $t \geq 0$, where s is in feet and t in seconds. Find (a) when the velocity is zero, (b) when the particle reverses direction, if it does, (c) the position of the particle when the velocity is zero, and (d) the distance traveled between t = 1 and t = 3 seconds.

****

(a) $v = \frac{ds}{dt} = 3t^2 - 2t - 8 \;(= 0$

$[3t+4][t-2] \;(= 0$

$\therefore t = -\frac{4}{3},\; t = 2 \quad \therefore V = 0$ when $t = 2$ ($-\frac{4}{3}$ is irrelevant as $t \geq 0$)

(b) The particle will reverse direction when the velocity changes sign - if it does. (Remember velocity is a vector and the sign, in 1-dimension here, carries the directional information.) The velocity function is just a parabola bending up. Clearly the velocity only changes sign at $t = 2$ (for $t \geq 0$).

V(t); 15, 12, 9, 6, 3, -3, -6, -9, -12; t; -1, 1/3, 2, 3; (3,13); A; B; $V(\frac{1}{3}, -8\frac{1}{3})$

(c.) When $V = 0$, $t = 2$

$\therefore S = 8 - 4 - 16 + 12 = 0$

$\therefore$ the particle is at the origin (of the straight line) at $t = 2$.

(d) at $t = 1$, $S = 1 - 1 - 8 + 12 = +4$

at $t = 3$, $S = 27 - 9 - 24 + 12 = +6$

Now from $t = 0$ to $t = 2$ the particle is traveling to the left, while for $t > 2$, the particle is traveling to the right

t=3; t=2; t=1; t=0; S; 0, 4, 6, 12; Line

By inspection the total distance traveled is $4 + 6 = 10$ feet.

Note: the distance traveled is Not just $|S(t=1) - S(t=3)|$ because of the change in direction. Also, observe that geometrically the area under V-vs-t curve, A + B, is the total distance traveled but the 'displacement' is $B - A = 6 - 4 = 2$ for this time interval.

2-4

Pulley A is pulled vertically down with a constant velocity v_o through the action of a time-varying force P(t) not shown in the figure. Determine the velocity and acceleration of the body B in terms of v_o, b and x. Neglect the weight and radius of the pulley.

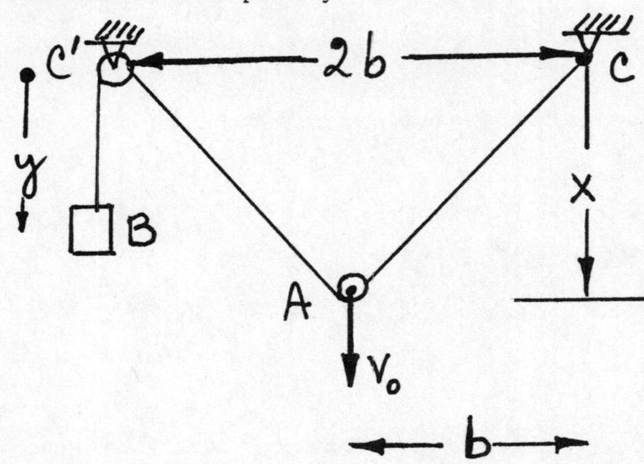

****

Denote the position of B by y. Then $AC = AC' = (x^2+b^2)^{1/2}$. The total length of the string is $y + 2(x^2+b^2)^{1/2} = \text{const}$.

Differentiating and substituting $\frac{dx}{dt} = V_o$ gives

$$\frac{dy}{dt} = -\frac{2xV_o}{(x^2+b^2)^{1/2}} \quad \text{or} \quad V_B = \frac{2xV_o}{(x^2+b^2)^{1/2}} \uparrow$$

A second differentiation yields

$$\frac{d^2y}{dt^2} = -2V_o\left[\frac{(x^2+b^2)^{1/2}V_o - x(x^2+b^2)^{-1/2}\,2xV_o}{x^2+b^2}\right]$$

Multiplying and dividing by $(x^2+b^2)^{1/2}$ and collecting terms gives

$$a_B = \frac{2V_o^2 b^2}{(x^2+b^2)^{3/2}} \uparrow$$

2-5

A movie stuntman is required to perform the following stunt. He is to hang from a cable 55 ft above a roadway and drop into a cushioned truck traveling down the road at a constant speed of 40 miles per hour. How far from the stuntman should the truck between the stuntman lets go of the cable ?

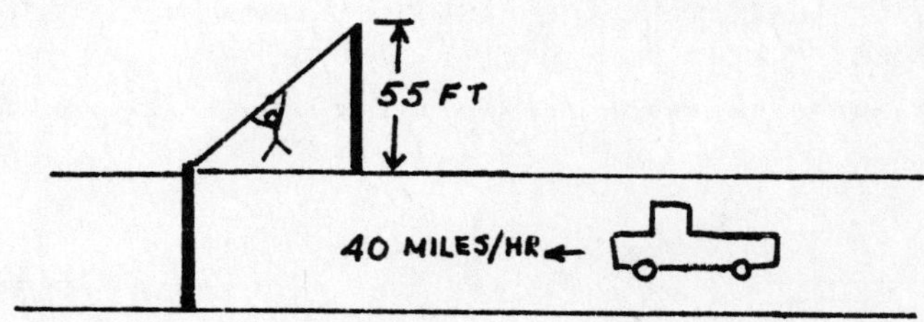

****

NEGLECT AIR RESISTANCE AND ASSUME THAT THE STUNTMAN FALLS AS A PARTICLE. LET $t=0$ BE THE TIME WHEN THE STUNTMAN RELEASES THE CABLE. HE WILL FALL WITH A CONSTANT ACCELERATION OF $g = 32.2$ FT/SEC² DOWNWARD. THUS

$$v = -gt$$

$$y = -gt^2/2$$

TO CALCULATE THE TIME FOR THE STUNTMAN TO REACH THE TRUCK WE SET

$$-55 \text{ FT} = -32.2 \text{ FT/SEC}^2 \; t^2/2$$

$$t = 1.85 \text{ SEC}$$

THE TRUCK IS MOVING WITH A CONSTANT VELOCITY OF 40 MILES/HR. IN 1.85 SECONDS, THE DISTANCE COVERED BY THE TRUCK IS

$$x = vt = \left(40 \frac{\text{MILES}}{\text{HR}}\right)\left(\frac{5280 \text{ FT}}{\text{MILE}}\right)\left(\frac{1 \text{ HR}}{3600 \text{ SEC}}\right)(1.85 \text{ SEC})$$

$$x = 108.5 \text{ FT}$$

THE STUNTMAN SHOULD RELEASE THE CABLE WHEN THE TRUCK IS 108.5 FT AWAY.

2-6

A motorcycle initially at rest accelerates from point A in an easterly direction according to the relation: $a = 4t^3 + t$ where t is the elapsed time in seconds and a is the acceleration in ft/sec^2. A second motorcycle initially at rest accelerates from point B in an easterly direction according to the relation: $a = 3t^3 + t$. Point B is 450 feet east of point A and both motorcycles begin travel at the same time. Assuming straight line motion, determine the distance travelled by motorcycle A just as it passes motorcycle B.

****

A, B, 450′

$\to X_A$, $\to a_A = 4t^3 + t$ $\qquad$ $\to X_B$, $\to a_B = 3t^3 + t$

$$t_A = t_B$$
$$X_A = X_B + 450$$

Motorcycle A:

$$a_A = \frac{dV_A}{dt} = 4t^3 + t \qquad V_A = \int_0^t (4t^3 + t)\,dt = t^4 + \frac{t^2}{2}$$

$$V_A = \frac{dX_A}{dt} = t^4 + \frac{t^2}{2} \qquad X_A = \int_0^t \left(t^4 + \frac{t^2}{2}\right) dt = \frac{t^5}{5} + \frac{t^3}{6}$$

Motorcycle B:

$$a_B = \frac{dV_B}{dt} = 3t^3 + t \qquad V_B = \int_0^t (3t^3 + t)\,dt = \frac{3t^4}{4} + \frac{t^2}{2}$$

$$V_B = \frac{dX_B}{dt} = \frac{3t^4}{4} + \frac{t^2}{2} \qquad X_B = \int_0^t \left(\frac{3t^4}{4} + \frac{t^2}{2}\right) dt = \frac{3t^5}{20} + \frac{t^3}{6}$$

Substituting: $X_A = X_B + 450$

$$\frac{t^5}{5} + \frac{t^3}{6} = \frac{3t^5}{20} + \frac{t^3}{6} + 450 \qquad t = 6.178 \text{ sec.}$$

$$X_A = \frac{t^5}{5} + \frac{t^3}{6} = \frac{(6.178)^5}{5} + \frac{(6.178)^3}{6}$$

$$\underline{X_A = 1839.3 \text{ ft.}}$$

2-7

Phil accelerates his car from rest to 50 mph (75 ft/sec) at a constant rate in 10 seconds. The car continues on at 50 mph for 5 seconds and then runs out of gas and coasts to a stop for 20 seconds. Phil makes Greg get out and push him back to the starting point at a constant speed of 4 mph (6 ft/sec).

A) Complete the s,v,a curves
B) How long does this entire caper take?
C) How far does Greg push?

****

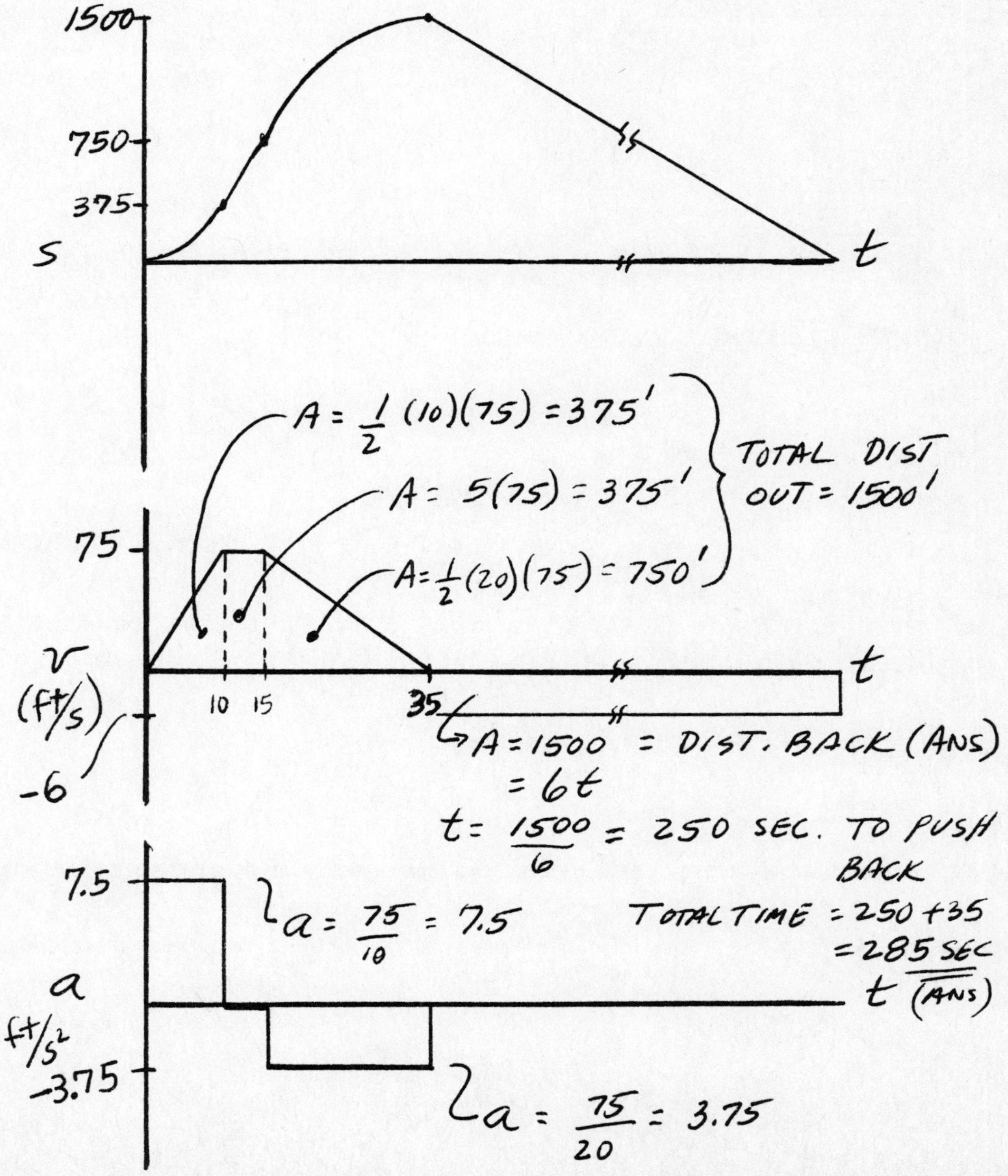

1500

750

375

s

t

$A = \frac{1}{2}(10)(75) = 375'$

$A = 5(75) = 375'$

$A = \frac{1}{2}(20)(75) = 750'$

TOTAL DIST OUT = 1500'

75

v (ft/s)

-6

10 15 35

t

$A = 1500$ = DIST. BACK (ANS)

$= 6t$

$t = \frac{1500}{6} = 250$ SEC. TO PUSH BACK

TOTAL TIME $= 250 + 35$

$= 285$ SEC (ANS)

7.5

a ft/s²

-3.75

t

$a = \frac{75}{10} = 7.5$

$a = \frac{75}{20} = 3.75$

2-8

What is the graphical relationship between displacement, velocity and acceleration ?

****

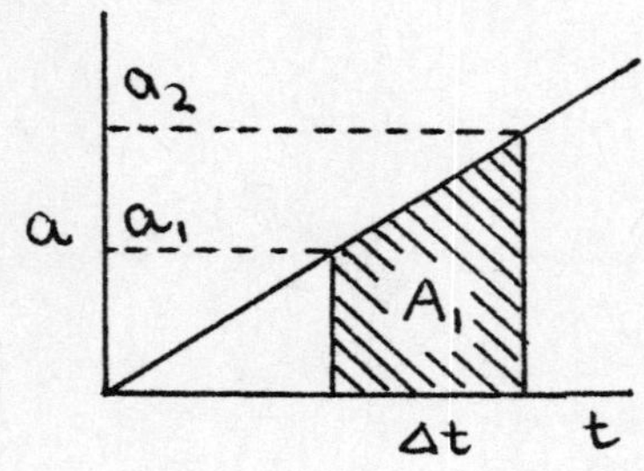

Area A_1 = Change in velocity in time Δt

$$= v_2 - v_1 = \Delta v$$

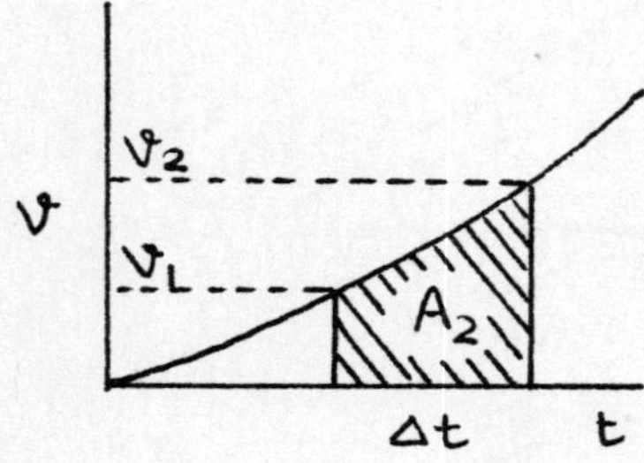

Area A_2 = Change in displacement in time Δt

$$= S_2 - S_1 = \Delta S$$

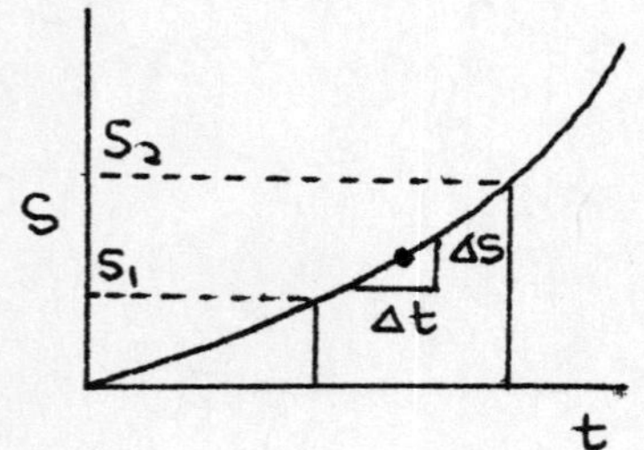

Slope of v-t curve = acceleration

Slope of S-t curve = velocity.

2-9

Rectilinear motion of a particle is expressed as the distance x (in meters) travelled in time t (in seconds), by the relationship

$$x = 2t^3 - t^2 + t + 4$$

Determine the velocity and acceleration of the particle at time t = 2 s.

****

Velocity $v = \frac{dx}{dt} = 6t^2 - 2t + 1$ m/s

Acceleration $a = \frac{dv}{dt} = 12t - 2$ m/s^2

at $t = 2$

$$v = 6 \times 2^2 - 2 \times 2 + 1 = \underline{21 \text{ m/s}}$$

$$a = 12 \times 2 - 2 = \underline{22 \text{ m/s}^2}$$

2-10

Rectilinear postion x of a particle at time t is given by

$$x = \sqrt{2t + 1}$$

Obtain a relationship between the velocity v and the displacement x of the particle. What is the acceleration at t = 1?

****

$$x^2 = 2t + 1$$

Differentiate with respect to t; $2x\frac{dx}{dt} = 2$

Now since $\frac{dx}{dt} = v$ we have,

$$\underline{xv = 1}$$

Differentiate again; $x\frac{dv}{dt} + \frac{dx}{dt} \cdot v = 0$

Since $\frac{dv}{dt} = a$ we have, $xa + v^2 = 0$

or, $a = -\frac{v^2}{x}$

At $t = 1$ $\quad x = \sqrt{2 \times 1 + 1} = \sqrt{3}$

$\quad v = \frac{1}{x} = \frac{1}{\sqrt{3}}$

substitute; $a = -\frac{1/3}{\sqrt{3}} = \underline{-\frac{1}{3\sqrt{3}}}$

2-11

The acceleration of a particle moving along a straight line is given by the law:

$a = -k\,x^n$

where $x > o$ is the displacement and k and n are constants ($n \neq -1$) (a) Determine the expression of the velocity v of the particle as a function of x, v_o and x_o (v_o and x_o being the initial velocity and displacement respectively) (b) Find the maximum displacement if k = 16, n=3, x_o = 2 m/s, v_o = 6 m/s.

****

Solution :

(a)

$$a = \frac{dv}{dt} = \frac{dv}{dx}\frac{dx}{dt} = v\,\frac{dv}{dx} = -k\,x^n$$

or

$$v\,dv = -k\,x^n\,dx$$

Integrating both sides:

$$\int_{v_o}^{v} v\,dv = \int_{x_o}^{x} -k\,x^n\,dx$$

$$\frac{v^2}{2} - \frac{v_o^2}{2} = -\frac{k}{n+1}\left(x^{n+1} - x_o^{n+1}\right)$$

or

$$v = \sqrt{v_o^2 - \frac{2k}{n+1}\left(x^{n+1} - x_o^{n+1}\right)}$$

(b) The maximum displacement :

$$\frac{k}{n+1}\,x^{n+1} = \frac{k}{n+1}\,x_o^{n+1} + \frac{v_o^2}{2} - \frac{v^2}{2}$$

The displacement is maximum when $v = 0$.

Thus ,

$$\frac{k}{n+1} x_{max}^{n+1} = \frac{k}{n+1} x_0^{n+1} + \frac{v_0^2}{2}$$

$$\frac{16}{4} x_{max}^4 = \frac{16}{4}(2)^4 + \frac{(6)^2}{2}$$

or

$$x_{max} = 2.13 \text{ m}$$

2-12

The acceleration of a particle is defined by the relation $a = 21-12x^2$, where a is expressed in m/s^2 and x in meters. The particle starts with no initial velocity at the position x = 0. Determine a) the velocity when x = 1.5 m, b) the position where the velocity is again zero, and c) the position where the velocity is maximum.

****

$$a = v\frac{dv}{dx} \qquad \int(21-12x^2)\,dx = \int v\,dv$$

$$21x - \frac{12x^3}{3}\Bigg|_{x=0}^{x} = \frac{v^2}{2} - \frac{v_0^2}{2} \quad (v_0 = 0)$$

$$21x - 4x^3 = \frac{v^2}{2}$$

a) $21(1.5) - 4(1.5)^3 = 18 \text{ m/s} = \frac{v^2}{2}$ $\qquad v_{x=1.5} = 6 \text{ m/s}$

b) $21x - 4x^3 = 0 \qquad x = 0, \; 21 - 4x^2 = 0$

$$x_{v=0} = \sqrt{21/4} = 2.29 \text{ m.}$$

c) $a = \frac{dv}{dx} = 0 \qquad 21 - 12x^2 = 0$

$$x_{v_{MAX}} = \sqrt{21/12} = 1.32 \text{ m}$$

2-13

In a test the muzzle velocity of a bullet fired from a rifle is determined by mounting two cardboard disks four feet apart on a long spindle which is rotated at a constant speed of 2500 rpm. The rifle is fired parallel to the spindle and the bullet passes through each disk in succession. If the bullet hole in the far disk is displaced 15 degrees with respect to the hole in the near disk, what is the muzzle velocity of the bullet?

****

Time required for disks to rotate thru 15° equals time required for bullet to travel 4 feet.

$$\text{Time} = \frac{1 \text{ min}}{2500 \text{ rev}} \times \frac{1 \text{ rev}}{360 \text{ deg}} \times \frac{60 \text{ sec}}{\text{min}} \times 15 \text{ deg} = .001 \text{ sec.}$$

$$\text{Velocity of bullet} = \frac{4 \text{ ft}}{.001 \text{ sec}} \qquad V_{BULLET} = 4000 \text{ ft/sec}$$

2-14

A particle moves in a straight line with an acceleration $a = 3t^2 - 4$. If a is given in ft/sec^2 and t in seconds, determine the position of the particle when t = 4 sec. Assume when t = 0, the particle is located 4 ft to the left of the origin, and when t = 3 sec, it is 10 ft to the left of the origin.

****

$\frac{d^2s}{dt^2} = a = 3t^2 - 4$, Integrating twice one obtains

$$s = \frac{1}{4}t^4 - 2t^2 + ct + c_1 \quad \text{------------------ (i)}$$

Substituting the given conditions in (i), one gets for $s = -4$ ft, when $t = 0$, and $s = -10$ ft, when $t = 3$ sec, the values $c = -2.75$, and $c_1 = -4$.

Therefore, $s = \frac{1}{4}t^4 - 2t^2 - 2.75t - 4$ ------ (ii)

Inserting $t = 4$ sec in (ii) one has

$$s = 17 \text{ ft to the right of origin}$$

2-15

Bill enters a metricated Interstate highway at kilometer marker 100 on his motorcycle. At that instant he is traveling at 20 kilometers per hour, and is accelerating at a constant rate of 4 meters per second squared. At the same instant he spots Sam on a cycle 500 meters ahead and gives chase. If Sam holds a constant speed of 70 kph, and Bill obeys the speed limit of 80 kph, how much time will it take him to catch up? Will he catch up in time to tell Sam to pull off at kilometer 105 for a beer?

****

$$\frac{1\ km}{hr} = \frac{1000\ m}{3600\ s} = 0.277\ m/s$$

So: Bill's initial speed is 5.56 m/s
Bill's top speed is 22.2 m/s
Sam's speed is 19.4 m/s

I prefer to use v-t curves

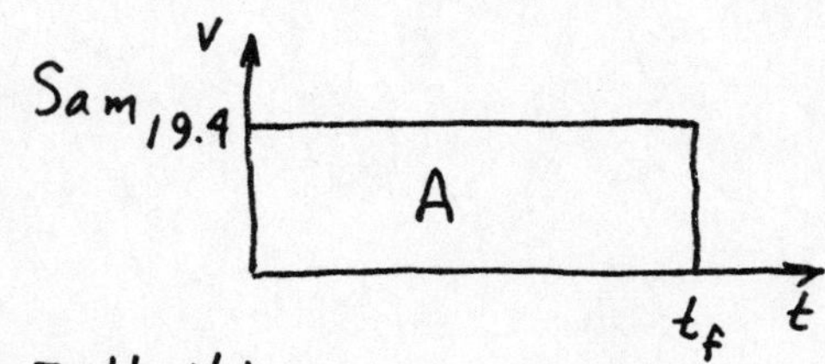

Sam travels distance A

$$A = 19.4\, t_f$$

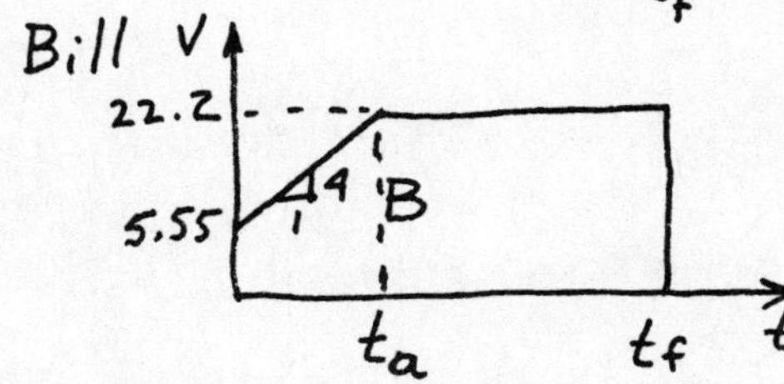

Bill travels distance B

$$B = 22.2\, t_f - \left(\tfrac{1}{2}\right)(22.2 - 5.55)\, t_a$$

$$t_a = \frac{(22.2 - 5.55)}{4} = 4.16\ s$$

Bill travels 500 m more than Sam so $B = A + 500$

$$22.2\, t_f - \frac{1}{2}\frac{(22.2 - 5.55)^2}{4} = 19.4\, t_f + 500$$

Answer $t_f = 191$ seconds

At that point Bill has travelled $B = 19.4(191) + 500 = 4205\,m$

Answer
YES! He catches him at km 104.2 and gets his beer.

2-16

A particle moves along a horizontal straight line with a constant acceleration. The velocity of the particle changes from 10 feet per second to the right, to 25 feet per second to the left during a time interval of 7 seconds. Determine: (a) displacement (b) total distance traveled, during the 7 seconds.

****

(a) SOLUTION ANALYTICALLY. Let $\longrightarrow$ be positive.

WHEN $t=0$; $V=+10$, $S=0$

$t=7$; $V=-25$, $S=S_1$

Let $a=K$, then $V=Kt+C_1$

$$S=\frac{Kt^2}{2}+C_1t+C_2$$

By initial conditions: $C_2=0$, $C_1=10$, $K=-5$.

then $S=-\frac{5t^2}{2}+10t$. When $t=7$,

$S_1=-5\left(\frac{49}{2}\right)+70 \qquad S_1=-52.5$ ft

but $V=0$ if $t=2$ sec, then $S=10$ ft.

TOTAL Distance $D=10+10+52.5=72.5$ ft.

(b) SOLUTION BY V-t diagram:

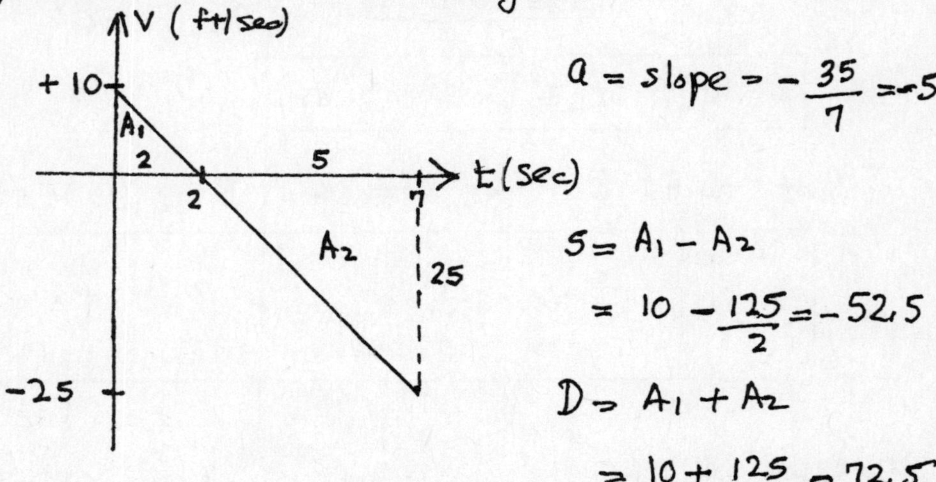

2-17

A particle travels in a straight line with a constant acceleration of 6 m/s^2. Its initial velocity is −48 m/s. Determine the total distance the particle travels in the first 12 seconds of its movement.

$V = 48\ m/s$ ← ○ → $a = 6\ m/s$ (CONSTANT)

AT $t = 0$

****

CHECK IF PARTICLE CHANGES DIRECTION ($V=0$) DURING THE 12 SECONDS.

$$V = V_0 + at \Rightarrow 0 = -48 + 6(t) \Rightarrow t = 8\ s$$

THE PARTICLE TRAVELS TO THE LEFT FOR THE FIRST 8 SECONDS, COMES MOMENTARILY TO REST, THEN MOVES TO THE RIGHT.

FOR FIRST 8 SECONDS: $X = V_0 t + \frac{1}{2}at^2 = (-48)(8) + \frac{1}{2}(6)(8)^2$

$X = -384 + 192 = -192\ m$

FOR LAST 4 SECONDS: $X = V_0 t + \frac{1}{2}at^2 = 0 + \frac{1}{2}(6)4^2$

$X = +48\ m$

THEREFORE THE DISTANCE TRAVELLED $= 192 + 48$

$= 240\ m$

Note: It may be easier to visualize the motion of the particle if the curves of acceleration, velocity, and displacement versus time are constructed.

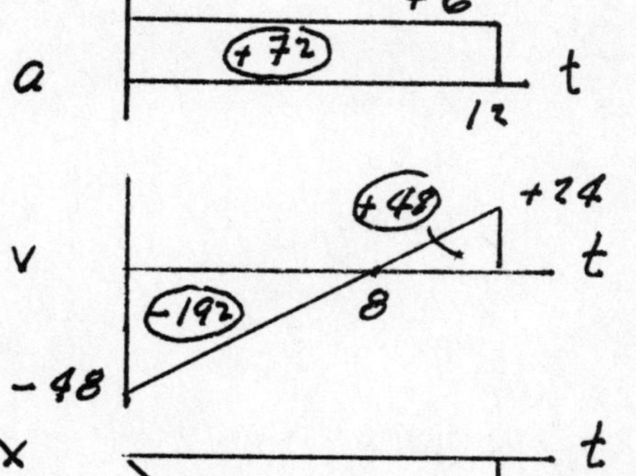

The area under the a-t curve (+72) equals the change in velocity. Therefore the velocity at t=12 is −48+72=+24 m/s. The velocity curve is linear and v=0 at t=8 seconds.

The change in displacement is equal to the area under the velocity curve, and from t=0 to t=8 is −192 m. From t=8 to t=12 the displacement of the particle is +48 m.

<u>Therefore the total distance travelled is 192+48= 240 m</u>. At t=12 the particle will be at −192+48 = −144 m. The motion can be shown as:

2-18

A point travels along the x axis according to the law $v = 2t^2 - 13t + 20$, where v is in feet per second, and t is in seconds. It starts from a point 10 feet to the right of the origin, and moves initially to the right. Find the distance it travels and its average velocity between t=0 and t=5.

****

To find distance and average velocity we need an expression for position

$$X = \int v\,dt = \int (2t^2 - 13t + 20)\,dt = \frac{2t^3}{3} - \frac{13}{2}t^2 + 20t + C$$

at $t=0$, $X=10 = \frac{2(0)^3}{3} - \frac{13}{2}(0)^2 + 20(0) + C$ so $C = 10$

To find distance we need to know if it changes direction. It does that when $V = 0 = 2t^2 - 13t + 20$

Using $t = \frac{-b \pm \sqrt{b^2 - 4ac}}{2a}$, we find turns at $t = 2.5$ and 4

t	0	2.5	4	5
X	10	29.79	28.67	30.83

$$\text{distance} = S = |X_{2.5} - X_0| + |X_4 - X_{2.5}| + |X_5 - X_4|$$

$$= 19.79 + 1.12 + 2.16$$

$$\boxed{S = 23.07 \text{ feet}}$$

$$V_{AVG} = \frac{\Delta X}{\Delta t} = \frac{X_5 - X_0}{5 - 0} = \frac{30.83 - 10}{5}$$

$$\boxed{V_{AVG} = 4.17 \text{ fps}}$$

2-19

How are acceleration velocity and displacement related ?

****

Velocity $v = \frac{ds}{dt}$ = Time rate of change of displacement

Average velocity $v_{avg} = \frac{\Delta s}{\Delta t}$

acceleration $a = \frac{dv}{dt}$ = Time rate of change of velocity

Average acceleration

$$a_{avg} = \frac{\Delta v}{dt}$$

acceleration $a = \frac{dv}{dt} = \frac{dv}{ds} \cdot \frac{ds}{dt} = \frac{dv}{ds} v$

2-20

A particle is moving in a horizontal direction in an electric field. The acceleration is 3t cm/sec^2. The initial velocity is 11 cm/sec. If the particle hits the plate with a velocity of 500 cm/sec, find when it hit the plate and how far it travelled.

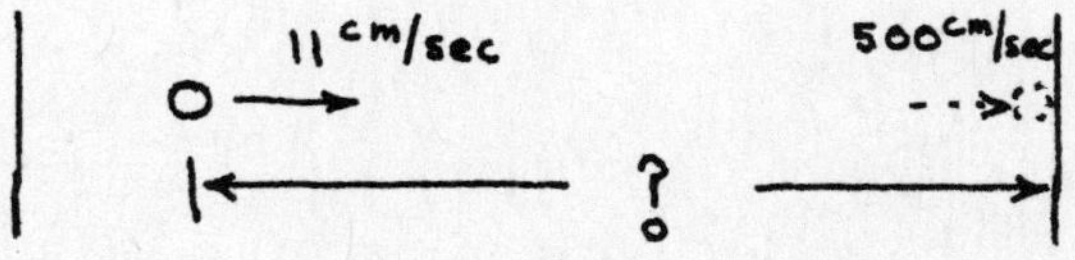

11 cm/sec 500 cm/sec

?

****

at $t = 0$, $v = 11$ cm/sec, $X = 0$ cm

$$v = \int_0^t a\,dt = \int_0^t 3t\,dt = \frac{3}{2}t^2 + V_0 \qquad \text{at } t=0,\ v=V_0=11$$

find the time when the particle hits the plate, i.e. the time for $v = 500$ cm/sec

$$v = 500 = 1.5t^2 + 11 \Rightarrow t = \sqrt{\frac{500-11}{1.5}} = 18.06 \text{ sec}$$

find the distance travelled in 18.06 sec.

$$X = \int_0^{18.06} v\,dt = \int_0^{18.06} [1.5t^2 + 11]\,dt = \left[\frac{1}{3}1.5t^3 + 11t + x_0\right]_0^{18.06}$$

at $t = 0$, $x = 0 = x_0$ $\quad\therefore\ X = 0.5(18.06)^3 + 11(18.06)$

$X = 3{,}144$ cm

TWO-DIMENSIONAL MOTION
RECTANGULAR CARTESIAN COORDINATES

2-21

A charged particle is accelerated by an electrical field. The acceleration is a constant 5 mm/sec^2 horizontally and a constant 3 mm/sec^2 vertically. The velocity at 13 sec is 52 mm/sec horizontally and 37 mm/sec vertically. (a) Find the velocity and position of the particle at zero time. (Put the coordinate system at the position of the particle at 13 sec.) (b) If the particle is 23 mm from the top plate, find when and where it hits the plate.

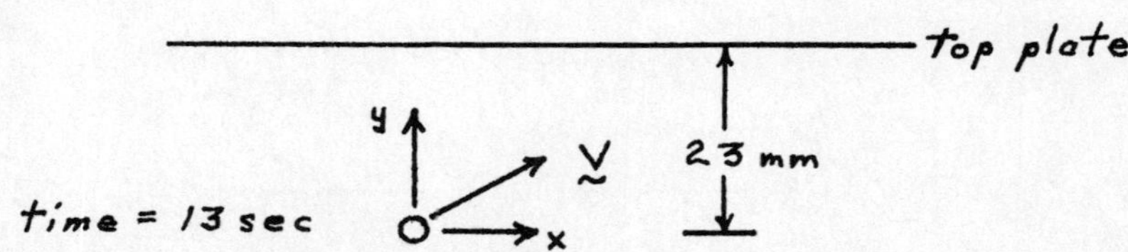

****

(a)

the acceleration vector: $\underset{\sim}{a} = 5\hat{i} + 3\hat{j}$

the velocity vector: $\underset{\sim}{v} = \int \underset{\sim}{a}\,dt = \int (5\hat{i} + 3\hat{j})dt = 5t\hat{i} + 3t\hat{j} + \underset{\sim}{v}_0$

at $t = 13$ sec $\underset{\sim}{v} = 52\hat{i} + 37\hat{j} = 5(13)\hat{i} + 3(13)\hat{j} + \underset{\sim}{v}_0$

$\therefore \underset{\sim}{v}_0 = -13\hat{i} - 2\hat{j}$ which is the velocity at $t = 0$

the position vector: $\underset{\sim}{r} = \int \underset{\sim}{v}\,dt = \int (5t\hat{i} + 3t\hat{j} - 13\hat{i} - 2\hat{j})\,dt$

$\underset{\sim}{r} = 2.5t^2\hat{i} + 1.5t^2\hat{j} - 13t\hat{i} - 2t\hat{j} + \underset{\sim}{r}_0$

at $t = 13$ sec $\underset{\sim}{r} = 0 = 2.5(13)^2\hat{i} + 1.5(13)^2\hat{j} - 13(13)\hat{i} - 2(13)\hat{j} + \underset{\sim}{r}_0$

$\therefore \underset{\sim}{r}_0 = -253.5\hat{i} - 227.5\hat{j}$ which is the position at $t=0$

(b)

$$\underset{\sim}{r} = [2.5t^2 - 13t - 253.5]\hat{i} + [1.5t^2 - 2t - 227.5]\hat{j}$$

when the particle hits the top plate $\underset{\sim}{r} = x\hat{i} + 23\hat{j}$

to find the time, use the $\hat{j}$ components:

$$23 = 1.5t^2 - 2t - 227.5$$

from the quadratic formula

$$t = \frac{2 \pm \sqrt{2^2 - 4(-250.5)(1.5)}}{2(1.5)} = 13.61 \text{ or } -12.27$$

since negative time is meaningless, $t = 13.61$ sec

to find the x-position, use the $\hat{i}$ components

$$x = 2.5(13.61)^2 - 13(13.61) - 253.5 = 32.65 \text{ mm}$$

$\therefore$ the position is $\underset{\sim}{r} = 32.65\hat{i} + 23\hat{j}$

2-22

A ball has an initial velocity of 20 m/s and moves in a vertical plane from A to C. Find the horizontal distance H between A and C.

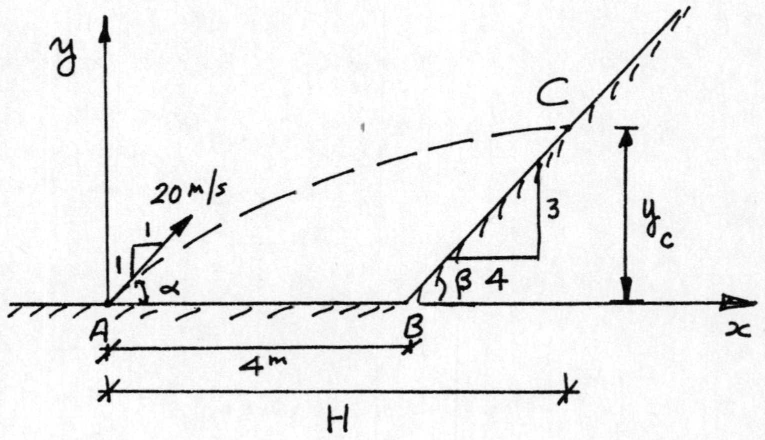

****

Solution :

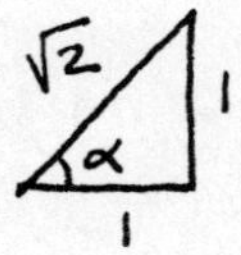

$$\cos\alpha = \frac{\sqrt{2}}{2} \quad ; \quad \sin\alpha = \frac{\sqrt{2}}{2}$$

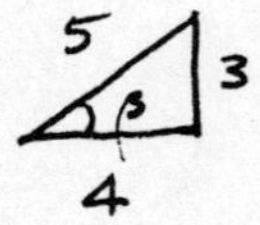

$$\cos\beta = \frac{4}{5} \quad ; \quad \sin\beta = \frac{3}{5} \quad ; \quad \tan\beta = \frac{3}{4}$$

$$x_0 = 0 , \quad y_0 = 0 , \quad g = 9.81\ m/s^2$$

$$v_{ox} = v_0 \cos\alpha = 20\,\frac{\sqrt{2}}{2} = 14.14\ m/s$$

$$v_{oy} = v_0 \sin\alpha = 20\,\frac{\sqrt{2}}{2} = 14.14\ m/s$$

$$x = x_0 + v_{ox} \cdot t$$

$$y = y_0 + v_{oy} \cdot t - \frac{1}{2} g t^2$$

Substituting :

$$x = 14.14\, t$$

$$y = 14.14\, t - \frac{9.81}{2} t^2$$

At C:

$$x_c = H \;;\; y_c = (H-4)\tan\beta = \frac{3}{4}(H-4)$$

Then :

$$H = 14.14\, t \quad \text{or} \quad t = \frac{H}{14.14}$$

$$\frac{3}{4}(H-4) = (14.14)\left(\frac{H}{14.14}\right) - \frac{9.81}{2}\left(\frac{H}{14.14}\right)^2$$

$$0.0245\, H^2 - 0.25\, H - 3 = 0$$

or

$$H = 17.29 \text{ m} .$$

2-23

A ball is thrown by a player from a position 2 m above the ground surface as shown with a velocity of 40 m/s inclined at an angle of 60^0 with the horizontal. Determine the maximum height the ball would attain.

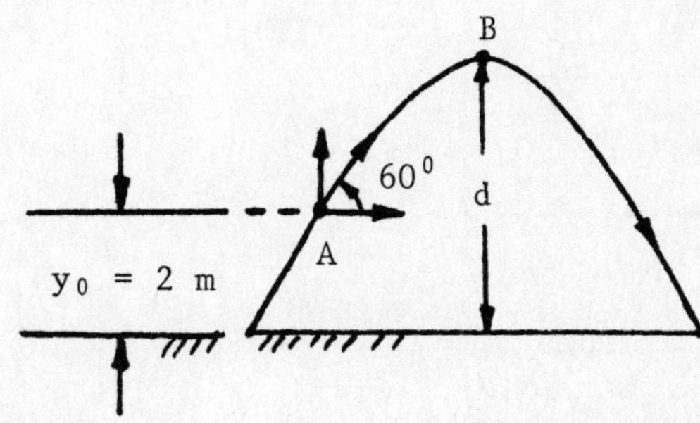

****

Vertical component of velocity of ball

$$(V_A)_y = 40 \sin 60^\circ = 34.64 \text{ m/s}$$

When the ball attains the heighest position B, the vertical component of velocity of ball is zero and is given as, $(V_B)_y^2 = 0 = (V_A)_y^2 - 2g(d - y_0)$

or, $(34.64)^2 - 2 \times 9.81(d - 2) = 0$

or, $d = 63.2$ m above the ground

2-24

A particle moves along the curve $y = 4 - 2x^2$ according to the equation $x = 0.5\ t$. Determine the velocity and acceleration of the particle when $t = 2$ seconds.

****

$$x = \tfrac{1}{2}t \qquad y = 4 - 2x^2$$

$$\dot{x} = \frac{dx}{dt} = \frac{1}{2} \qquad \dot{y} = \frac{dy}{dt} = -4x\dot{x}$$

$$\ddot{x} = \frac{d^2x}{dt^2} = 0 \qquad \ddot{y} = \frac{d^2y}{dt^2} = -4\dot{x}^2 - 4x\ddot{x}$$

$$\bar{V} = \dot{x}\bar{i} + \dot{y}\bar{j} = \tfrac{1}{2}\bar{i} + (-4)(\tfrac{1}{2}t)(\tfrac{1}{2})\bar{j} = \tfrac{1}{2}\bar{i} - t\bar{j}$$

$$\boxed{\text{at } t = 2,\ \bar{V} = \tfrac{1}{2}\bar{i} - 2\bar{j}}$$

$$\bar{a} = \ddot{x}\bar{i} + \ddot{y}\bar{j} = 0\bar{i} + \left[(-4)(\tfrac{1}{2})^2 - 4(\tfrac{1}{2}t)(0)\right]\bar{j}$$

$$\boxed{\text{at } t = 2,\ \bar{a} = -\bar{j}}$$

2-25

A golfer is 150 yards from the flag of an elevated green. Determine the initial velocity of the ball if the ball is to strike the base of the flag stick.

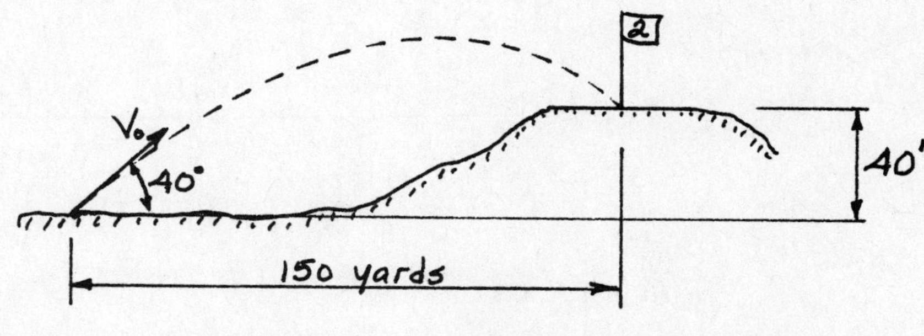

****

Horiz. Motion:

$$x = x_0 + V_0 t + \tfrac{1}{2} a t^2$$

$$450 = 0 + (V_0 \cos 40)t + 0$$

$$V_0 t = \frac{450}{\cos 40}$$

Vertical Motion:

$$y = y_0 + V_0 t + \tfrac{1}{2} a t^2$$

$$40 = 0 + (V_0 \sin 40)t + \tfrac{1}{2}(-32.2)t^2$$

$$40 = V_0 t \sin 40 - 16.1 t^2$$

substituting: $40 = \frac{450}{\cos 40} \cdot \sin 40 - 16.1 t^2$

$$t = 4.58 \text{ sec.}$$

$$V_0 = \frac{450}{t \cos 40} = \frac{450}{4.58 \cos 40}$$

$$V_0 = 128.3 \text{ ft/sec}$$

2-26

A projectile is fired up the incline at the angle and velocity shown. Find the distance d on the incline to the place where the projectile lands.

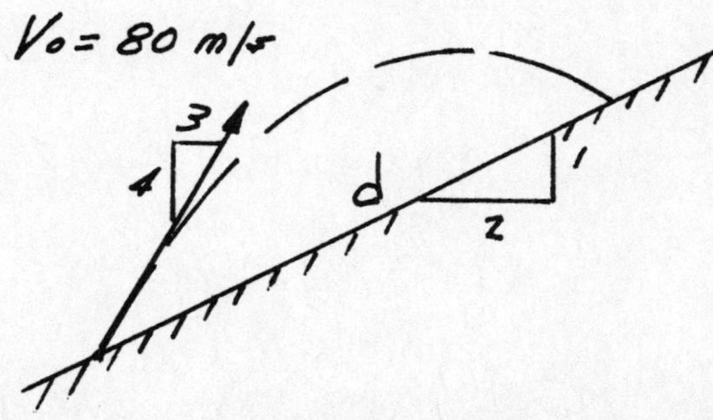

****

MOTION IN X-DIRECTION (CONSTANT VELOCITY)

① $x = V_{0x} t = 48t \Rightarrow t = \frac{x}{48}$

MOTION IN Y-DIRECTION (CONSTANT ACCELERATION)

② $y = V_{0y} t - \frac{1}{2}(g) t^2$

$y = 64t - \frac{1}{2}(9.81) t^2$

③ $y = \frac{1}{2} x$ (FROM SLOPE OF PLANE)

d, $\sqrt{5}$, 1, y, 2, x

$V_0 = 80$ m/s, 5, 4, 3

$V_{0y} = \frac{4}{5}(80) = 64$ m/s

$V_{0x} = \frac{3}{5}(80) = 48$ m/s

③ IN ② $\frac{1}{2} x = 64t - \frac{1}{2}(9.81) t^2$

① IN ② $\frac{1}{2} x = 64\left(\frac{x}{48}\right) - \frac{1}{2}(9.81)\left(\frac{x}{48}\right)^2$

$x = 391.4$ m AND $d = \frac{\sqrt{5}}{2} x = 438$ m

2-27

A man wants to throw a ball onto the flat roof of a building 35 ft. high. Standing 30 ft. from the wall, he throws the ball with a velocity of 60 fps at an angle of 60° with the horizontal, releasing it at the same level as the building base. How far from the edge of the roof will the ball fall? Check that the ball will clear the edge of the roof.

****

$V_{x_0} = 60 \cos 60 = 30$ fps = CONSTANT

$V_{y_0} = 60 \sin 60 = 52$ fps

THE TIME REQD. TO GET TO THE EDGE OF THE BUILDING (30 ft) AT 30 fps = 1 SEC.

THE HEIGHT OF THE BALL AT THAT TIME IS

$y_{t=1} = y_0^{\,0} + V_{0y} t + \frac{1}{2} a t^2 = 52(1) + \frac{1}{2}(-32.2)(1)^2$

$= 52 - 16.1 = 35.9$ ft > 35 ft. SO THE BALL CLEARS THE EDGE OF THE BUILDING.

AT $y = 35'$ $\quad 35 = 52t - \frac{1}{2}(32.2)t^2$

$t = .95$ SEC, 2.28 SEC.

$x = x_0 + vt = -30' + 30(2.28) = 38.4$ ft FROM EDGE OF BUILDING.

2-28

A ball is thrown from a 400 foot high building with an initial velocity of 100 ft/sec in a direction down and to the right of $\tan^{-1} 3/4$. Determine the velocity of the ball (in terms of *x* and *y* components) when it strikes the level ground.

****

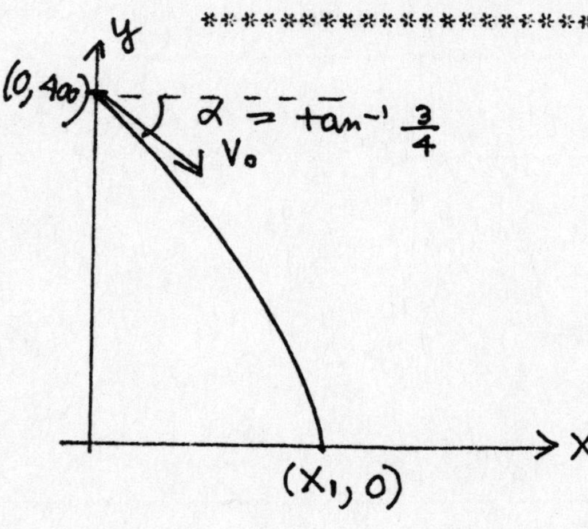

$a_x = 0$

$V_x = \frac{4}{5} \cdot 100 = 80$

$x = 80t$

$a_y = -g$

$V_y = -gt - \frac{3}{5} \cdot 100$

$= -gt - 60$

$y = -\frac{gt^2}{2} - 60t + 400$

Substituting, $16.1t^2 + 60t - 400 = 0$ when $y = 0$

BY QUADRATIC FORMULA, $t = 3.46$ sec.

then $V_x = 80$ ft/sec

$V_y = -32.2(3.46) - 60 = -171.3$ ft/sec.

2-29

A projectile is launched as shown. A) Show your origin and +X and +Y direction. B)Will it hit building B? C) If so, where will it strike on the building, or, if not, where will the projectile land?

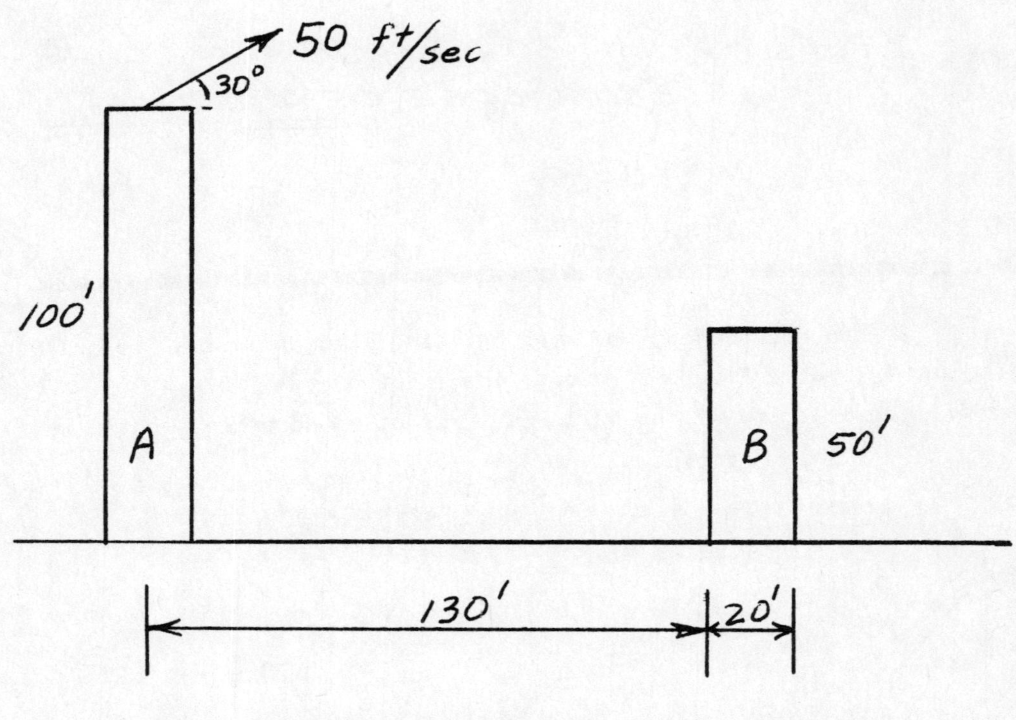

****

A) ORIGIN AT LAUNCH POINT; $\overset{+x}{\rightarrow}$, $\uparrow +y$

$$s = v_0 t + \frac{1}{2} a t^2$$
$$v^2 = v_0^2 + 2as$$
$$v = v_0 + at$$

X DIR.	Y DIR.
$s = 130$ ft.	$s = ?$
$v_0 = 43.3$ ft/s	$v_0 = 25$ ft/s
$a = 0$	$a = -32.2$ ft/s²
$t = ?$	$t = ?$

$s_x = 130 = 43.3t$; $t = 3.00$ sec

$$s_y = 25(3.00) - \frac{32.2}{2}(3.00)^2 = -69.9'$$

IT WILL STRIKE THE BUILDING B 30.1' UP ON THE LEFT SIDE.

2-30

A projectile is fired from the top of a 40 foot high building at 100 feet per second. IF the angle of initial velocity is 15° below the horizon, determine the velocity just prior to striking the ground.

****

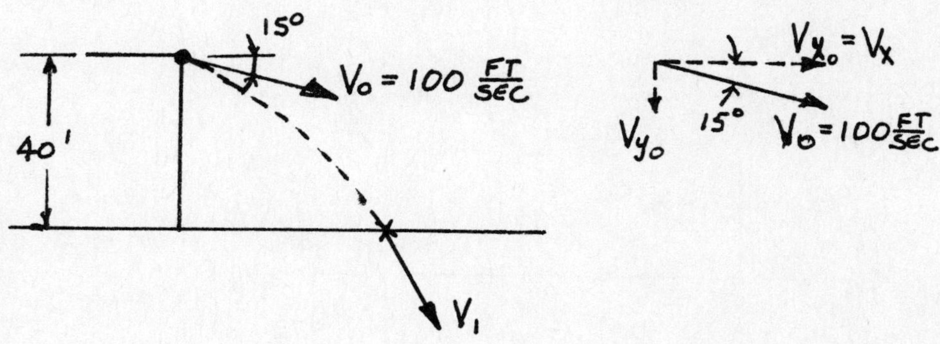

$$V_{x_o} = V_x = \text{Constant} = \left(100 \frac{FT}{SEC}\right) \cos 15^\circ = 96.59 \frac{FT}{SEC} \rightarrow$$

$$V_{y_o} = \left(100 \frac{FT}{SEC}\right) \sin 15^\circ = 25.88 \frac{FT}{SEC} \downarrow$$

$$a_y = 32.2 \frac{FT}{SEC^2} \downarrow \quad \& \quad S_y = 40\ FT \downarrow$$

$$(V_{y_1})^2 = (V_{y_o})^2 + 2 a_y S_y$$

$$(V_{y_1})^2 = \left(-25.88 \frac{FT}{SEC}\right)^2 + 2\left(-32.2 \frac{FT}{SEC^2}\right)(-40\ FT)$$

$$(V_{y_1})^2 = 669.87 \frac{FT^2}{SEC^2} + 2576 \frac{FT^2}{SEC^2} = 3245.87 \frac{FT^2}{SEC^2}$$

$$V_{y_1} = 56.97 \frac{FT}{SEC} \downarrow \quad \& \quad V_{x_1} = 96.59 \frac{FT}{SEC} \rightarrow \text{(From Above)}$$

$$V_1 = \sqrt{\left(96.59 \frac{FT}{SEC}\right)^2 + \left(56.97 \frac{FT}{SEC}\right)^2}$$

$$V_1 = 112.14 \frac{FT}{SEC}$$

96.59

56.97

2-31

The nozzle of a garden hose is held at a height 1.5 m above a football field. The water jet is discharged with a velocity of 5 m/s. What is the radius x reached by the water stream if the nozzle is held upward at 30° to the horizontal?

****

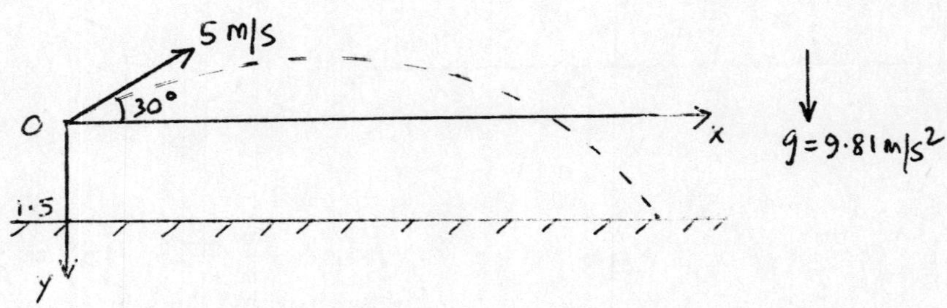

Chose the coordinates x and y as shown. Vertical motion with acceleration g, is given by

$$y = u_y t + \frac{1}{2} g t^2$$

with

Initial velocity $u_y = -5 \sin 30° = -2.5$ m/s

When a water droplet reaches the field at x, the vertical distance travelled is $y = 1.5$ m. Hence,

$$1.5 = -2.5t + \frac{1}{2} \times 9.81 t^2 \quad \text{or} \quad 9.81t^2 - 5t - 3 = 0$$

Solving, $$t = \frac{5 \pm \sqrt{5^2 + 4 \times 9.81 \times 3}}{2 \times 9.81} = 0.86 \text{ s}$$ {ignore negative root.

Horizontal motion has constant velocity $5 \cos 30°$ m/s. Hence,

$$x = 5 \cos 30° \times 0.86 \text{ m}$$

$$= \underline{3.72 \text{ m}}$$

2-32

Hank Aaron is trying to hit a home run. The ball leaves his bat (which is 5 ft. high) with an unknown speed v and at an angle of 30° to the ground. The fence is 130 yards from home plate and is 10 feet high. Find the speed v to just make a home run (i.e. to touch the top of the fence).

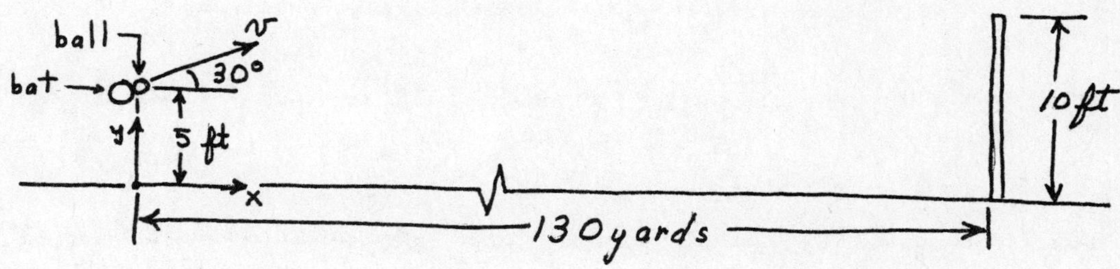

****

the acceleration vector: $\underset{\sim}{a} = -32.2\,\hat{j}$ ft/sec^2

the velocity vector: $\underset{\sim}{v} = \int \underset{\sim}{a}\,dt = -32.2t\,\hat{j} + \underset{\sim}{v}_0$

the position vector: $\underset{\sim}{r} = \int \underset{\sim}{v}\,dt = -16.1t^2\,\hat{j} + \underset{\sim}{v}_0 t + \underset{\sim}{r}_0$

at $t = 0$, $\underset{\sim}{v} = \underset{\sim}{v}_0 = v(\cos 30^\circ\,\hat{i} + \sin 30^\circ\,\hat{j})$

and $\underset{\sim}{r} = \underset{\sim}{r}_0 = 5\,\hat{j}$

when the ball touches the top of the fence, $t = t_{HR}$

and $\underset{\sim}{r} = 130\,\text{yd} \times 3\frac{\text{ft}}{\text{yd}}\,\hat{i} + 10\,\text{ft}\,\hat{j} = 390\,\hat{i} + 10\,\hat{j}$

$\therefore\; 390\,\hat{i} + 10\,\hat{j} = -16.1(t_{HR})^2\,\hat{j} + v(\cos 30^\circ\,\hat{i} + \sin 30^\circ\,\hat{j})t_{HR} + 5\,\hat{j}$

breaking into $\hat{i}$ and $\hat{j}$ components

$\hat{i}:\quad 390 = v\cos 30^\circ\, t_{HR} \Rightarrow t_{HR} = \frac{390}{\cos 30^\circ}\,\frac{1}{v} = 450.3\frac{1}{v}$

$\hat{j}:\quad 10 = -16.1(t_{HR})^2 + v\sin 30^\circ\, t_{HR} + 5$

$10 - 5 = -16.1\left(450.3\frac{1}{v}\right)^2 + v\left(\frac{1}{2}\right)\left(450.3\frac{1}{v}\right)$

$v = 121.8$ ft/sec

2-33

A particle moving in the xy plane has x and y velocity components given by $v_x = 4t - 1$ and $v_y = 2\left(\frac{ft}{sec}\right)$. The particle passes through (2,3) when t = 1 second. (a) Find the equation of the particles trajectory in terms of x and y and (b) sketch a graph of the motion of this particle.

****

$v_x = \frac{dx}{dt} = 4t - 1 \Rightarrow x = 2t^2 - t + C_1$ but when $t = 1$, $x = 2$

$\therefore 2 = 2 - 1 + C_1 \Rightarrow C_1 = 1$

$\therefore x(t) = 2t^2 - t + 1 \quad (*)$

Similarly, $v_y = \frac{dy}{dt} = 2 \Rightarrow y = 2t + C_2$, but when $t = 1$, $y = 3$

$\therefore 3 = 2 + C_2 \Rightarrow C_2 = 1$

$\therefore y(t) = 2t + 1$

Next we wish to eliminate the parameter t:

noting that $t = \frac{y-1}{2}$ from the last equation, we insert this in (*):

$$x = 2\left(\frac{y-1}{2}\right)^2 - \left(\frac{y-1}{2}\right) + 1$$

$$x = \frac{2(y^2 - 2y + 1)}{4} - \frac{(y-1)}{2} + 1 = \frac{y^2 - 2y + 1 - (y-1) + 2}{2}$$

$$\therefore 2x = y^2 - 2y + 1 - y + 1 + 2$$

or $2x = y^2 - 3y + 4$ — a parabola opening to the right. When $t = 0 \Rightarrow x = 1, y = 1$

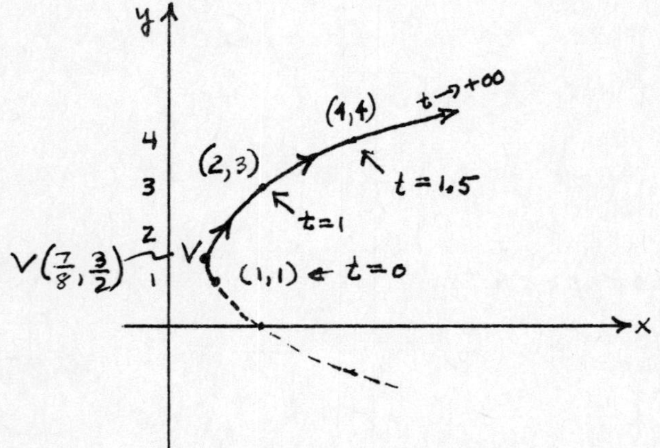

2-34

A nozzle discharges a stream of water into a room with an initial velocity of 40 ft/sec from a height of 2 ft with an angle of 10°.The height and length of the room are 8 ft and 20 ft,respectively.Where does this stream of water hit the other wall ?

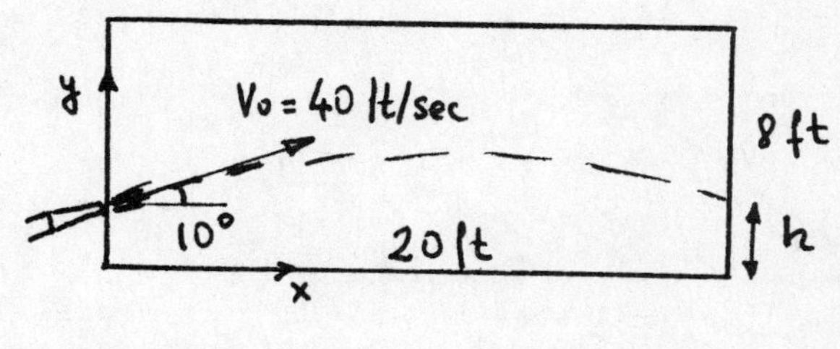

****

The coordinate system can be chosen as shown in the figure :

$$\therefore \quad x = x_0 + V_{x_0} \cdot t = x_0 + V_0 \cdot \cos\alpha \cdot t$$

$$20 = 0 + 40 \cdot \cos 10^\circ \cdot t \quad \rightarrow \quad t = 0.508 \text{ sec.}$$

$$y = y_0 - \frac{1}{2} g t^2 + V_{y_0} \cdot t$$

$$= y_0 - \frac{1}{2} g \cdot t^2 + V_0 \cdot \sin\alpha \cdot t$$

$$= 2 - \frac{1}{2} \times 32.2 \times 0.508^2 + 40 \times \sin 10^\circ \times 0.508 = 1.37 \text{ ft} = h$$

CHECK : Does the stream hit the top surface?

$$y = y_0 - \frac{1}{2} g t^2 + V_0 \sin 10^\circ \cdot t$$

$$8 = 2 - \frac{1}{2} \times 32.2 \times t^2 + 40 \times \sin 10^\circ \cdot t$$

$t_{1,2} \rightarrow$ complex numbers. $\therefore$ it does not hit the top surface.

2-35

A man working on the deck of a ship moves a distance of 15 m from A to B, as shown in the accompanying illustration. If the ship moves forward a distance of 100 m during the same time period, calculate the linear displacement as well as the net forward movement of the man.

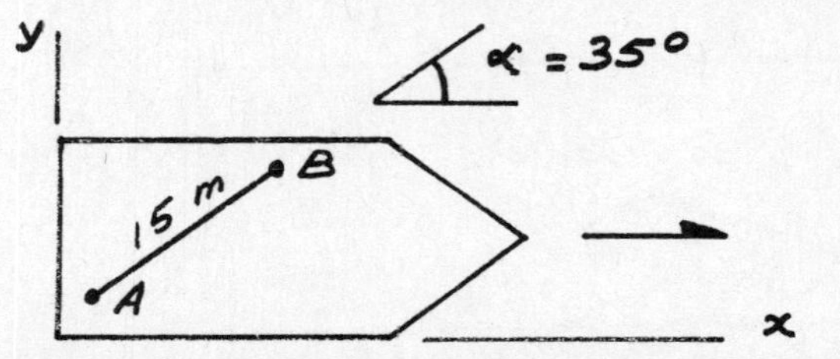

****

Solution

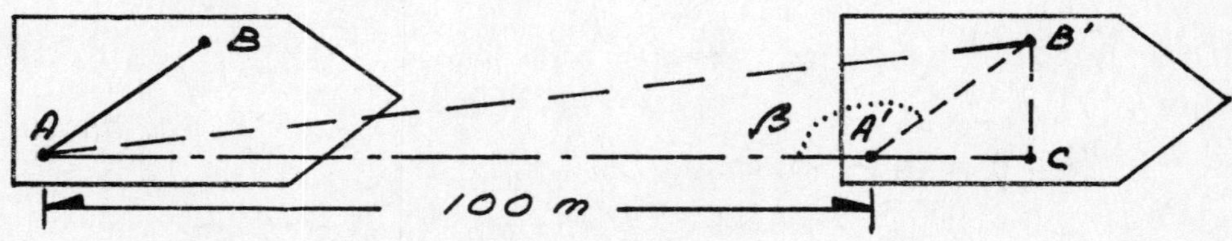

From trignometry, $\beta = \angle AA'B' = 180° - 35° = 145°$

Linear displacement of man, AB' is given by law of cosines : $(AB')^2 = (AA')^2 + (A'B')^2 - 2(AA')(A'B')\cos\beta$

$\therefore AB' = \sqrt{(100\,m)^2 + (15\,m)^2 - 2(100\,m)(15\,m)(-0.819)}$

$= 112.614$ m Ans.

Net forward movement is given by distance AC.

$\frac{B'C}{A'B'} = \sin 35°$; $B'C = (15\,m)(0.573) = 8.603$ m

From trignometry, $AC = \sqrt{(AB')^2 - (B'C)^2}$

$\therefore AC = \sqrt{(112.614\,m)^2 - (8.603\,m)^2} = 112.284\,m$ Ans.

CHECK : Alternative method for calculating the distance AC : $AC = AA' + AB\cos\alpha$

$\therefore AC = 100\,m + [(15\,m)(0.819)] = 112.287\,m$ Ans.

Note the small difference (0.003 m) between the two methods in computation.

2-36

The Mudville Hens are trailing the visiting team 2-1 with 2 outs in the bottom of the ninth. A runner is on first base as mighty Casey steps to the plate. The count quickly becomes three balls and two strikes as the tension mounts. The pitcher gives Casey a waist high (3 ft above the ground) fastball on the outside corner. Casey gives a mighty swing and at the crack of the bat everyone knows that it is gone. Indeed, 4 seconds later the ball hits the upper deck, 43 ft high, 400 ft away in right centerfield. The game is over and Mudville wins 3-2.

a. What was the velocity of the ball as it left the bat ?
b. How far would the ball have traveled had it not hit the stands ?

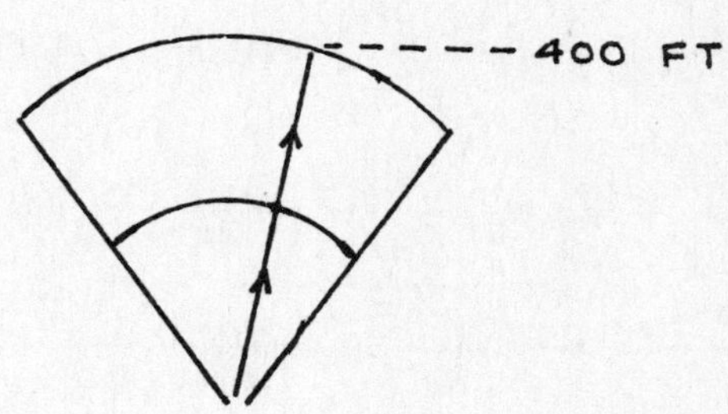

****

WE ASSUME THE FOLLOWING:

1) AIR RESISTANCE IS NEGLECTED
2) THE BALL CAN BE TREATED AS A PARTICLE
(THESE ARE NOT REALLY VERY GOOD ASSUMPTIONS)

LET x BE A COORDINATE ALONG THE GROUND FOLLOWING THE PATH OF THE BALL. LET y BE THE VERTICAL COORDINATE. THE PROBLEM IS A CLASSIC CASE OF PROJECTILE MOTION. THUS

$$\vec{a} = -g\vec{j}$$
$$\vec{v} = V_{0x}\vec{i} + (V_{0y} - gt)\vec{j}$$
$$\vec{r} = (V_{0x}t + x_0)\vec{i} + (V_{0y}t - \frac{gt^2}{2} + y_0)\vec{j}$$

IF WE SET THE ORIGIN OF THE COORDINATE SYSTEM AT CASEY'S FEET THEN

$$x_0 = 0 \quad , \quad y_0 = 3 \text{ FT}$$

AT $t = 4$ SEC , $\vec{r} = (400\vec{i} + 43\vec{j})$ FT

THUS

$$(400\vec{i} + 43\vec{j})\text{ FT} = [V_{0x}(4\text{ SEC})]\vec{i} + [V_{0y}(4\text{sec}) - \frac{(32.2\text{ FT/SEC}^2)(4\text{sec}^2)}{2} + 3\text{FT}]\vec{j}$$

OR

$$400\ FT = V_{ox}(4\ SEC)$$
$$V_{ox} = 100\ FT/SEC$$
$$43\ FT = 4\ V_{oy} - 257.6 + 3$$
$$V_{oy} = 74.4\ FT/SEC$$

THUS

$$\vec{V}_o = (100\vec{i} + 74.4\vec{j})\ FT/SEC$$

b) THE BALL WILL LAND WHEN $y = 0$ FT. THUS TO FIND THE TIME WE SOLVE

$$0 = (74.4\ FT/SEC)t - \frac{32.2\ FT/SEC^2}{2}t^2 + 3\ FT$$

$$t = \frac{-74.4 \pm \sqrt{(74.4)^2 + 4(16.1)(3}}{-2(16.1)} = 4.66\ SEC, -0.04\ SEC$$

THUS THE BALL WILL LAND 4.66 SEC. AFTER IT LEAVES THE BAT. AT THIS TIME

$$x = (100\ FT/SEC)(4.66\ SEC) = 466\ FT$$

TWO-DIMENSIONAL MOTION PLANE POLAR COORDINATES

2-37

Rod AB rotates about point A. Its motion is defined by $\theta = 4t^3$. θ is in radians and t is in seconds. At t = 0.5 seconds, the acceleration of end B will be:

a) 9 m/s^2 b) 27 m/s^2 c) 36 m/s^2 d) 45 m/s^2 e) 63 m/s^2

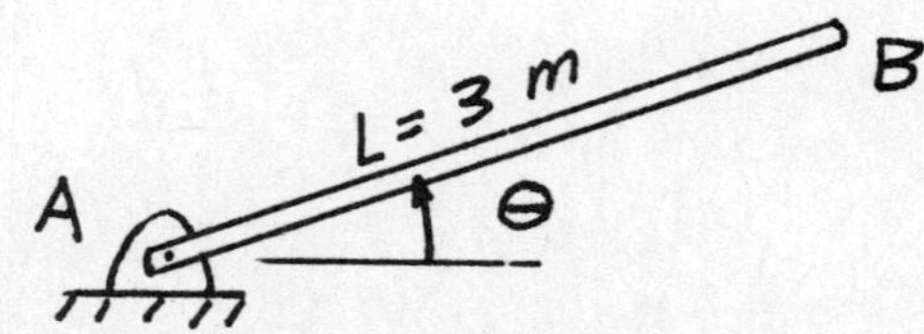

****

$$\theta = 4t^3$$

$$\omega = \frac{d\theta}{dt} = 12t^2, \quad \text{at } t = 0.5\ ;\ \omega = 12(0.5)^2 = 3 \text{ rad/s}$$

$$\alpha = \frac{d\omega}{dt} = 24t, \quad \text{at } t = 0.5;\ \alpha = 24(0.5) = 12 \text{ rad/s}^2$$

$$a_N = r\omega^2 = 3(3)^2 = 27 \text{ m/s}^2$$

$$a_T = r\alpha = 3(12) = 36 \text{ m/s}^2$$

$$\underset{\sim}{a} = \underset{\sim}{a}_N + \underset{\sim}{a}_T \qquad a = \sqrt{a_N^2 + a_T^2} = \sqrt{27^2 + 36^2} = 45 \text{ m/s}^2$$

THE ANSWER IS d

2-38

A particle is travelling along an Archimedian Spiral, which is given by the equation $r = \theta$. If the particle is at the point, $\theta = \frac{7\pi}{8}$, and $\dot{\theta} =$ 3 rad/sec and $\ddot{\theta} = 5$ rad/sec^2, find $\underset{\sim}{v}$ and $\underset{\sim}{a}$ in polar coordinates.

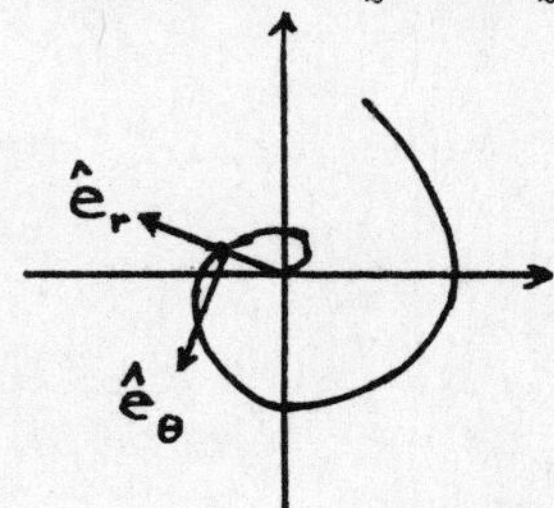

****

$$r = \theta = \frac{7\pi}{8}, \quad \dot{r} = \dot{\theta} = 3, \quad \ddot{r} = \ddot{\theta} = 5$$

$$\underset{\sim}{v} = \dot{r}\,\hat{e}_r + r\dot{\theta}\,\hat{e}_\theta = 3\hat{e}_r + \frac{7\pi}{8} 3\hat{e}_\theta = 3\hat{e}_r + 8.25\hat{e}_\theta$$

$$\underset{\sim}{a} = (\ddot{r} - r\dot{\theta}^2)\hat{e}_r + (r\ddot{\theta} + 2\dot{r}\dot{\theta})\hat{e}_\theta$$

$$\underset{\sim}{a} = \left(5 - \frac{7\pi}{8} 3^2\right)\hat{e}_r + \left(\frac{7\pi}{8} 5 + 2(3)(3)\right)\hat{e}_\theta$$

$$\underset{\sim}{a} = -19.7\hat{e}_r + 31.7\hat{e}_\theta$$

2-39

A particle moves radially outward along the spoke of a wheel with velocity V_o, while the wheel rotates with constant angular velocity ω_o about a shaft supported by bearings and going through the center of the wheel. Express the velocity and acceleration of the particle at any time in polar coordinates. Assume that $\theta = 0$, $r = 0$ when $t = 0$.

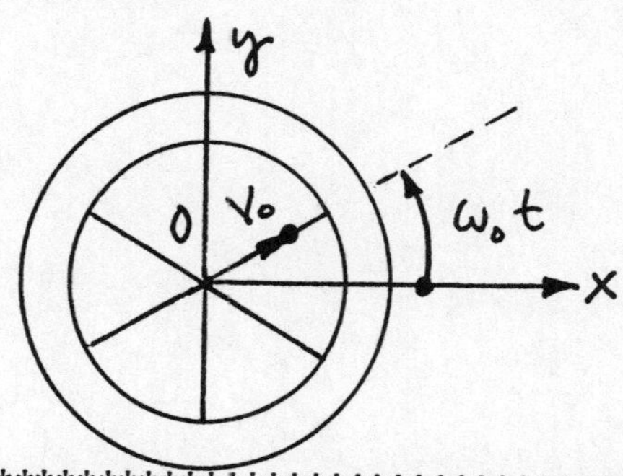

****

In the fixed xoy coordinate system the velocity of the particle is given by

$$\vec{V} = \dot{r}\,\vec{e}_r + r\dot{\theta}\,\vec{e}_\theta = V_r\,\vec{e}_r + V_\theta\,\vec{e}_\theta$$

where r and θ are the polar coordinates, and $\vec{e}_r$ and $\vec{e}_\theta$ are unit vectors in the radial and transverse directions.

Integrating $\dot{r} = V_o$, leads to $r = V_o t$. Also $\dot{\theta} = \omega_o$ yields $\theta = \omega_o t$, in view of initial conditions. Substituting

$$\vec{V} = V_o\,\vec{e}_r + V_o t\,\omega_o\,\vec{e}_\theta$$

Hence, the speed is $|\vec{V}| = V_o(1+\omega_o^2 t^2)^{1/2}$.

The angle it makes relative to the radius vector is $\arctan(V_\theta/V_r) = \tan^{-1}(\omega_o t)$

The absolute angle is $\omega_o t + \tan^{-1}(\omega_o t)$.

The acceleration is given by

$$\vec{a} = (\ddot{r} - r\dot{\theta}^2)\vec{e}_r + (2\dot{r}\dot{\theta} + r\ddot{\theta})\vec{e}_\theta$$

with $\ddot{r} = 0$ and $\ddot{\theta} = 0$, $\vec{a} = -V_0\omega_0^2 t\vec{e}_r + 2V_0\omega_0\vec{e}_\theta$

The angle the acceleration makes with the radius vector is $\theta = \arctan(a_\theta/a_r)$

$\theta = \arctan(-2/\omega_0 t)$. Add $\omega_0 t$ to obtain the absolute angle from the x axis.

2-40

A particle is travelling upwards along a path $y = (x-2)^3$. Express $\hat{e}_r$ and $\hat{e}_\theta$ (unit vectors of polar coordinates) in terms of $\hat{i}$ and $\hat{j}$ at the position $y = 4$ on the path.

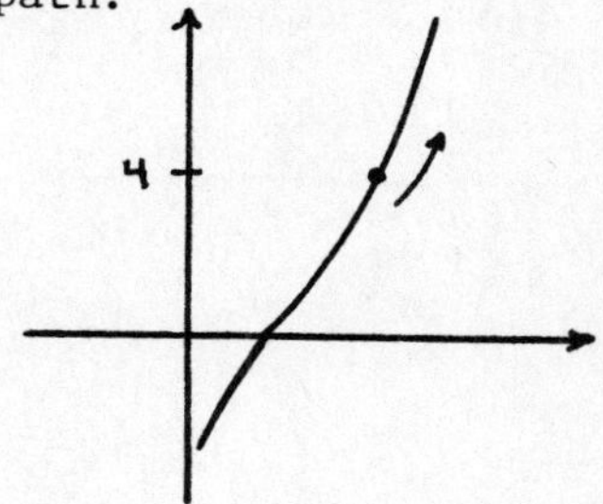

****

$x = y^{1/3} + 2 = (4)^{1/3} + 2 = 3.587$

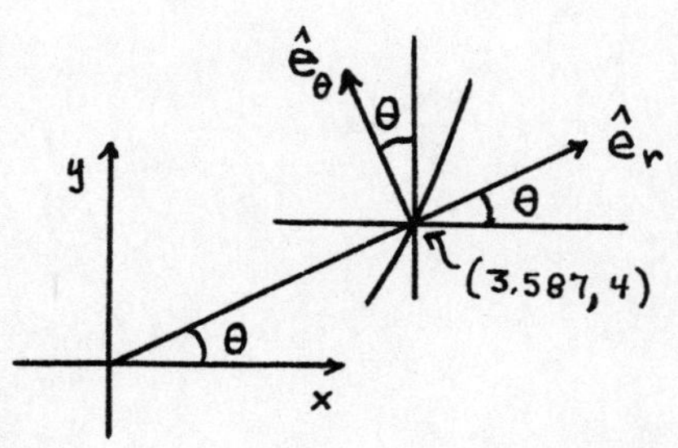

$\tan\theta = \dfrac{4}{3.587} \Rightarrow \theta = 48.12°$

$\hat{e}_r = \cos\theta\,\hat{i} + \sin\theta\,\hat{j}$

$\hat{e}_r = 0.668\,\hat{i} + 0.745\,\hat{j}$

$\hat{e}_\theta = -\sin\theta\,\hat{i} + \cos\theta\,\hat{j}$

$\hat{e}_\theta = -0.745\,\hat{i} + 0.668\,\hat{j}$

2-41

The parametric equations of a particle in polar coordinates are given by:

$$r = a_0 \sin\theta \quad \text{and} \quad \theta = \lambda \frac{\pi}{n} t$$

Find the expressions of velocity and acceleration at any given time t.

****

Solution :

$$r = a_0 \sin \lambda \frac{\pi}{n} t \quad ; \quad \theta = \lambda \frac{\pi}{n} t$$

$$\dot{r} = a_0 \lambda \frac{\pi}{n} \cos \lambda \frac{\pi}{n} t \quad ; \quad \dot{\theta} = \lambda \frac{\pi}{n}$$

$$\ddot{r} = -a_0 \lambda^2 \frac{\pi^2}{n^2} \sin \lambda \frac{\pi}{n} t \quad ; \quad \ddot{\theta} = 0$$

The velocity :

$$v_r = \dot{r} = a_0 \lambda \frac{\pi}{n} \cos \lambda \frac{\pi}{n} t$$

$$v_\theta = r\dot{\theta} = a_0 \lambda \frac{\pi}{n} \sin \lambda \frac{\pi}{n} t$$

$$v = \sqrt{v_r^2 + v_\theta^2} = \sqrt{a_0^2 \lambda^2 \frac{\pi^2}{n^2} \left(\cos^2 \lambda \frac{\pi}{n} t + \sin^2 \lambda \frac{\pi}{n} t\right)}$$

$$v = a_0 \lambda \frac{\pi}{n}$$

The acceleration :

$$a_r = \ddot{r} - r\dot{\theta}^2 = -a_0 \lambda^2 \frac{\pi^2}{n^2} \sin \lambda \frac{\pi}{n} t - a_0 \lambda^2 \frac{\pi^2}{n^2} \sin \lambda \frac{\pi}{n} t$$

$$= -2 a_0 \lambda^2 \frac{\pi^2}{n^2} \sin \lambda \frac{\pi}{n} t$$

$$a_\theta = r\ddot{\theta} + 2\dot{r}\dot{\theta} = 0 + 2 a_0 \lambda^2 \frac{\pi^2}{n^2} \cos \lambda \frac{\pi}{n} t$$

$$a = \sqrt{a_r^2 + a_\theta^2} = \sqrt{4 a_0^2 \lambda^4 \frac{\pi^4}{n^4} \left(\sin^2 \lambda \frac{\pi}{n} t + \cos^2 \lambda \frac{\pi}{n} t\right)}$$

or $a = 2a_0 \lambda^2 \frac{\pi^2}{n^2}$

The magnitudes of velocity and acceleration are independent of time.

2-42

The motion of a particle is defined by the following equations:

$$r = 2 \sin \pi t \qquad \theta = \frac{1}{2} \pi^2 t$$

Determine the velocity and acceleration of the particle when t = 0.50. (r is in meters, θ is in radians, and t is in seconds.)

****

$$\theta = \tfrac{1}{2}\pi^2 t \qquad r = 2 \sin \pi t$$

$$\frac{d\theta}{dt} \equiv \dot{\theta} = \frac{1}{2}\pi^2 \qquad \frac{dr}{dt} \equiv \dot{r} = 2\pi \cos \pi t$$

$$\frac{d^2\theta}{dt^2} \equiv \ddot{\theta} = 0 \qquad \frac{d^2 r}{dt^2} \equiv \ddot{r} = -2\pi^2 \sin \pi t$$

$$\bar{V} = \frac{d\bar{r}}{dt} \equiv \dot{\bar{r}} = \dot{r}\bar{e}_r + r\dot{\theta}\bar{e}_\theta$$

$$= (2\pi \cos \pi t)\bar{e}_r + \left[(2 \sin \pi t)(\tfrac{1}{2}\pi^2)\right]\bar{e}_\theta$$

at $t = 0.5$, $\bar{V} = \pi^2 \bar{e}_\theta$ or **Answer** $\bar{V} = 9.87\,\bar{e}_\theta$

$$\bar{a} = \dot{\bar{V}} = \frac{d^2\bar{r}}{dt^2} = \ddot{\bar{r}} = (\ddot{r} - r\dot{\theta}^2)\bar{e}_r + (2\dot{r}\dot{\theta} + r\ddot{\theta})\bar{e}_\theta$$

$$= \left[(-2\pi^2 \sin \pi t) - (2 \sin \pi t)(\tfrac{1}{2}\pi^2)^2\right]\bar{e}_r$$

$$+ \left[2(2\pi \cos \pi t)(\tfrac{1}{2}\pi^2) + (2 \sin \pi t)(0)\right]\bar{e}_\theta$$

substituting $t = 0.5$ **Answer** $\bar{a} = -68.4\,\bar{e}_r$

2-43

A particle is sliding in a slot on an arm that is rotating about a fixed point O. The distance between the particle and O, as a function of time, is given by

$$r(t) = (3 + \cos 2t) \text{ in}$$

where t is in seconds. The arm is rotating about O at a constant rate of 0.3 rad/sec counterclockwise starting at t=0 from the horizontal position.

a. Use radial and transverse coordinates to express the velocity and acceleration of the particle when $\theta = 30°$.
b. Express the velocity and acceleration of the particle in Cartesian coordinates when $\theta = 30°$.

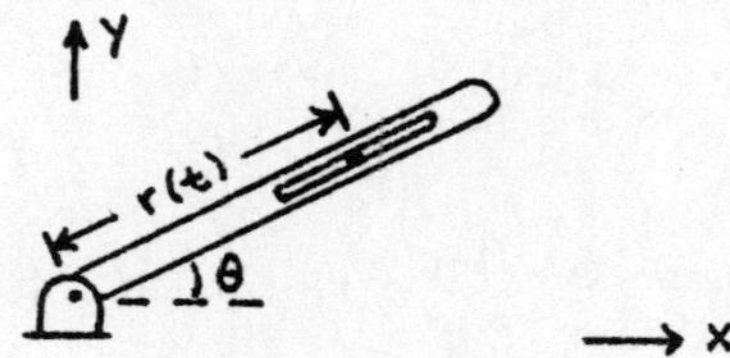

****

a) LET $\vec{i}_r$ BE A UNIT VECTOR DIRECTED AWAY FROM O ALONG THE ARM. LET $\vec{i}_\theta$ BE A UNIT VECTOR PERPENDICULAR TO $\vec{i}_r$.

IN THESE COORDINATES

$$\vec{V} = \dot{r}\vec{i}_r + r\dot{\theta}\vec{i}_\theta$$

$$\vec{a} = (\ddot{r} - r\dot{\theta}^2)\vec{i}_r + (r\ddot{\theta} + 2\dot{r}\dot{\theta})\vec{i}_\theta$$

$$\theta = \int \dot{\theta}\,dt = \dot{\theta}t + \theta_0 \qquad \text{FOR A CONSTANT } \dot{\theta}$$

$$\theta(0) = 0 \longrightarrow \theta_0 = 0 \qquad \theta(t) = \dot{\theta}t$$

FOR $\theta = 30° = \pi/6 \text{ RAD} = (.3 \text{ RAD/SEC})t \longrightarrow t = 1.745 \text{ SEC}$

THEN

$$r(t) = (3 + \cos 2t)\text{ IN} = (3 + \cos 2(1.745))\text{ in} = 2.06 \text{ IN}$$

$$\dot{r}(t) = -2\sin 2t \text{ IN/SEC} = -2\sin 2(1.745) \text{ IN/SEC} = -0.682 \text{ IN/SEC}$$

$$\ddot{r}(t) = -4\cos 2t \text{ IN/SEC}^2 = 3.76 \text{ IN/SEC}^2$$

THEN

$$\vec{V} = -0.682 \text{ IN/SEC}\,\vec{i}_r + (2.06 \text{ IN})(0.3 \text{ RAD/SEC})\vec{i}_\theta$$

$$= (-0.682\,\vec{i}_r + 0.618\,\vec{i}_\theta) \text{ IN/SEC}$$

$$\vec{a} = [3.76 \text{ in/sec}^2 - 2.06 \text{ IN}(.3 \text{ RAD/SEC})^2]\vec{i}_r + [(2.06 \text{ in})(0 \text{ RAD/SEC}^2) + 2(-0.682 \text{ IN/SEC})(.3 \text{ RAD/SEC})]\vec{i}_\theta$$

$$= (3.57\,\vec{i}_r - 0.409\,\vec{i}_\theta) \text{ IN/SEC}^2$$

b) SIMPLE TRANSFORMATIONS CAN BE MADE BETWEEN $\vec{i}_r$ AND $\vec{i}_\theta$ AND $\vec{i}$ AND $\vec{j}$. CONSIDER THE FOLLOWING DIAGRAM

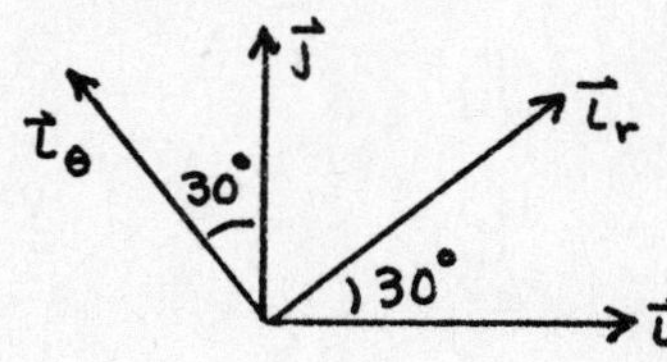

FROM LAWS OF VECTOR ADDITION

$$\vec{i} = \vec{i}_r \cos 30° - \vec{i}_\theta \sin 30°$$
$$\vec{j} = \vec{i}_r \sin 30° + \vec{i}_\theta \cos 30°$$

OR

$$\vec{i}_r = \vec{i} \cos 30° + \vec{j} \sin 30° = \frac{\sqrt{3}}{2}\vec{i} + \frac{1}{2}\vec{j}$$
$$\vec{i}_\theta = -\vec{i} \sin 30° + \vec{j} \cos 30° = -\frac{1}{2}\vec{i} + \frac{\sqrt{3}}{2}\vec{j}$$

THEN

$$\vec{V} = [-0.682(\sqrt{3}/2\,\vec{i} + 1/2\,\vec{j}) + 0.618(-1/2\,\vec{i} + \sqrt{3}/2\,\vec{j})] \text{ IN/SEC}$$
$$= (-0.9\,\vec{i} + 0.194\,\vec{j}) \text{ IN/SEC}$$

$$\vec{a} = [3.57(\sqrt{3}/2\,\vec{i} + 1/2\,\vec{j}) - 0.409(-1/2\,\vec{i} + \sqrt{3}/2\,\vec{j})] \text{ IN/SEC}^2$$
$$= (3.30\,\vec{i} + 1.43\,\vec{j}) \text{ IN/SEC}^2$$

2-44

The motion of a particle is given by $r=20t^2+1-2t$ and $\theta = 3t^2-1+t$ where r is in meters, θ in radians and t in seconds. Determine the time when a) $V_r=4$ m/sec, b) $a_\theta=5$ m/sec^2, c) $V_r = V_\theta/r$.

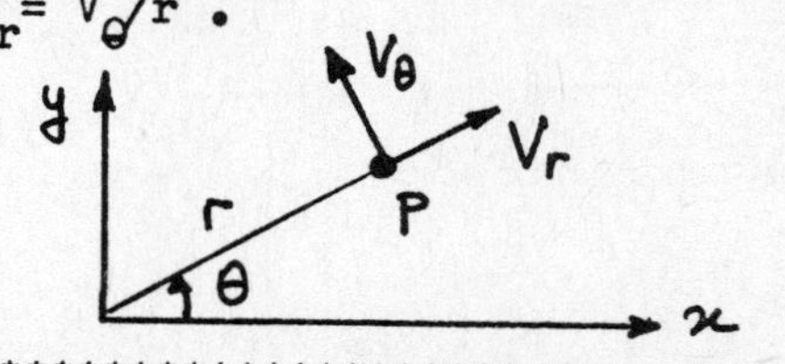

****

The motion has been expressed in cylindrical (polar) coordinates, where velocity vector is given by $\vec{V} = V_r\vec{i}_r + V_\theta\vec{i}_\theta = \dot{r}\vec{i}_r + r\dot{\theta}\vec{i}_\theta$,

acceleration is given by $\vec{a} = \vec{a}_r\vec{i}_r + a_\theta\vec{i}_\theta = (\ddot{r} - r\dot{\theta}^2)\vec{i}_r + (r\ddot{\theta} + 2\dot{r}\dot{\theta})\vec{i}_\theta$

a) $V_r = \dot{r} = 40t - 2 = 4$

$$40t = 6 \rightarrow t = 0.15 \text{ sec.}$$

b) $a_\theta = r\ddot{\theta} + 2\dot{r}\dot{\theta}$ $\quad \theta' = 6t+1 \quad , \theta'' = 6$

$\therefore \quad a_\theta = (20t^2-2t+1)\,6 + 2\,(40t-2)(6t+1) = 5$

$$600t^2 + 44t - 3 = 0$$

$$t_{1,2} = \frac{-44 \pm (44 - 4 \times 600 \times (-3))^{1/2}}{2 \times 600} \rightarrow t_1 = 0.034 \text{ sec} \qquad t_2 \rightarrow \text{negative}.$$

c) $V_r = V_\theta / r \rightarrow \dot{r} = r\theta'/r \rightarrow \dot{r} = \theta'$

$$(40t-2) = (6t+1)$$

$$34t - 3 = 0 \rightarrow t = 0.088 \text{ sec.}$$

2-45

A rocket is fired vertically from a launching pad at A. Its flight is tracked by radar from point A. Find an expression for the velocity of the rocket in terms of d, θ and $\dot{\theta}$.

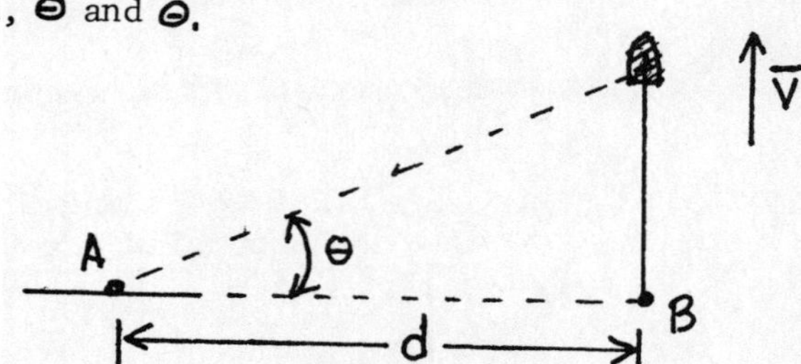

****

USING AXES (θ, r) $\qquad r = d \sec\theta$

$$V_r = \dot{r} = d \sec\theta \tan\theta\, \dot{\theta}$$

$$V_\theta = r\dot{\theta} = d \sec\theta\, \dot{\theta}$$

then $V^2 = V_r^2 + V_\theta^2$

$$= d^2 \sec^2\theta \tan^2\theta\, \dot{\theta}^2 + d^2 \sec^2\theta\, \dot{\theta}^2$$

$$V^2 = d^2 \sec^2\theta \cdot \dot{\theta}^2 \cdot \sec^2\theta$$

and $V = d \sec^2\theta\, \dot{\theta}$

2-46

If a particle moves along the path r = 2θ (meters). The angle, θ, is given by $\theta(t) = t^2$ (radians). (a) Find the velocity and the acceleration of the particle when $\theta = \pi/3$ radians.

****

Note: when $\theta = \frac{\pi}{3} = t^2 \Rightarrow t = \sqrt{\pi/3} \cong 1.02 \text{ sec.}$

Since $r = 2\theta = 2t^2$, $\therefore \dot{r} = 4t$, $\ddot{r} = 4$, $\dot{\theta} = 2t$, $\ddot{\theta} = 2$

now $\vec{V} = \dot{r}\, i_r + r\dot{\theta}\, i_\theta + \dot{z} K$

$= 4t\, i_r + 2t^2(2t)\, i_\theta$

$\vec{V} = 4t\, i_r + 4t^3\, i_\theta$

$\vec{V}\Big|_{t=1.02} = 4.08\, i_r + 4.24\, i_\theta$

$\therefore |\vec{V}| = \sqrt{4.08^2 + 4.24^2} = 5.88$

$\therefore \vec{V} = 5.88 \text{ m/sec}$, $\alpha = 1.29$ rad.

$\alpha = \frac{\pi}{6} + \beta$, $\tan\beta = \frac{4.08}{4.24} = .96226$

$\Rightarrow \beta = .766$ rad

$\therefore \alpha = .523 + .766 = 1.29$ rad

$\vec{a} = (\ddot{r} - r\dot{\theta}^2)\, i_r + (r\ddot{\theta} + 2\dot{r}\dot{\theta})\, i_\theta$

$= (4 - 2t^2 \cdot (4t^2))\, i_r + (2t^2 \cdot 2 + 2(4t)(2t))\, i_\theta$

$\vec{a} = (4 - 8t^4)\, i_r + (4t^2 + 16t^2)\, i_\theta$

$\vec{a}\Big|_{t=1.02} = -4.66\, i_r + 20.8\, i_\theta \Rightarrow |\vec{a}| = \sqrt{(-4.66)^2 + (20.8)^2} = 21.3 \text{ m/sec}^2$

$\therefore \vec{a} = 21.3 \text{ m/sec}^2$, $\alpha = .303$ (rad)

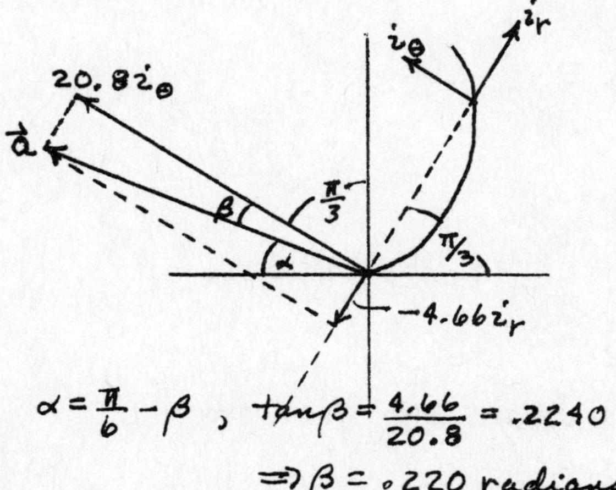

$\alpha = \frac{\pi}{6} - \beta$, $\tan\beta = \frac{4.66}{20.8} = .2240$

$\Rightarrow \beta = .220$ radians

$\therefore \alpha = .523 - .220 = .303$ radians

2-47

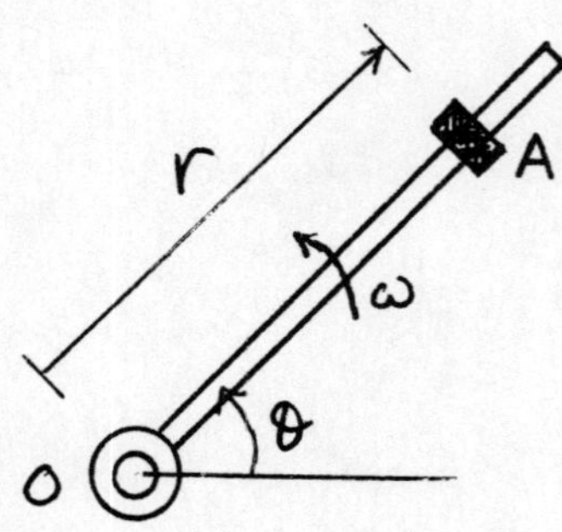

A rod rotates about O in a plane at an angular velocity $\omega = 3t^2 - 2t$. A ring A slides along the rod, and its distance from O is given by $r = 10t^2 - 5t$.

(a) Determine the angle of rotation θ of the rod when t = 2.
(b) Determine the velocity and acceleration of the ring when t = 2.

****

$$\frac{d\theta}{dt} = \omega = 3t^2 - 2t$$

Integrate; $\theta = t^3 - t^2 + C$

Using $\theta = 0$ at $t = 0 \Rightarrow C = 0$. Hence $\theta = t^3 - t^2$

(a) at $t = 2$ $\quad \theta = 2^3 - 2^2 = \underline{4}$

(b) In polar co-ordinates

radial velocity component $V_r = \frac{dr}{dt} = 20t - 5$

transverse velocity component $V_\theta = \omega r$

$$= (3t^2 - 2t)(10t^2 - 5t)$$

At $t = 2$: $\quad V_r = 20 \times 2 - 5 = 35$

$$V_\theta = (3 \times 2^2 - 2 \times 2)(10 \times 2^2 - 5 \times 2) = 240$$

Velocity magnitude $V = \sqrt{V_r^2 + V_\theta^2} = \sqrt{35^2 + 240^2} = \underline{242.5}$

Velocity direction α made to the rod is given by

$$\tan\alpha = \frac{V_\theta}{V_r} = \frac{240}{35} = 6.86$$

or,

$$\alpha = \underline{81.7^\circ}$$

radial acceleration component $a_r = \frac{d^2r}{dt^2} - \omega^2 r$

transverse acceleration component $a_\theta = \frac{d\omega}{dt} r$

Note: $\frac{d^2r}{dt^2} = 20$ and $\frac{d\omega}{dt} = 6t-2$

Hence

$$a_r = 20 - (3t^2-2t)^2(10t^2-5t)$$

$$a_\theta = (6t-2)(10t^2-5t)$$

at $t=2$: $a_r = -1900$

$$a_\theta = 300$$

magnitude $a = \sqrt{a_r^2+a_\theta^2} = \sqrt{1900^2+300^2} = \underline{1923.5}$

direction $\alpha' \Rightarrow \tan\alpha' = \frac{300}{1900}$

or $\alpha' = \underline{9^\circ}$ as shown in figure.

300
a
α'
1900
θ

2-48

A boy is flying a kite. At a particular instant the kite is 100 meters off the ground and is rising vertically at 2m/sec. If the kite is 150 meters from the boy at this instant, how fast is the kite string being let out?

****

V = 2 m/sec
KITE
r = 150 m
h = 100 m
θ
Boy
a

$$a = \sqrt{(150)^2 - (100)^2} = 111.8 \text{ m}$$

$$\cos\theta = \frac{a}{r} = \frac{111.8}{r}$$

$$r = 111.8 \sec\theta$$

$$\dot{r} = 111.8 \tan\theta \sec\theta\, \dot{\theta}$$

$$\tan\theta = \frac{h}{a} \qquad h = 111.8 \tan\theta \qquad \dot{h} = 111.8 \sec^2\theta\, \dot{\theta}$$

$$\dot{\theta} = \frac{\dot{h}}{111.8 \sec^2\theta} = \frac{\dot{h}\cos^2\theta}{111.8} = \frac{(2\text{ m/sec})\left(\frac{111.8\text{ m}}{150\text{ m}}\right)^2}{111.8\text{ m}}$$

$$\dot{\theta} = 0.009938 \text{ rad/sec}$$

Substituting: $\dot{r} = 111.8\text{ m}\left(\frac{100\text{ m}}{111.8\text{ m}}\right)\left(\frac{150\text{ m}}{111.8\text{ m}}\right)\left(\frac{.009938}{\text{sec}}\right)$

$$\underline{\underline{\dot{r} = 1.33 \text{ m/sec}}}$$

2-49

Find the velocity and acceleration vectors in terms of the unit vectors u_r and u_θ if $r = a(1 + s\ in\ t)$ and $\theta = 1 - e^{-t}$.

****

First we recall that $\vec{v} = \frac{dr}{dt}\vec{u}_r + r\frac{d\theta}{dt}\vec{u}_\theta$

$$\vec{v} = (a\cos t)\vec{u}_r + a(1+\sin t)e^{-t}\vec{u}_\theta$$

then we recall that

$$\vec{a} = [\ddot{r} - r\dot{\theta}^2]\vec{u}_r + [r\ddot{\theta} + 2\dot{r}\dot{\theta}]\vec{u}_\theta$$

$$= [-a\sin t - a(1+\sin t)(e^{-t})^2]\vec{u}_r + [a(1+\sin t)(-e^{-t}) + (2a\cos t)e^{-t}]\vec{u}_\theta$$

$$\therefore\ \vec{a} = -a[\sin t + (1+\sin t)e^{-2t}]\vec{u}_r + ae^{-t}[2a\cos t - 1 - \sin t]\vec{u}_\theta$$

2-50

A ski-jump track is designed as a circular arc with its center at O and a radius of 65 m, as shown in the accompanying illustration. The subtended angle of the arc between the start (S) and the take-off (T) points is 50°, such that the take-off acceleration is defined by $a_t = g \sin \phi$, where g is the acceleration due to gravity. Assuming no losses, calculate the take-off speed of this track.

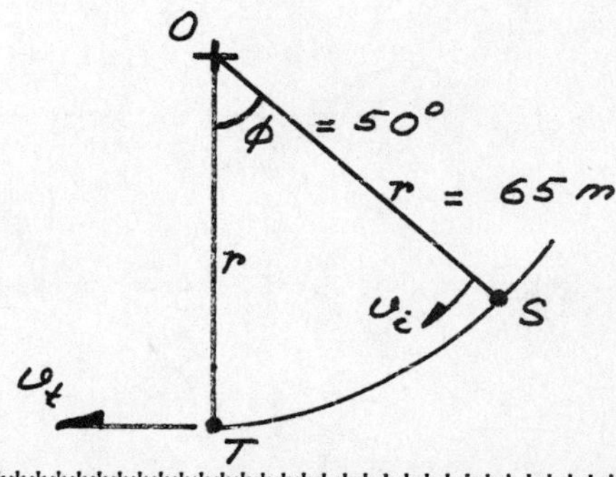

****

Solution

From the given conditions,

$$a_t = g \cdot \sin\phi = \frac{dv}{dt} \; ; \quad dv = g \cdot \sin\phi \; dt$$

Also, using transverse scalar components of acceleration, we have

$$g \cdot \sin\phi = \frac{r\dot{\phi}}{dt} = \frac{r\,d\dot{\phi}}{d\phi} \cdot \frac{d\phi}{dt} = r\,\omega \frac{d\omega}{d\phi}$$

OR,

$$\omega\, d\omega - \frac{g}{r} \sin\phi \; d\phi = 0$$

By integrating this equation, we get

$$\frac{\omega^2}{2} + \frac{g}{r} \cos\phi = c \quad (\text{a constant})$$

At the starting point, where $\phi = 0$,

$$v_i = 0 \quad \text{and} \quad \omega = 0$$

Hence, $c = \frac{g}{r}$ and $\omega = \sqrt{\frac{2g}{r}(1 - \cos\phi)}$

By substituting the values of g, r, and ϕ, we get

$$\omega = \sqrt{\frac{2(9.81\ m/s^2)}{65\ m}(1 - \cos 50^{\circ})} = 0.328\ rad/s$$

Hence,

$$v_t = r\,\omega$$

$$= (65\,m)(0.328\ rad/s) = 21.346\ m/s \quad \underline{\text{Ans.}}$$

2-51

An object A moves at a constant speed of v_a = 25 m/s along a spiral path such that r = 15/ϕ m, where ϕ is in radians. Calculate (a) the magnitudes of v_r and v_ϕ as functions of ϕ, and (b) the values of v_r and v_ϕ when ϕ = 1.5 rad.

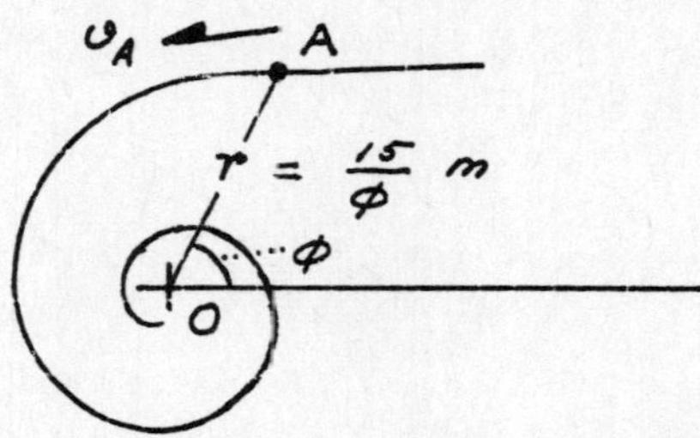

****

<u>Solution</u>

$$r = \frac{15}{\phi} \quad ; \quad \dot{r} = -\frac{15}{\phi^2}\dot{\phi} \qquad \cdots\cdots (1)$$

Now, $v^2 = \dot{r}^2 + (r\dot{\phi})^2$

or, $(25 \text{ m/s})^2 = \left(-\frac{15}{\phi^2}\dot{\phi}\right)^2 + \left(\frac{15}{\phi}\dot{\phi}\right)^2$

$$= \frac{15^2}{\phi^4}(1+\phi^2)\dot{\phi}^2$$

or, $\dot{\phi} = \sqrt{\frac{(25 \text{ m/s})^2 \phi^4}{(15 \text{ m/s})^2(1+\phi^2)}} = \frac{1.666\,\phi^2}{\sqrt{1+\phi^2}}$

(a) By substituting the value of $\dot{\phi}$ in Eq. (1), we get

$v_r = \dot{r} = \left(-\frac{15}{\phi^2}\right)\left(\frac{1.666\,\phi^2}{\sqrt{1+\phi^2}}\right) = \frac{-25}{\sqrt{1+\phi^2}}$ **Ans.**

and $v_\phi = r\dot{\phi} = \left(\frac{15}{\phi}\right)\left(\frac{1.666\,\phi^2}{\sqrt{1+\phi^2}}\right) = \frac{25\,\phi}{\sqrt{1+\phi^2}}$ **Ans.**

(b) For ϕ = 1.5 rad, we have

$v_r = \frac{-25}{\sqrt{1+(1.5)^2}} = -13.873$ m/s **Ans.**

and $v_\phi = \frac{25\,(1.5)}{\sqrt{1+(1.5)^2}} = 20.810$ m/s **Ans.**

TWO-DIMENSIONAL MOTION: TANGENTIAL AND NORMAL COMPONENTS

2-52

A projectile is fired with velocity 50 m/s at an angle of 60° to the horizontal. Neglecting the aerodynamic resistance, determine the radius of curvature ρ of the trajectory at its highest point.

****

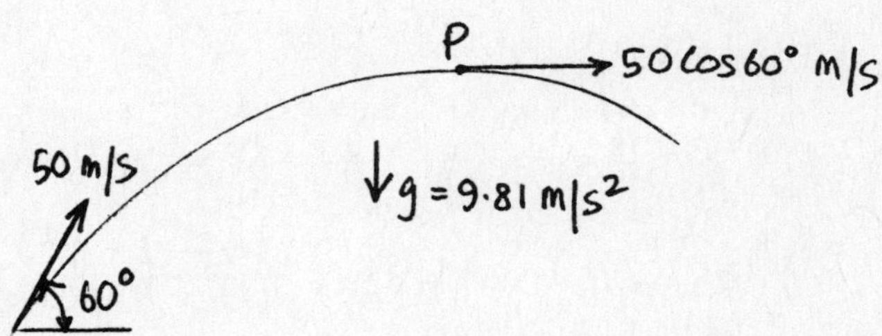

Horizontal velocity component $u_x = 50 \cos 60° = 25$ m/s

This remains constant throughout the trajectory.

At the highest point P the velocity is horizontal

Using normal - tangential co-ordinate system,

Normal (inward) acceleration component $a_n = \dfrac{v^2}{\rho}$

with v = velocity magnitude

At P: $a_n = g = 9.81$ m/s² and $v = u_x = 25$ m/s

Hence $9.81 = \dfrac{25^2}{\rho}$ or $\rho = \dfrac{25^2}{9.81} = \underline{63.7 \text{ m}}$

2-53

A particle is moving along a curved path as shown. At point A the radius of curvature of the path is 50 m and the particle is moving with a speed of 50 m/sec. The particle's speed is increasing at a constant rate of 0.5 m/sec^2.

a. What is the acceleration of the particle at A ?
b. The distance along the curve from A to B is 100 m. The radius of curvature of the curve at B is 25m. What is the particles velocity when it is at B ?
c. What is the acceleration of the particle when it is at B ? B ?

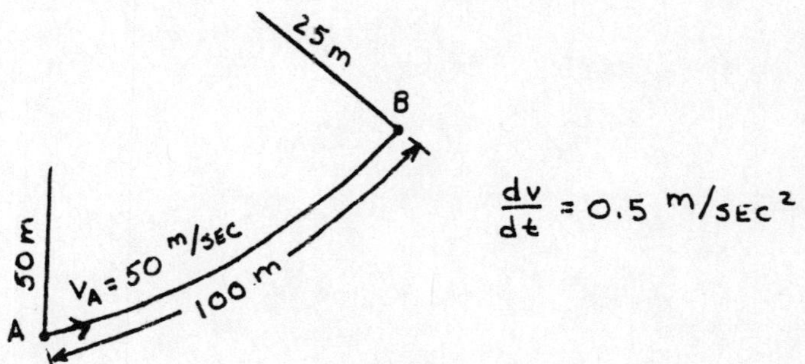

$$\frac{dv}{dt} = 0.5 \text{ m/SEC}^2$$

****

a)

$$\vec{a}_A = \frac{dv}{dt}\,\vec{\iota}_t + \frac{v^2}{\rho}\,\vec{\iota}_n$$

$$\vec{a}_A = 0.5 \text{ m/sec}^2\,\vec{\iota}_t + \frac{(50 \text{ m/sec})^2}{50 \text{ m}}\,\vec{\iota}_n$$

$$\vec{a}_A = 0.5 \text{ m/sec}^2\,\vec{\iota}_t + 50 \text{ m/sec}^2\,\vec{\iota}_n$$

b) LET S BE A COORDINATE THAT MEASURES THE DISTANCE TRAVELED BY THE PARTICLE FROM POINT A. BY DEFINITION THE INSTANTANEOUS VELOCITY OF THE PARTICLE IS $v = ds/dt$. RECALL

$$\frac{dv}{dt} = 0.5 \text{ m/sec}^2$$

USING THE CHAIN RULE

$$\frac{dv}{dt} = \frac{dv}{ds}\frac{ds}{dt} = v\frac{dv}{ds} = 0.5 \text{ m/sec}^2$$

INTEGRATING BOTH SIDES WITH RESPECT TO S AS S VARIES FROM 0 TO 100 m YIELDS

$$\int_{v=v_A}^{v=v_B} v\frac{dv}{ds}\,ds = \int_{s=0}^{s=100} 0.5 \frac{m}{sec^2}\,ds$$

$$\left.\frac{v^2}{2}\right|_{v_A}^{v_B} = 0.5 \frac{m}{sec^2}\, s\,\Big|_0^{100 \text{ m}}$$

$$\frac{V_B^2}{2} - \frac{V_A^2}{2} = 0.5\ \text{m/sec}^2\ (100\ \text{m})$$

$$V_B = 2\sqrt{50\ \frac{\text{m}^2}{\text{sec}^2} + \frac{1}{2}\left(50\ \frac{\text{m}}{\text{sec}}\right)^2} = 72.1\ \text{m/sec}$$

c) $\vec{a}_B = \frac{dv}{dt}\vec{\iota}_{tB} + \frac{V_B^2}{\rho}\vec{\iota}_{nB}$

$$\vec{a}_B = 0.5\ \frac{\text{m}}{\text{sec}^2}\vec{\iota}_{tB} + \frac{(72.1\ \text{m/sec})^2}{25\ \text{m}}\vec{\iota}_{nB}$$

$$\vec{a}_B = 0.5\ \frac{\text{m}}{\text{sec}^2}\vec{\iota}_{tB} + 208\ \frac{\text{m}}{\text{sec}^2}\vec{\iota}_{nB}$$

2-54

At a certain instant, a particle traveling a circular path is in the position shown. Its instantaneous acceleration is 30 feet per second squared to the right. Find its instantaneous velocity. The circle is 6 feet in diameter.

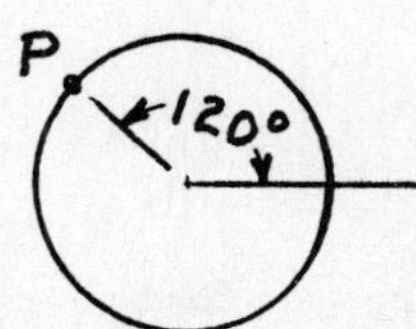

****

Resolve acceleration into normal and tangential parts.

$a_T = 30\cos 30 = 15\sqrt{3}$

$a_N = 30\cos 60 = 15$

In normal-tangential system

$$\bar{a} = \dot{V}\bar{e}_T + \frac{V^2}{r}\bar{e}_N$$

so $\frac{V^2}{r} = a_N = 15$, $r = 3$

$$V^2 = ra_N = 45$$

Answer
$V = 6.71$ fps

2-55

A point moves on a circular path of radius 10 feet in such a way that $v = 2t^2 + 4$ where v is in ft/sec and t is in seconds.

(a) What are the normal and tangential components of acceleration at a time t of 4 seconds?

(b) How much distance has been traveled during those 4 seconds?

****

(a) If $V = 2t^2 + 4$, $a_t = dv/dt = 4t$

For $t = 4$, $V = 36$ and $a_t = 16$ ft/sec^2

$$a_n = V^2/r = 36^2/10 = 129.6 \text{ ft/sec}^2$$

(b) $V = ds/dt$ or $ds = (2t^2 + 4)\,dt$

$$s = \int_0^4 (2t^2 + 4)\,dt = 58.67 \text{ ft}$$

2-56

A pilot testing an airplane at 2000 mph wishes to test various items while experiencing a normal component of 10 g's. Find the radius of the circular path which would allow the pilot to do this.

****

On a circular path the acceleration is a_{normal}.

Hence, $a_n = 10g = \dfrac{v^2}{\rho}$ $\therefore \rho = \dfrac{v^2}{10g}$

$v = 2000 \dfrac{Mi}{Hr} \cdot \dfrac{1 \; Hr}{3600 \; Sec}$

$v = \dfrac{5}{9} \dfrac{Mi}{Sec}$

$$\therefore \rho = \frac{\left(\frac{5}{9}\frac{Mi}{sec}\right)^2}{10\cdot(32.2)\frac{ft}{sec^2}\cdot\frac{1}{5{,}280}\frac{Mi}{ft}}$$

$$= \frac{\frac{25}{81}\frac{Mi^2}{Sec^2}}{\frac{322}{5{,}280}\frac{Mi}{Sec^2}} = \frac{25}{81}\cdot\frac{5{,}280}{322} \text{ Mi}$$

$$\therefore \rho = 5.06 \text{ Mi}$$

2-57

A particle is traveling upwards along a path $y = (x-2)^3$. When $y = 4$, the speed of the particle is 11 m/sec and the rate of change of its speed is 3.6 m/sec². Find the velocity and acceleration of the particle in terms of path components, $\hat{e}_t$ and $\hat{e}_n$.

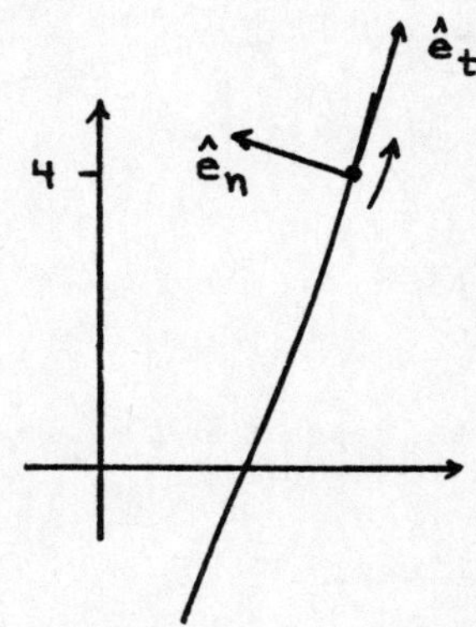

****

$$\dot{s} = 11 \text{ m/sec}, \quad \ddot{s} = 3.6 \text{ m/sec}^2$$

$$\underset{\sim}{v} = \dot{s}\,\hat{e}_t + 0\,\hat{e}_n = 11\,\hat{e}_t \text{ m/sec}$$

$$\underset{\sim}{a} = \ddot{s}\,\hat{e}_t + \frac{1}{R}(\dot{s})^2\,\hat{e}_n$$

$$R = \frac{\left[1 + (dy/dx)^{1.}\right]^{1.5}}{\left|d^2y/dx^2\right|}$$

$$y = (x-2)^3 \Rightarrow x = y^{1/3} + 2 = 4^{1/3} + 2 = 3.587 \text{ m}$$

$$dy/dx = 3(x-2)^2 = 3(3.587-2)^2 = 7.556$$

$$d^2y/dx^2 = 6(x-2) = 6(3.587-2) = 9.522$$

$$R = \frac{\left[1 + (7.556)^2\right]^{1.5}}{9.522} = 46.50 \text{ m}$$

$$\underset{\sim}{a} = 3.6\,\hat{e}_t + \frac{1}{46.50}(11)^2\,\hat{e}_n = 3.6\,\hat{e}_t + 2.602\,\hat{e}_n$$

2-58

A car enters a curved track which has a radius of 1500 ft at a steady speed of 45 m.p.h. If the driver then accelerates at 3 ft/s^2 for a period of 10s, calculate the normal and tangential components of acceleration at the end of the acceleration period.

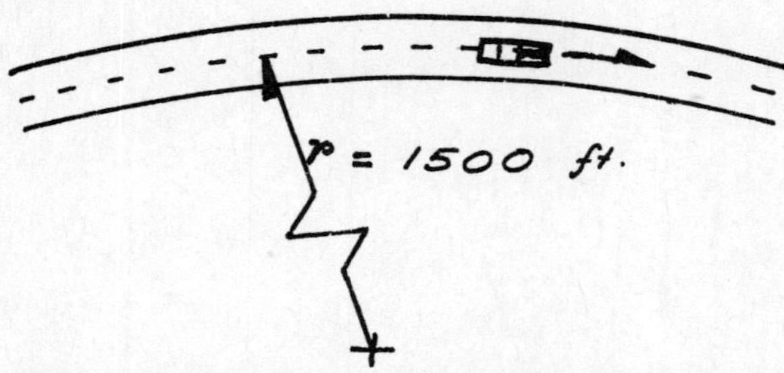

****

<u>Solution</u>

Velocity of car after 10 s :

$$v_f = v_i + at = v_i + \dot{v}_i t$$

$$= \frac{(45 \text{ mph})(5280'/\text{mile})}{(3600 \text{ sec/hr})} + (3 \text{ ft/s}^2)(10 \text{ s})$$

$$= 96 \text{ ft/s}$$

Acceleration :

$$\bar{a} = \dot{v}_i \bar{a}_t + \frac{v_f^2}{\rho} \bar{a}_n$$

$$= (3 \text{ ft/s}^2)(\bar{a}_t) + \frac{(96 \text{ ft/s}^2)^2}{(1500 \text{ ft})} (\bar{a}_n)$$

$$= (3\,\bar{a}_t + 6.144\,\bar{a}_n) \text{ ft/s}^2$$ <u>Ans.</u>

2-59

A nozzle discharges a stream of water in the direction shown with an initial velocity of 25 m/sec. Determine the radius of curvature of the stream
(a) as it leaves the nozzle at A
(b) at the maximum height B of the stream.

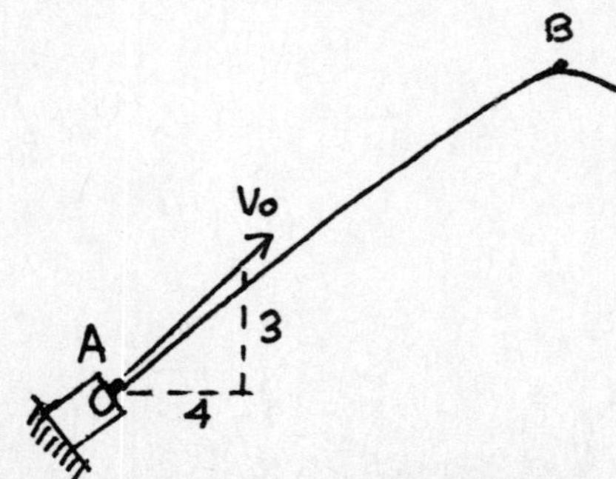

****

At A,

(a)

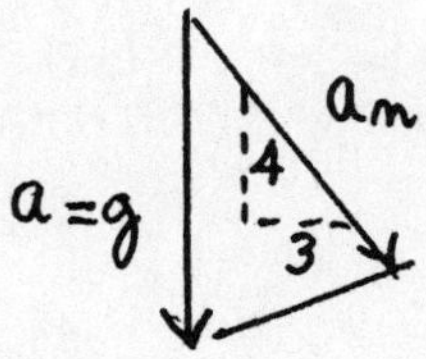

$a_m = \frac{4}{5} g = \frac{25^2}{\rho_A}$

SOLVING $\rho_A = 79.6$ meters

(b) At B, $V = (V_0)_x = 20$ m/s , $a_m = a = g$

OR $a_m = g = \frac{20^2}{\rho_B}$; $\rho_B = 40.8$ meters.

2-60

Particle A travels on a circular path with a constant speed of 6 ft/sec..
(a) Find the acceleration of A in the position shown knowing the vertical distance of 1.8 feet is increasing.
(b) Determine the horizontal and vertical components of acceleration in that position.

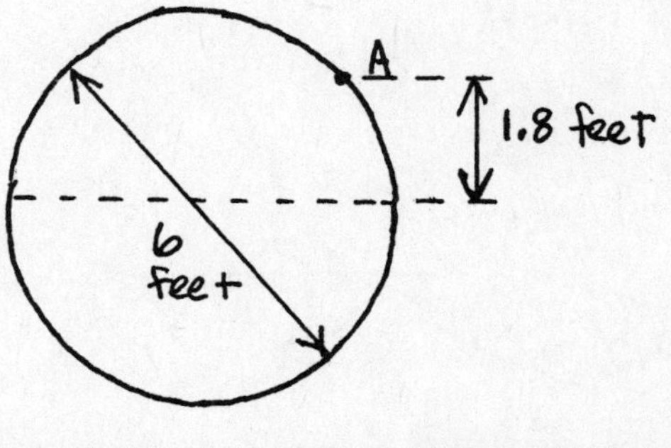

****

Since V = constant, $a_T = 0$

(a) $a = a_N = \frac{V^2}{r} = \frac{6^2}{3} = 12$ ft/sec^2

a_N is directed

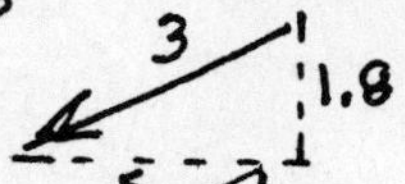

OR

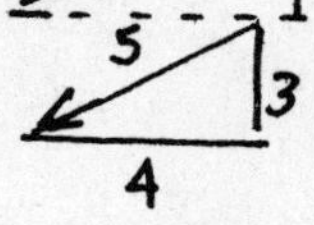

(b) $a_x = 4/5 \cdot 12 = 9.6$ ft/sec^2 ←

$a_y = 3/5 \cdot 12 = 7.2$ ft/sec^2 ↓

2-61

A particle is traveling upward along the path $y = x^2$. Find the acceleration and velocity at x = 5 ft., if $\dot{s} = 3$ ft/sec and $\ddot{s} = 7$ ft/sec^2 at this point. Express in both path components, $\hat{e}_T$ and $\hat{e}_N$, and in rectangular components, $\hat{i}.\hat{j}$.

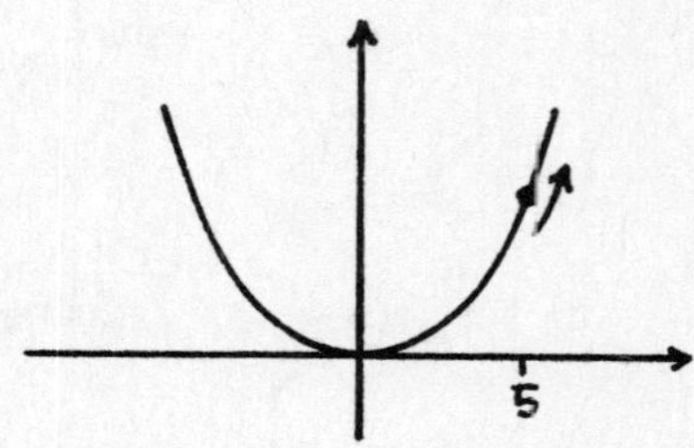

****

$$\underset{\sim}{v} = \dot{s}\,\hat{e}_T = 3\,\hat{e}_T$$

$$\underset{\sim}{a} = \ddot{s}\,\hat{e}_T + \frac{1}{R}(\dot{s})^2\,\hat{e}_N \qquad R = \frac{\left[1 + (dy/dx)^2\right]^{1.5}}{\left|d^2y/dx^2\right|}$$

$$R = \frac{[1 + 10^2]^{1.5}}{2} = 507.5 \text{ ft.} \qquad \frac{dy}{dx} = 2x = 2(5) = 10$$

$$\underset{\sim}{a} = 7\hat{e}_T + \frac{1}{507.5}(3)^2\hat{e}_N \qquad \frac{d^2y}{dx^2} = 2$$

$$\underset{\sim}{a} = 7\hat{e}_T + 0.01773\,\hat{e}_N$$

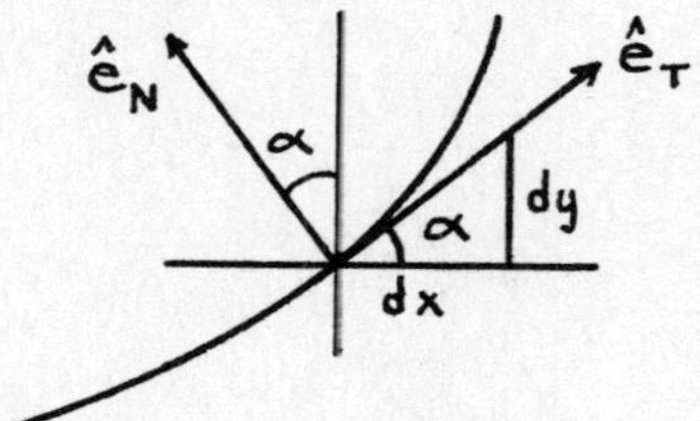

$$\tan\alpha = dy/dx = 10 \text{ at } x = 5 \qquad \therefore \alpha = 84.29°$$

$$\hat{e}_T = \cos\alpha\,\hat{i} + \sin\alpha\,\hat{j} = 0.09949\hat{i} + 0.9950\hat{j}$$

$$\hat{e}_N = -\sin\alpha\,\hat{i} + \cos\alpha\,\hat{j} = -0.9950\,\hat{i} + 0.09949\,\hat{j}$$

$$\underset{\sim}{v} = 3\,\hat{e}_T = 0.2985\,\hat{i} + 2.985\,\hat{j}$$

$$\underset{\sim}{a} = 7\,\hat{e}_T + 0.01773\,\hat{e}_N$$

$$\underset{\sim}{a} = 7[0.09949\hat{i} + 0.9950\hat{j}] + 0.01773[-0.9950\hat{i} + .09949\hat{j}]$$

$\underset{\sim}{a} = 0.6788\,\hat{i} + 6.967\,\hat{j}$

2-62

A point P moves in the x-y plane in a curved path. At the instant shown P has the x and y components of velocity and acceleration given. Find the radius of curvature of the path at this instant.

$a_y = 0$, $V_y = 8$ m/s, $V_x = 10$ m/s, P, $a_x = 6$ m/s

****

FIND VELOCITY (MAGNITUDE AND DIRECTION)

$$V = \sqrt{V_x^2 + V_y^2} = \sqrt{10^2 + 8^2} = \sqrt{164} = 12.81 \text{ m/s}$$

V, θ, 8, 10

$$\theta = \tan^{-1}\frac{8}{10} = 38.66^\circ$$

VELOCITY IS TANGENT TO THE PATH, AND SO IS THE TANGENTIAL ACCELERATION a_T, WHILE a_N, THE NORMAL ACCELERATION, IS PERPENDICULAR TO V AND a_T.

FIND a_N

a_T, $V = \sqrt{164}$, 38.66°, P, PATH, $a_x = a = 6$, a_N

$$a_N = 6 \sin 38.66^\circ$$

$$a_N = 3.748 \text{ m/s}^2$$

FROM $a_N = \dfrac{V^2}{\rho}$, $\rho = \dfrac{V^2}{a_N} = \dfrac{164}{3.748} = 43.8 \text{ m}$

2-63

A train leaves station S at time t=o and is traveling along the curved track shown in the figure. The distance s traveled by the train is given by $s=0.12\ t^2$, where s is measured in meters and the time t in seconds. After 10 seconds the train reaches joint B where the total acceleration a is known to be 0.25 m/s^2. Determine the radius of curvature ρ of the curved track at B.

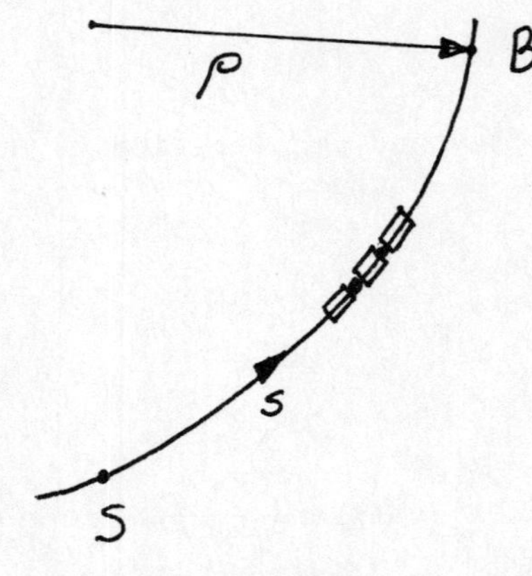

****

Speed of the train at B :

$$v = \frac{ds}{dt} = 0.24t = 0.24(10) = 2.4 \text{ m/s}$$

Tangential component of acceleration at B:

$$a_t = \frac{dv}{dt} = 0.24 \text{ m/s}^2$$

The normal component of acceleration :

$$a = \sqrt{a_n^2 + a_t^2} \quad , \text{ then}$$

$$a_n = \sqrt{a^2 - a_t^2} = \sqrt{(0.25)^2 - (0.24)^2} = 0.07 \text{ m/s}^2$$

The radius of curvature at B ;

$$a_n = \frac{v^2}{\rho} \quad ; \quad \rho = \frac{v^2}{a_n} = \frac{(2.4)^2}{0.07} = 82.29 \text{ m}$$

2-64

A bead is forced to travel along the curved path as shown at a constant speed of 8 ft/sec. Determine the acceleration of the bead at the point P where x = 2 ft.

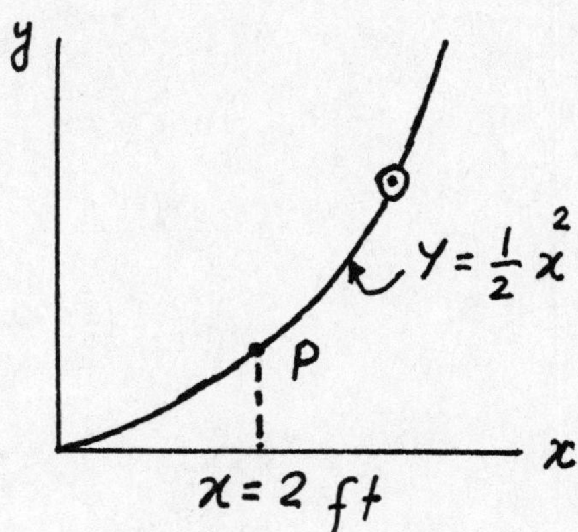

****

Acceleration at $P = \vec{a}_p = \frac{dv}{dt}\vec{u}_t + \frac{v^2}{\rho}\vec{u}_n$

Where $\vec{u}_t$, $\vec{u}_n$ are tangential and normal unit vectors, respectively.

V = constant $\frac{dv}{dt} = 0$, $\quad y = \frac{1}{2}x^2$, $\frac{dy}{dx} = x$

$\frac{d^2y}{dx^2} = 1$, Also, $\rho = \frac{\left[1 + \left(\frac{dy}{dx}\right)^2\right]^{3/2}}{d^2y/dx^2}$

or $\rho = \frac{\left[1 + (2)^2\right]^{3/2}}{1} = 11.18$ ft

Therefore, $\vec{a}_p = 0 + \frac{8 \times 8}{11.18}\vec{u}_n = \left\{5.7\,\vec{u}_n\right\} \frac{ft}{sec^2}$

2-65

A motorist enters a curve at a speed of 55 miles per hour. He applies his brakes and reduces his speed at a rate of 5 feet per second squared. The radius of the curve is 600 feet. What is his total acceleration when his speed is 45 miles per hour?

****

Using normal-tangential system

$$\bar{a} = \dot{V}\,\bar{e}_T + \frac{V^2}{r}\,\bar{e}_n$$

$$\dot{V} = -5 \text{ fps}^2 , \quad r = 600 \text{ ft}$$

$$V = 45 \text{ mph} \left(\frac{88 \text{ fps}}{60 \text{ mph}}\right) = 66 \text{ fps}$$

$$\bar{a} = -5\,\bar{e}_T + \frac{(66)^2}{600}\,\bar{e}_N$$

$$\boxed{\bar{a} = -5\,\bar{e}_T + 7.26\,\bar{e}_N \text{ fps}^2 \quad \text{or} \quad a = 8.82 \text{ fps}^2}$$

TWO-DIMENSIONAL MOTION: MOTION RELATIVE TO TRANSLATING REFERENCE FRAMES

2-66

Vehicle A starts its motion from rest at t = 0 with an acceleration of 10 ft/sec^2, when a second vehicle B passes it with a velocity of 20 ft/sec and an acceleration of 12 ft/sec^2. Determine the velocity of vehicle B relative to the velocity of vehicle A when t = 5 sec. Assume both vehicles are travelling in the same direction.

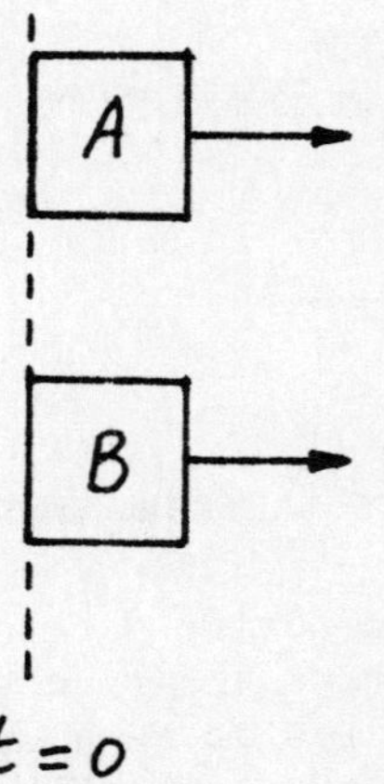

****

$$v_B = v_A + v_{B/A}$$

velocity of vehicle A $= v_A = (v_A)_0 + at = 10t$

velocity of vehicle B $= v_B = (v_B)_0 + at = 20 + 12t$

Relative velocity of vehicle B with respect to vehicle A at $t = 5$ sec, $= v_{B/A} = v_B - v_A$

or, $v_{B/A} = 20 + 12t - 10t = 20 + 2t = 30$ ft/sec

2-67

If rain drops fall vertically at the speed 5 m/s and a woman walks at the speed 2m/s, what is the best angle to hold the umbrella?

By what factor is the rain intensity increased when she walks, in comparison to if she were to stand still in the rain?

****

Velocity of rain relative to earth $V_{R/E} = 5$ m/s ↓

Velocity of woman relative to earth $V_{W/E} = 2$ m/s ←

$V_{R/E}$ $V_{R/W}$ θ $V_{W/E}$

But $\underline{V}_{R/E} = \underline{V}_{R/W} + \underline{V}_{W/E}$

Hence

velocity of rain relative to woman $\underline{V}_{R/W} = \underline{V}_{R/E} - \underline{V}_{W/E}$

From vector triangle $\tan\theta = \frac{5}{2} = 2.5$ or $\theta = 68.2°$

The best angle to hold the umbrella is in the direction of $\underline{V}_{R/W}$; $\underline{68.2°}$ to the horizontal.

Now, the intensity ratio of the rain $= \frac{V_{R/W}}{V_{R/E}} = \frac{1}{\sin\theta}$

$$= \frac{1}{\sin 68.2°} = \underline{1.08}$$

2-68

A man walks from point A at a speed of 4 mph. Three minutes later a man walks from point B at a speed of 3.5 mph. How far apart are the men after the first man has walked 4000 feet?

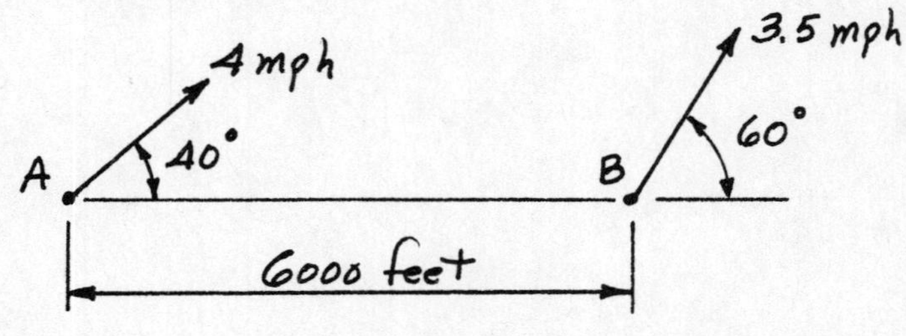

****

$$v_A = 4\,\frac{mi}{hr} \times \frac{1\,hr}{3600\,sec} \times \frac{5280\,ft}{mi} = 5.87\,ft/sec$$

$$v_B = 3.5\,\frac{mi}{hr} \times \frac{1\,hr}{3600\,sec} \times \frac{5280\,ft}{mi} = 5.13\,ft/sec$$

$$t_A = \frac{s_A}{v_A} = \frac{4000\,ft}{5.87\,ft/sec} = 681.4\,sec.$$

$$t_B = (681.4 - 180)\,sec = 501.4\,sec$$

$$s_B = v_B \cdot t_B = 5.13\,\frac{ft}{sec} \cdot 501.4\,sec = 2572.2\,ft.$$

$$s_B = s_A + s_{B/A}$$

<u>x-direction</u>: $2572.2\cos 60 + 6000 = 4000\cos 40 + (s_{B/A})_x$

$$(s_{B/A})_x = 4221.9\,ft.$$

<u>y-direction</u>: $2572.2\sin 60 = 4000\sin 40 + (s_{B/A})_y$

$$(s_{B/A})_y = -343.6\,ft.$$

$$s_{B/A} = \sqrt{(4221.9)^2 + (-343.6)^2} \qquad \underline{\underline{s_{B/A} = 4235.9\,ft.}}$$

2-69

Two particles are moving with the velocities as shown in the figure. What is the relative velocity between the particles ?

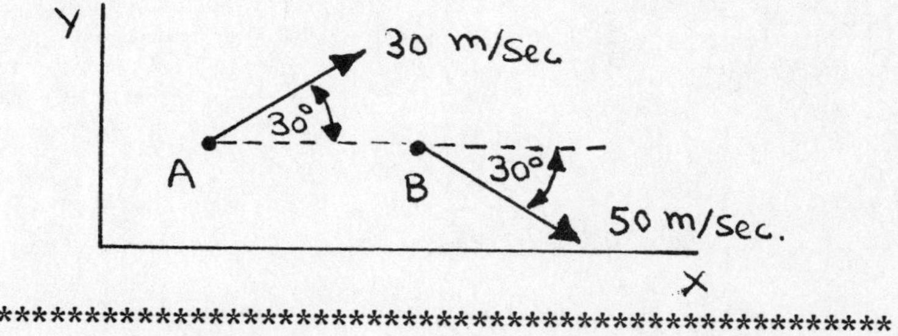

****

Relative velocity of A with respect to B

$$\underline{v_{AB}} = \underline{v_A} - \underline{v_B}$$

$$v_{AB_x} = v_{A_x} - v_{B_x}$$

$$= 30 \cos 30° - 50 \cos 30°$$

$$= -17.32 \text{ m/sec.}$$

v_{AB_x}, $-\underline{v_B}$, v_{AB_y}, θ, $\underline{v_{AB}}$, $\underline{v_A}$

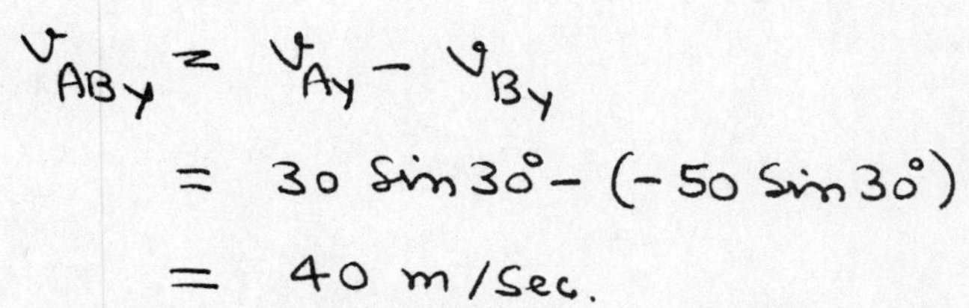

$$v_{AB_y} = v_{A_y} - v_{B_y}$$

$$= 30 \sin 30° - (-50 \sin 30°)$$

$$= 40 \text{ m/sec.}$$

$$v_{AB} = \sqrt{v_{AB_x}^2 + v_{AB_y}^2} = \sqrt{40^2 + (17.32)^2} = 43.59 \text{ m/sec.}$$

$$\tan\theta = \frac{v_{AB_x}}{v_{AB_y}} = \frac{17.32}{40} = 0.433 \,, \quad \theta = 23.4°$$

$v_{BA} = 43.59$ m/sec but opposite in direction

2-70

A pilot has a choice of flying to his destination 100 miles due east of his present position by flying his plane, capable of an air speed of 200 mph, at 3,000 feet where he faces a 20 mph headwind or at 6,000 feet where the wind velocity is 40 mph 20° east of north. What altitude should be chosen to make the best ground speed, what will this speed be, and what is the pilot's heading?

****

Clearly, in the first case, he would have a ground speed of 200 − 20 = 180 MPH.

In the second case, $\vec{V}_P = \vec{V}_{P/A} + \vec{V}_A$ ($\vec{V}_A$ = Velocity of the air)

($\vec{V}_P$: Velocity of the plane as measured on the ground)

$V_{P/A} = 200$, $40 = V_A$, α, β, $110°$, $20°$, V_P, $70°$

$$\therefore \frac{\sin 110°}{200} = \frac{\sin\alpha}{40} \Rightarrow \sin\alpha = \frac{40}{200}\sin 110° = .1879$$

$$\Rightarrow \alpha = 10.83°$$

$$\therefore \beta = 180° - 110° - 10.83°$$

$$\beta = 59.17°$$

$$\therefore \frac{\sin\beta}{V_P} = \frac{\sin 110°}{200}$$

$$\therefore V_P = 200\frac{\sin\beta}{\sin 110°} = 200\frac{\sin 59.17°}{\sin 110°}$$

$$V_P = 182.8 \text{ MPH}$$

Conclusion: The pilot should fly at 6,000 ft (2nd case), attaining a ground speed of 182.8 MPH, and the pilots heading would be 10.83° North of East.

2-71

A young couple of equal height device a game standing on the top of a ramp. The boy runs down the ramp at a constant speed of 10 m.p.h.; at the instant he has travelled 15 ft, the girl throws an apple horizontally off the side of the slope, which the boy catches further down the slope. Calculate (a) the time required by the boy B to catch the apple, (b) the speed at which the girl G throws the apple, and (c) the speed at which the apple reaches the boy's hand.

G v_i ; B ; 15 ft. ; v_B = 10 m.p.h. ; 35°

****

Solution

v_i ; y ; 35° ; x

As the boy and the girl are of equal height, assume their hands are also at equal height from the ground at the time of throwing and catching the apple.

$$10 \text{ mph} = \frac{(10 \text{ mph})(5280'/\text{mile})}{(3600 \text{ sec/hr})} = 14.666 \text{ ft/s.}$$

$\xrightarrow{+}$ $s = s_i + v_i t$

or $x = 0 + v_i t$ ...... (1)

$+\downarrow$ $s = s_i + v_i t + \frac{1}{2} a_c t^2$

$y = 0 + 0 + \frac{1}{2}(32.2 \text{ ft/s}^2)\, t^2$

$= (16.1 \text{ ft/s}^2)\, t^2$ .... (2)

$\frac{y}{x} = \tan 35° = 0.7$; $y = 0.7\,x$ .... (3)

Also, $s' = (15 \text{ ft}) + (14.666 \text{ ft})\, t$

or $x = [(15) + (14.666\, t)] \text{ ft} \cos 35°$

$= (12.285 + 12.011\, t)$ ft ...... (4)

and $y = 0.7[(12.285 + 12.011\, t)]$ ft

$= (8.599 + 8.408\, t)$ ft ...... (5)

(a) Time t: Substituting Eq. (5) in Eq. (2), we get

$$[(8.599) + (8.408\, t)] \text{ ft} = (16.1 \text{ ft/s}^2)\, t^2$$

or $16.1\, t^2 - 8.408\, t - 8.599 = 0$

Solving the quadratic equation, we have

$$t = \frac{-(-8.408) \pm \sqrt{(-8.408)^2 - \{4(16.1)(-8.599)\}}}{2\,(16.1)}$$

$= 1.037$ s <u>Ans</u>. (Ignoring the negative value of t)

(b) speed of throw: By substituting t in Eq. (4), we get

$$x = [(12.285) + (12.011)(1.037)] \text{ ft} = 24.740 \text{ ft}$$

and in Eq. (5), $y = (0.7)(24.740) \text{ ft} = 17.318 \text{ ft}$

Hence, for v_i, substituting x in Eq. (1), we get

$$24.740 \text{ ft} = v_i (1.037 \text{ s})$$

$$\text{or, } v_i = \frac{24.740 \text{ ft}}{1.037 \text{ s}} = 23.857 \text{ ft/s} \quad \underline{\text{Ans.}}$$

(c) speed of apple v_a reaching boy:

$$(v_a)_x = 23.857 \text{ ft/s} \longrightarrow$$

$$(v_a)_y = (v_a)_{y_i} + a_c t$$

$$= 0 + (32.2 \text{ ft/s}^2)(1.037 \text{ s}) = 33.391 \text{ ft/s} \downarrow$$

$$V_a = V_B + V_{a/B}$$

$$\text{or, } \underset{\longrightarrow}{(v_a)_x} + \underset{\downarrow}{(v_a)_y} = \underset{35^\circ}{V_B} + \underset{\longrightarrow}{(V_{a/B})_x} + \underset{\downarrow}{(V_{a/B})_y}$$

Considering the horizontal terms,

$$\overset{+}{\rightarrow} \; 23.857 \text{ ft/s} = 14.666 \text{ ft/s} \cos 35^\circ + (V_{a/B})_x$$

$$\text{or } (V_{a/B})_x = [(23.857) - (14.666)(0.819)] \text{ ft/s}$$

$$= 11.846 \text{ ft/s} \longrightarrow$$

Similarly, considering the vertical terms,

$$+\downarrow \; 33.391 \text{ ft/s} = 14.666 \text{ ft/s} \sin 35^\circ + (V_{a/B})_y$$

$$\text{or } (V_{a/B})_y = [(33.391) - (14.666)(0.573)] \text{ ft/s}$$

$$= 24.979 \text{ ft/s} \downarrow$$

$$\therefore \; V_{a/B} = \sqrt{(11.846 \text{ ft/s})^2 + (24.979 \text{ ft/s})^2}$$

$$= 27.645 \text{ ft/s} \quad \underline{\text{Ans.}}$$

2-72

A rocket is fired with an initial velocity of 700 m/sec and an angle of 50° to hit an escaping airplane.The airplane flies with a speed of 350 m/sec and an acceleration of 0 m/sec^2 at an altitude of 3000 m when the rocket is fired.Calculate the relative velocity and acceleration of the rocket with respect to the airplane when they hit.

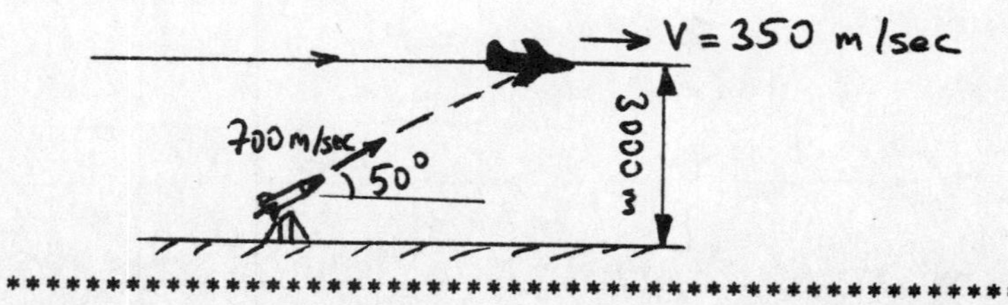

****

For rocket : $V_{0x} = V_0 . \cos\alpha = 700 \times \cos 50° = 449.95$ m/sec

$V_{0y} = V_0 . \sin\alpha = 700 \times \sin 50° = 536.23$ m/sec

at 3000 m : $V_y^2 = V_{y0}^2 - 2.g.\Delta y = 536.23^2 - 2 \times 9.81 \times 3000$

$V_y = 478.2$ m/sec , $V_x = V_{x0} = 449.95$ m/sec .

Relative velocity : $\vec{V}_R = \vec{V}_P + \vec{V}_{R/P}$

$\vec{V}_{R/P} = \vec{V}_R - \vec{V}_P = (449.95\vec{i} + 478.2\vec{j}) - 350\vec{i}$

$= 99.95\vec{i} + 478.2\vec{j}$

Relative acceleration : $\vec{a}_R = \vec{a}_P + \vec{a}_{R/P}$

$\vec{a}_{R/P} = \vec{a}_R - \vec{a}_P = -9.81\vec{j}$

THREE DIMENSIONAL MOTION RECTANGULAR CARTESIAN COORDINATES

2-73

The position vector of a particle is given by:

$$\underset{\sim}{r} = 3t^2 \underset{\sim}{i} + 2t \underset{\sim}{j} + (50 - t^2) \underset{\sim}{k} \qquad [r]: m \quad , [t]: sec$$

Determine the magnitudes of velocity and acceleration at t = 3 sec.

****

Solution :

$x = 3t^2$; $y = 2t$; $z = 50 - t^2$

$\dot{x} = 6t$; $\dot{y} = 2$; $\dot{z} = -2t$

$\ddot{x} = 6$; $\ddot{y} = 0$; $\ddot{z} = -2$

At $t = 3$ sec we have:

$$\dot{x} = 6(3) = 18 \text{ m/s} \; ; \quad \dot{y} = 2 \text{ m/s} \; ; \quad \dot{z} = -2(3) = -6 \text{ m/s}$$

$$\ddot{x} = 6 \text{ m/s}^2 \; ; \quad \ddot{y} = 0 \; ; \quad \ddot{z} = -2 \text{ m/s}^2$$

The velocity:

$$v = \sqrt{\dot{x}^2 + \dot{y}^2 + \dot{z}^2} = \sqrt{(18)^2 + (2)^2 + (-6)^2} = 19.08 \text{ m/s}$$

The acceleration:

$$a = \sqrt{\ddot{x}^2 + \ddot{y}^2 + \ddot{z}^2} = \sqrt{(6)^2 + (0)^2 + (-2)^2} = 6.32 \text{ m/s}^2$$

2-74

The motion of a particle is defined by the position vector $\vec{r} = t\,i + 2t^3 j + \sqrt{3}t^2 K$ (t in seconds, r in meters). Describe the surface this curve is on. (Find the equation of this surface. Find the path length as t varies from 0 to 1.)

****

By inspection $x = t$, $y = 2t^3$, $z = \sqrt{3}\, t^2$

We note $\dfrac{y}{x} = \dfrac{2t^3}{t} = 2t^2 = 2 \cdot \dfrac{z}{\sqrt{3}}$

or $y = \dfrac{2\sqrt{3}}{3} x z$ (this is a hyperbolic paraboloid)

$$ds = \sqrt{dx^2 + dy^2 + dz^2}$$

$$= \sqrt{(dt)^2 + (6t^2)^2 dt^2 + 12t^2 dt^2}$$

$$ds = \sqrt{1 + 36t^4 + 12t^2}\, dt = \sqrt{(6t^2+1)^2}\, dt$$

$$\therefore ds = (6t^2+1)\,dt \rightarrow S = \int ds = \int_0^1 (6t^2+1)\,dt = \frac{6t^3}{3} + t \Big]_0^1$$

$$\therefore S = 2t^3 + t\Big]_0^1 = 3 - 0 = 3 \text{ (meters)}$$

2-75

A ball is thrown with a velocity v_o = 13 m/s from the origin 0 of the coordinate system in the direction of A (3, 4, 12) in the gravitational field. Find the coordinates of the point $B(x_B, y_B)$ where it will land.

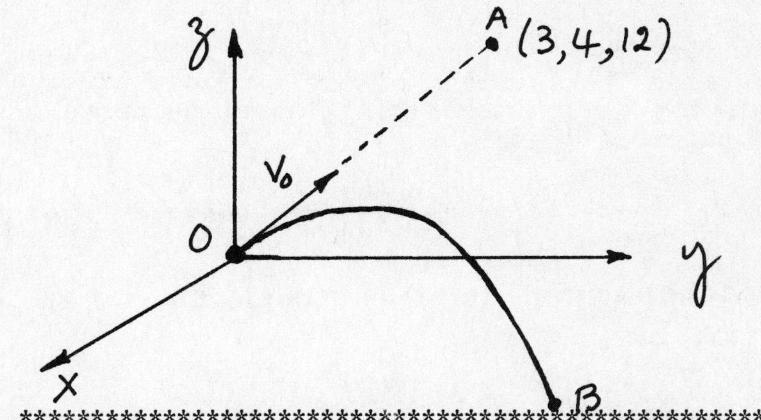

****

The direction cosines are $\cos\alpha = 3/OA$, $\cos\beta = 4/OA$, $\cos\gamma = 12/OA$, where $OA = 13$m

The velocities are $V_x = 3V_o/13$, $V_y = 4V_o/13$ and $V_z = \frac{12V_o}{13} - gt$. Integration gives the coordinates as $x = 3V_ot/13$, $y = 4V_ot/13$ $z = \frac{12V_ot}{13} - \frac{gt^2}{2}$. The ball hits the xy plane when $z = 0$. This happens when $t_B = \frac{24V_o}{13g}$ (excluding $t = 0$, when motion starts).

Substituting t_B into x and y yields $x_B = 72V_o^2/169g$, $y_B = 96V_o^2/169g$.

Using $g = 9.81$ m/s^2, and $V_o = 13$ m/s leads to $\underline{x_B = 7.34\text{m}}$, $\underline{y_B = 9.79\text{m}}$.

2-76

At t=0 a particle is at rest at the point (1,0,-2) when a force is applied to the particle to cause a time dependent acceleration of

$$\vec{a} = (3\,\vec{i} + \sin(2t)\,\vec{j} + \cos(2t)\,\vec{k})\ \text{in/sec}^2$$

a. Determine the velocity of the particle at t= π sec .
b. Determine the location of the particle at t= π sec .
c. Find an equation describing the space curve in which the particle travels.

****

a) THE ACCELERATION IS THE TIME DERIVATIVE OF THE VELOCITY. IF

$$\vec{v} = v_x\vec{i} + v_y\vec{j} + v_z\vec{k}$$

AND

$$\vec{a} = a_x\vec{i} + a_y\vec{j} + a_z\vec{k}$$

THEN

$$a_x = \frac{dv_x}{dt} \qquad a_y = \frac{dv_y}{dt} \qquad a_z = \frac{dv_z}{dt}$$

$$v_x = \int a_x\,dt \qquad v_y = \int a_y\,dt \qquad v_z = \int a_z\,dt$$

$$v_x = \int 3\,dt \qquad v_y = \int \sin 2t\,dt \qquad v_z = \int \cos 2t\,dt$$

$$V_x = 3t + C_x \qquad v_y = -\tfrac{1}{2}\cos 2t + C_y \qquad V_z = \tfrac{1}{2}\sin 2t + C_z$$

WHERE C_x, C_y, AND C_z ARE CONSTANTS OF INTEGRATION TO BE DETERMINED FROM THE INITIAL CONDITIONS

$$V_x = V_y = V_z = 0 \qquad \text{AT } t = 0$$

THUS

$$C_x = 0 \qquad C_y = \tfrac{1}{2} \qquad C_z = 0$$

AND

$$\vec{v} = 3t\,\vec{i} + \tfrac{1}{2}(1 - \cos 2t)\,\vec{j} + \tfrac{1}{2}\sin 2t\,\vec{k}$$

AT $t = \pi$ SEC $\qquad \vec{v} = 3\pi\,\vec{i}$ IN/SEC

b) THE VELOCITY IS THE TIME DERIVATIVE OF THE POSITION VECTOR. IF (x,y,z) DEFINES THE POSITION OF THE PARTICLE THEN

$$V_x = \frac{dx}{dt} \qquad V_y = \frac{dy}{dt} \qquad V_z = \frac{dz}{dt}$$

$$x = \int V_x\,dt \qquad y = \int V_y\,dt \qquad z = \int V_z\,dt$$

$x = \int 3t\,dt$ $\qquad y = \int \frac{1}{2}(1-\cos 2t)dt$ $\qquad z = \int \frac{1}{2}\sin 2t\,dt$

$x = \frac{3t^2}{2} + d_x$ $\qquad y = \frac{1}{2}t - \frac{1}{4}\sin 2t + d_y$ $\qquad z = -\frac{1}{4}\cos 2t + d_z$

WHERE d_x, d_y, AND d_z ARE CONSTANTS OF INTEGRATION TO BE DETERMINED FROM THE INITIAL CONDITIONS

$x = 1$ IN $\quad y = 0 \quad z = -2$ IN $\quad$ AT $t = 0$

THUS

$d_x = 1 \qquad d_y = 0 \qquad d_z = 1/4$

THUS

$$\vec{r}(t) = \left[\frac{3t^2}{2}\vec{i} + \left(\tfrac{1}{2}t - \tfrac{1}{4}\sin 2t\right)\vec{j} + \tfrac{1}{4}(1-\cos 2t)\vec{k}\right] \text{ IN}$$

AT $t = \pi$ SEC $\qquad \vec{r}(\pi) = \left(\frac{3\pi^2}{2}\vec{i} + \frac{1}{2}\pi\vec{j}\right)$ IN

c) TIME t CAN BE VIEWED AS A PARAMETER IN THE EQUATIONS

$x = \frac{3t^2}{2} \qquad y = \frac{1}{2}t - \frac{1}{4}\sin 2t \qquad z = \frac{1}{4}(1-\cos 2t)$

$t = \left(\frac{2x}{3}\right)^{1/2} \qquad \left[y - \frac{1}{2}\left(\frac{2x}{3}\right)^{1/2}\right] = -1/4 \sin 2t \qquad z - 1/4 = -1/4 \cos 2t$

SQUARING THE ABOVE EQUATIONS AND ADDING YIELDS

$$\left[y - \tfrac{1}{2}\left(\tfrac{2x}{3}\right)^{1/2}\right]^2 + \left(z - \tfrac{1}{4}\right)^2 = \frac{1}{16}$$

THE PARTICLE TRAVELS ON THE SURFACE DEFINED BY THE ABOVE EQUATION.

THREE DIMENSIONAL MOTION
TANGENTIAL AND NORMAL COMPONENTS

2-77

A particle travels along the surface of a cylinder with a constant velocity of 10 ft/sec. The path of the particle is at 45 degrees to the horinzontal plane. What is the normal and tangential acceleration experienced by the particle ?

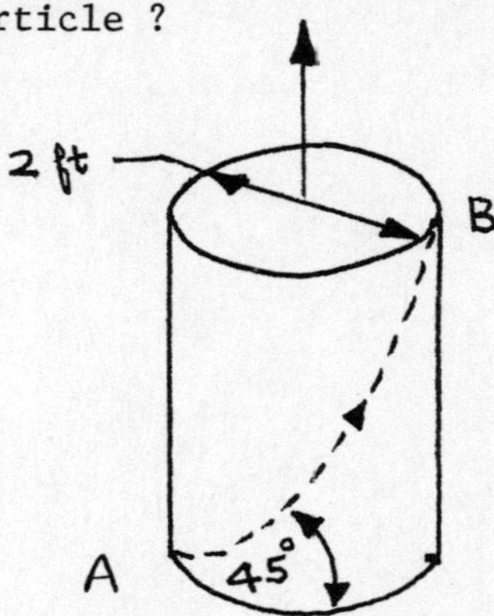

****

Effective radius of the path of the particle is $r = \frac{a}{\cos^2\alpha}$ $\quad a$ = cylinder rad.

α = Helix angle

$$r = \frac{1}{\cos^2 45^\circ} = 2 \text{ ft}$$

Normal acceleration of the particle

$$a_n = \frac{v^2}{r} = \frac{10^2}{2} = \frac{100}{2} = 50 \text{ ft/sec}^2$$

Normal acceleration is always directed towards the center axis of the cylinder.

Tangential acceleration $= \frac{dv}{dt} = 0$

(since v = const.)

2-78

A particle moves along a circular helical path at speed v. The distance from the axis of the helix to the path is a and the path makes an angle α with the plane perpendicular to the axis.

In terms of a, α, and v(t):

(a) what are the cylindrical coordinate components of velocity and acceleration?

(b) what are the tangential and normal components of acceleration/

(c) what is the radius of curvature of the helical path?

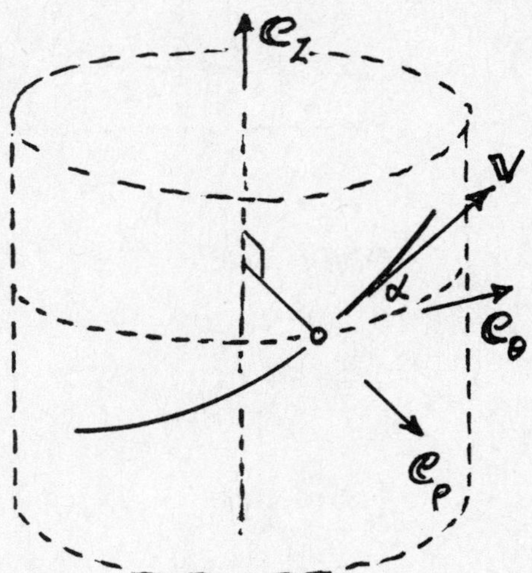

****

a) From the geometry of the diagram,

$$\mathbf{v} = v\cos\alpha\,\mathbf{e}_\theta + v\sin\alpha\,\mathbf{e}_z$$

Comparing with the cylindrical coordinate formulas for velocity,

$$\mathbf{v} = \dot{\rho}\,\mathbf{e}_\rho + \rho\dot{\phi}\,\mathbf{e}_\phi + \dot{z}\,\mathbf{e}_z$$

we find

$$\dot{\rho} = 0, \qquad \dot{\phi} = \frac{v\cos\alpha}{a}, \qquad \dot{z} = v\sin\alpha$$

Using these results in the cylindrical coordinate formulas for acceleration,

$$\mathbf{a} = (\ddot{\rho} - \rho\dot{\phi}^2)\mathbf{e}_\rho + (\rho\ddot{\phi} + 2\dot{\rho}\dot{\phi})\mathbf{e}_\phi + \ddot{z}\,\mathbf{e}_z$$

we find

$$\mathbf{a} = -a\left(\frac{v\cos\alpha}{a}\right)^2\mathbf{e}_\rho + \left(a\,\frac{\dot{v}\cos\alpha}{a}\right)\mathbf{e}_\phi + \dot{v}\sin\alpha\,\mathbf{e}_z$$

$$\mathbf{a} = -\frac{v^2\cos^2\alpha}{a}\mathbf{e}_\rho + \dot{v}\cos\alpha\,\mathbf{e}_\phi + \dot{v}\sin\alpha\,\mathbf{e}_z$$

b) From the diagram, the unit tangent vector is

$$\mathbf{e}_t = \mathbf{e}_\phi\cos\alpha + \mathbf{e}_z\sin\alpha$$

so that the above result may be rewritten as

$$\mathbf{a} = \dot{v}\,\mathbf{e}_t - \frac{v^2\cos^2\alpha}{a}\mathbf{e}_\rho$$

c)
Comparing with the general relationship for tangential and normal components,

$$\mathbf{a} = \dot{v}\,\mathbf{e}_t + \frac{v^2}{\rho_c}\mathbf{e}_n$$

we find that $\mathbf{e}_n = -\mathbf{e}_\rho$, and

$$\rho_c = \frac{a}{\cos^2\alpha}$$

2-79

A stunt driver in a motorcycle is travelling at a speed of 190 km/h in a circular track with a diameter of 500 m. If the driver accelerates the motorcycle in order to leave the track on a straight line some 25° beforehand so that $a_t = 2.5\ m/s^2$, calculate the magnitude and graphically indicate the direction of the net acceleration at that point.

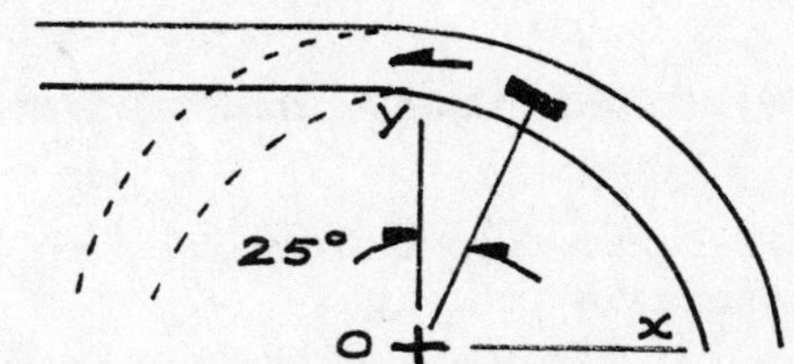

****

Solution

$$v = \frac{(190\ km/h)(1000\ m)}{(3600\ s)} = 52.777\ m/s$$

By expressing the given tangential acceleration in vector form, we get

$$a_t = 2.5\ m/s^2\,(-\cos 25^\circ i + \sin 25^\circ j)$$
$$= (-2.265\ i + 1.056\ j)\ m/s^2$$

The normal acceleration, acting towards the center O of the track, can be expressed as

$$a_n = \frac{v^2}{r}\ ;\ \text{and in vector form as}$$
$$a_n = \frac{(52.777\ m/s)^2}{(500/2\ m)}(-\cos 65^\circ i - \sin 65^\circ j)$$
$$= (-4.708\ i - 10.097\ j)\ m/s^2$$

Hence, the net acceleration in vector form is

$$a = a_t + a_n$$
$$= (-2.265\ i + 1.056 j) + (-4.708\ i - 10.097\ j)$$
$$= -6.973\ i - 9.041\ j\ \ m/s^2$$

and the magnitude is given by

$$a = \sqrt{(6.973)^2 + (9.041)^2}$$
$$= 11.417\ \ m/s^2 \quad \text{Ans.}$$

The direction of the net acceleration is shown in the diagram alongside.

THREE DIMENSIONAL MOTION: MOTION RELATIVE TO TRANSLATING REFERENCE FRAMES

2-80

Find the relative velocity between the two translating frames of references with velocities as shown in the figure.

$\underline{v}_B = -5i - 2j + 3k$

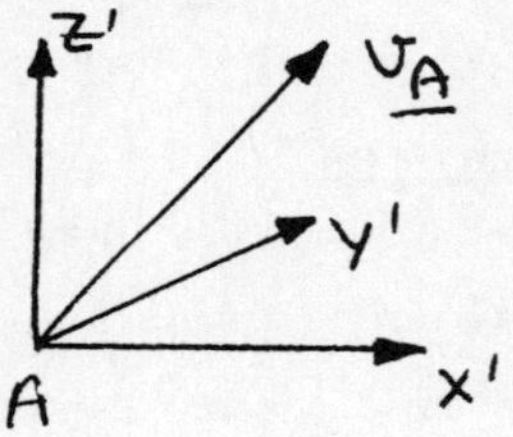

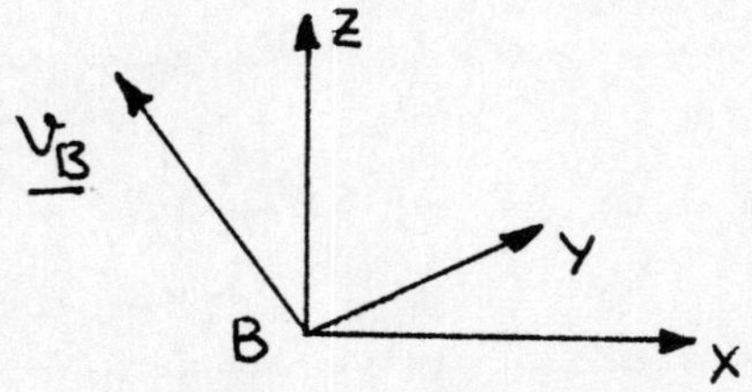

$\underline{v}_A = 2i + 4j + 6k$

****

Relative velocity of frame of reference A with respect to frame of reference B

$$\underline{v}_{AB} = \underline{v}_A - \underline{v}_B$$

$$\underline{v}_{AB} = (2i + 4j + 6k) - (-5i - 2j + 3k)$$

$$= 7i + 6j + 3k$$

$$v_{AB} = \sqrt{7^2 + 6^2 + 3^2} = 9.7 \text{ ft/sec.}$$

$v_{BA} =$ 9.7 ft/sec but opposite in direction.

3
KINEMATICS OF RIGID BODIES

ROTATION ABOUT A FIXED POINT

3-1

A rigid body rotates with a constant angular velocity $\omega = 3i + j - 2k$. Calculate (a) the speed of a particle P located on the body at the instant it passes through the point $4i + 4j$, and (b) the value of ω if the speed at point $4i + 4j$ is $9/\sqrt{14}$ m/s.

****

Solution

$$\vec{a} = \vec{3i} + \vec{j} - \vec{2k}$$

Hence, the unit vector of P parallel to the axis of rotation is given by

$$\frac{\vec{a}}{|\vec{a}|} = \frac{1}{\sqrt{14}}(3\vec{i} + \vec{j} - 2\vec{k})$$

(a) The position vector of P is given by

$$\vec{r} = \vec{4i} + \vec{4j}$$

Hence,

$$\vec{V} = \vec{\omega} \times \vec{r}$$

$$= \pm \frac{\omega}{\sqrt{14}} \begin{vmatrix} \vec{i} & \vec{j} & \vec{k} \\ 3 & 1 & -2 \\ 4 & 4 & 0 \end{vmatrix}$$

$$= \pm \frac{\omega}{\sqrt{14}} (\vec{8i} - \vec{8j} + \vec{8k})$$

$$\text{Magnitude of } V = \frac{\omega}{\sqrt{14}} \sqrt{8^2 + 8^2 + 8^2}$$

$$= 3.70\,\omega \text{ m/s} \quad \underline{\underline{\text{Ans.}}}$$

(b) For $\omega = 9/\sqrt{14}$ m/s, we get

$$V = (3.70)\, 9/\sqrt{14} = 8.899 \text{ m/s} \quad \underline{\underline{\text{Ans.}}}$$

3-2

Bar AB (when in the position shown) has an angular velocity of 10 rad/sec counterclockwise and an angular acceleration of 3 rad/sec^2 clockwise. Two seconds later, the angular acceleration changes to 4 rad/sec^2 clockwise and continues until the bar turns through a total angle of 34 radians.

Determine:

(a) The total time elapsed during the motion.

(b) The acceleration of point A at the end of the time (Total time).

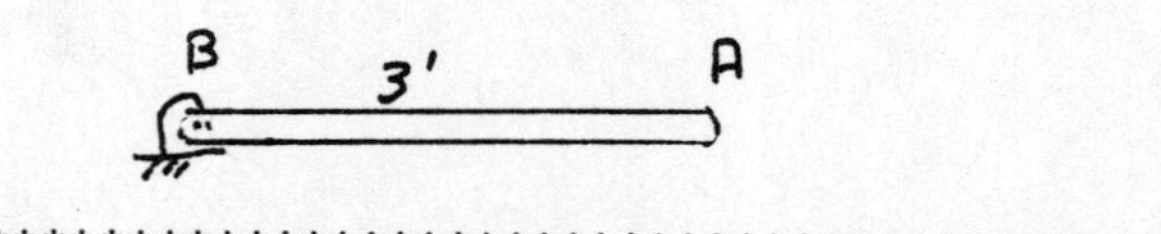

****

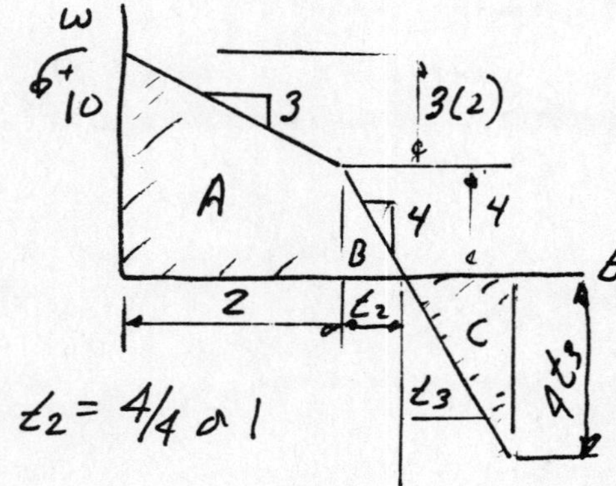

Draw ω-t Diagram as shown

Area A = 2(4) + ½ 2(6) or 14

Area B = ½(1)4 or 2

Total ∠ = Area (A + B + C) = 34

14 + 2 + C = 34   Area C = 18

$t_2 = 4/4$ or 1

$18 = \frac{1}{2} 4 t_3 t_3$ or $t_3 = 3$

total time = 2 + 1 + 3
= 6 Sec

$\omega_f = 4t_3$ or 12)

Acceleration at end of time is $r\alpha \rightarrow r\omega^2$

$3(4) \rightarrow 3(12)^2$

Since that only gives the magnitude we need to get position of bar at end of time.

Angular displacement is A + B − C (Areas)

or 14 + 2 − 18 or 2 rad)

2 rad = 114.6°

24.6°   114.6°

a = 3(4) $3(12)^2$ or 432.2 ft/sec^2 67°

3-3

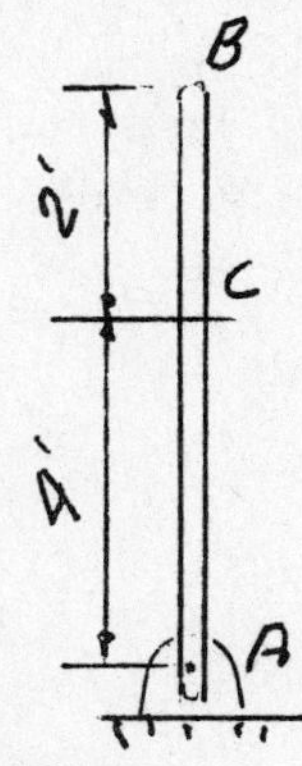

A member AB rotating in a plane has an initial angular velocity of 4 rad/sec clockwise. The angular velocity changes uniformly to 6 rad/sec counterclockwise, then it remains constant for 2 seconds. The total angle turned through is 25 radians. Determine (if the member is in the position shown 3 seconds after the velocity was 4):

(a) The acceleration of point B, 3 seconds after the angular velocity was 4 rad/sec.

(b) The acceleration of point C at the end of the time interval.

****

Using an ω-t diagram

Total ∠ turned = Total Area under ω-t

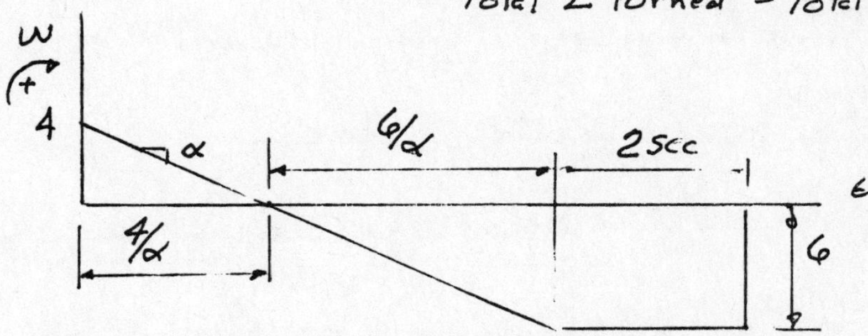

$\frac{1}{2}\,4\,\frac{4}{\alpha} + \frac{1}{2}\,6\,\frac{6}{\alpha} + 2(6) = 25$

$\alpha = \frac{8+18}{13}$ or $\alpha = 2\ rad/sec^2$ ↺

$a_B = \overleftarrow{r\alpha} \rightarrow r\omega^2\downarrow$ or $a_B = \overleftarrow{6(2)} \rightarrow 6(2)^2\downarrow$

$= \overleftarrow{12} \rightarrow 24\downarrow$

$a_B = 26.8\ ft/sec^2$ (direction: 1, 2)

Part b

Must Locate bar to get the direction at end of time.

Since position shown is 3 sec, find Area under curve after 3 Sec. (shown shaded)

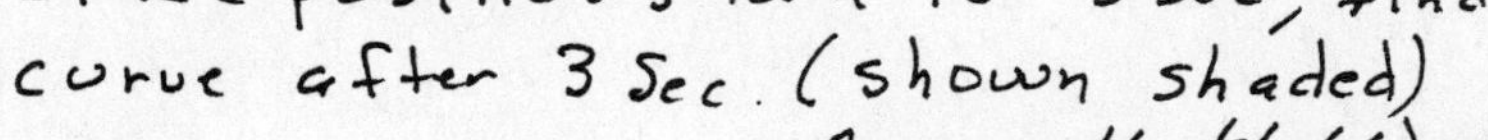

Area $= \frac{1}{2}\,\frac{6}{2}(6) - \frac{1}{2}(1)(2) + 2(6)$

$= 20$ rad ↻ or $1145.9°$ ↻

or 3 rev + 65.9°

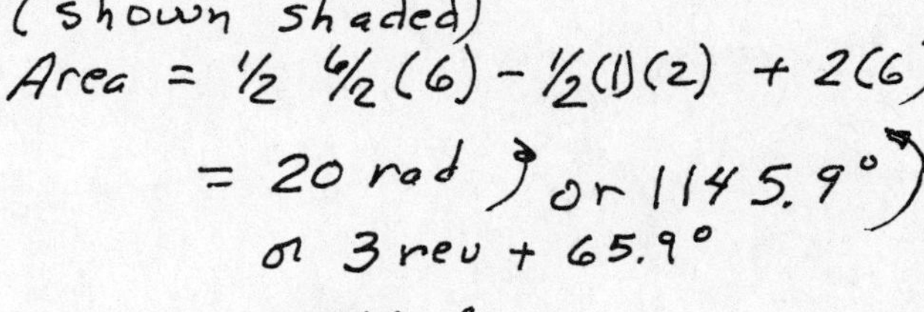

$a_{C_T} = 0$

$a_{C_N} = 4(6)^2$

$= 144\ ft/sec^2$

3-4

At the instant shown mass A has a velocity of 2 ft/s and an acceleration of 0.5 ft/s² both directed downward. Find the velocity and acceleration of points B and C located on the disk.

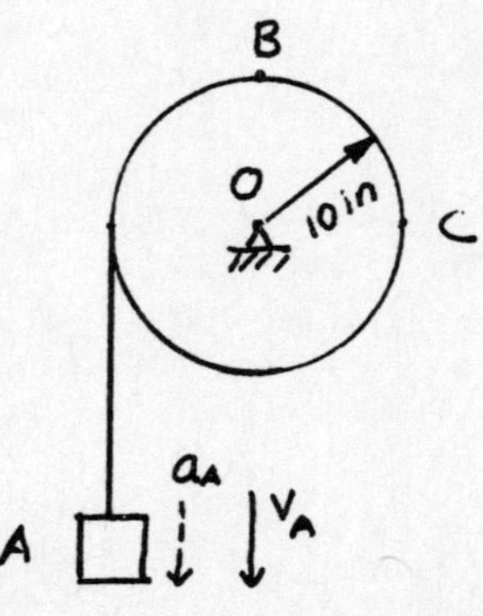

****

ω : angular velocity

α : angular acceleration

$v_A = 2\,ft/s = 24\ in/s \downarrow$

$a_A = 0.5\,ft/s = 6\ in/s^2 \downarrow$

$$\omega = \frac{v_A}{r_A} = \frac{24}{10} = 2.4\ rad/s$$

$$\alpha = \frac{(a_A)_t}{r_A} = \frac{6}{10} = 0.6\ rad/s^2$$

The velocities :

$v_B = \omega r_B = 2.4\,(10) = 24\ in/s \leftarrow$

$v_C = \omega r_C = 2.4\,(10) = 24\ in/s \uparrow$

The accelerations:

$$(a_B)_t = \alpha r_B = 0.6(10) = 6 \text{ in/s}^2 \leftarrow$$

$$(a_B)_n = \omega^2 r_B = (2.4)^2(10) = 57.6 \text{ in/s}^2 \downarrow$$

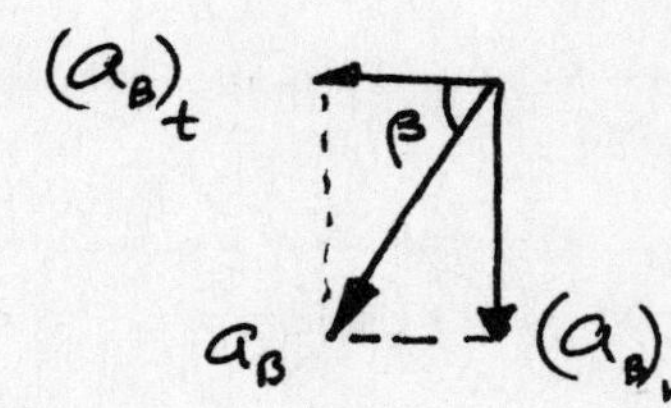

$$a_B = \sqrt{(a_B)_t^2 + (a_B)_n^2} = \sqrt{6^2 + (57.6)^2} = 57.91 \text{ in/s}^2 \quad 84^\circ$$

$$\tan\beta = \frac{(a_B)_n}{(a_B)_t} = \frac{57.6}{6} = 9.6 \quad \text{or} \quad \beta = 84^\circ$$

$$(a_c)_t = \alpha r_c = (0.6)(10) = 6 \text{ in/s}^2 \uparrow$$

$$(a_c)_n = \omega^2 r_c = (2.4)^2(10) = 57.6 \text{ in/s}^2 \leftarrow$$

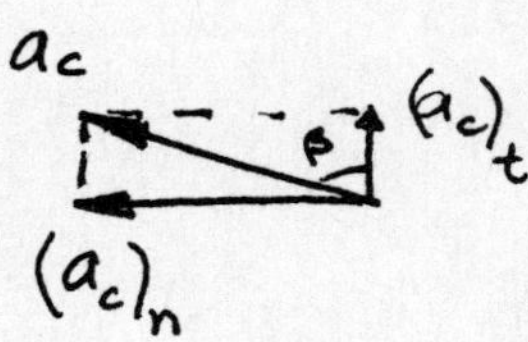

$$a_c = \sqrt{(a_c)_t^2 + (a_c)_n^2} = \sqrt{6^2 + (57.6)^2} = 57.91 \text{ in/s}^2 \quad 84^\circ$$

3-5

A fly wheel 2 feet in radius is brought uniformly from rest up to an angular velocity of 300 rpm in 30 sec. Find the velocity and the acceleration of a point p on the edge 2 seconds after starting from rest.

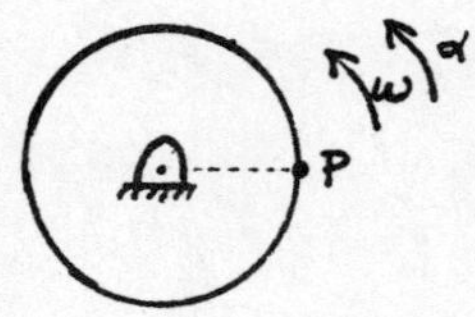

****

Since we have uniformly accelerated rotation:

$$\omega = \omega_0 + \alpha t \Rightarrow \alpha = \frac{\omega - \omega_0}{t} = \frac{300\,\frac{rev}{min}}{30\,sec} = \frac{300\frac{rev}{min}\cdot 2\pi\frac{rad}{rev}\cdot\frac{1}{60}\frac{min}{sec}}{30\,sec}$$

$$\alpha = \frac{2\pi}{30}\cdot\frac{300}{60}\,\frac{rad}{sec^2}$$

$$\alpha = \frac{\pi}{3}\,\frac{rad}{sec^2}\ \circlearrowleft$$

Now we can determine the angular velocity at any time t:

$\omega(t) = \omega_0 + \alpha t$, at $t = 2\,sec$

$\omega(2) = 0 + \frac{\pi}{3}(2) \cong 2.09\,\frac{rad}{sec}$

At this instant, $v_p = r\omega = 2(2.09)$

$\therefore \vec{V}_p = 4.18\,\frac{ft}{sec}\ \uparrow$

The magnitude of the normal acceleration of P is:

$$a_n = r\omega^2 = 2(2.09)^2 = 8.736\,\frac{ft}{sec^2},\ \therefore \vec{a}_n = 8.736 \leftarrow \left(\frac{ft}{sec^2}\right)$$

The magnitude of the tangential acceleration of P is:

$$a_t = r\alpha = 2\left(\frac{\pi}{3}\right) = 2.09\,\frac{ft}{sec^2},\ \therefore \vec{a}_t = 2.09 \uparrow \left(\frac{ft}{sec^2}\right)$$

The magnitude of the total acceleration is:

$$a = \sqrt{a_n^2 + a_t^2} = \sqrt{8.736^2 + (2.09)^2} \cong 8.99\,\frac{ft}{sec^2}$$

a, $a_t = 2.09$, O, α, $a_n = 8.736$

$$\tan\alpha = \frac{2.09}{8.736} = .2391$$

$$\Rightarrow \alpha = 13.45^\circ$$

$$\therefore \vec{a} = 8.99\,\frac{ft}{sec^2},\ \nwarrow 13.45^\circ$$

3-6

An electric motor drives pulley A at a speed of 600 rpm. The horizontal portion of the belt connecting pulleys B (8-inch dia) and C (6-inch dia) has a velocity of 30 ft/sec. The hub at D (2-inch dia) is securely connected to pulley B. Determine the diameter of pulley A. Assume no slippage in the belts.

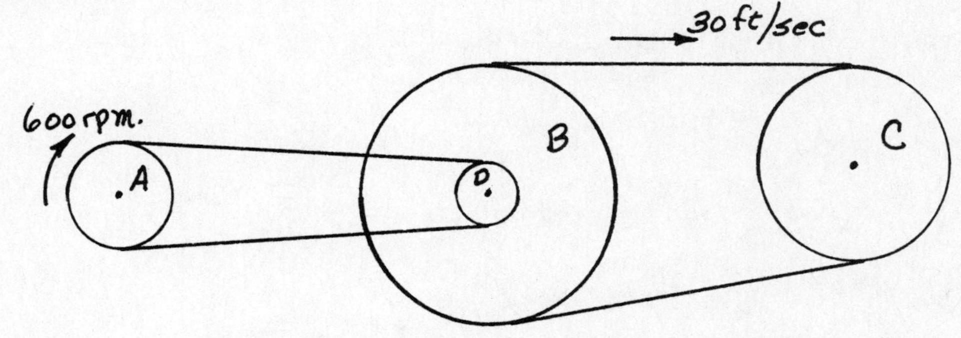

****

Pulley B: $V = r\omega \qquad \omega_B = \frac{V}{r} = \frac{30\ ft/sec}{\frac{4}{12}\ ft} = 90\ rad/sec.$

Hub D: $\omega_D = \omega_B = 90\ rad/sec$

$V_D = r\omega_D = (1\ in)(90\ rad/sec) = 90\ in/sec$

Pulley A: $V_A = V_D = 90\ in/sec$

$$r_A = \frac{V_A}{\omega_A} = \frac{90\ in/sec}{600\frac{rev}{min} \times \frac{1\ min}{60\ sec} \times \frac{2\pi\ rad}{rev}} = 1.43\ in.$$

$d_A = 2r_A = 2(1.43\ in) \qquad \underline{\underline{d_A = 2.86\ in.}}$

3-7

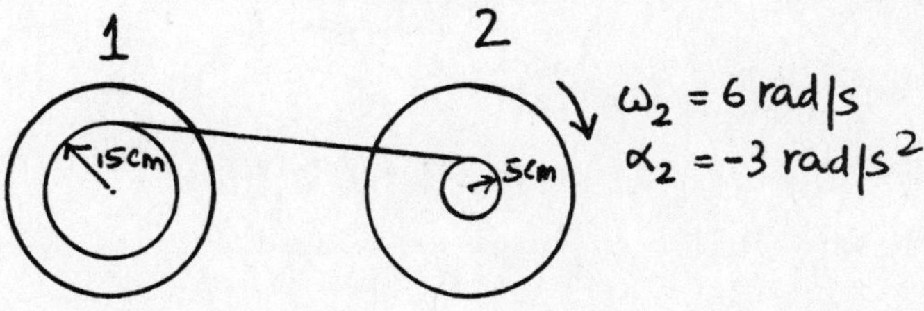

Schematic for a magnetic-tape-drive mechanism of a main-frame computer is shown in the figure. At a given instant spool radii in reels 1 and 2 are 15 cm and 5 cm respectively. If the angular speed ω_2 of reel 2 is 6 rad/s and the angular acceleration α_2 is -3 rad/s^2 in the same direction, determine the angular speed ω_1 and angular acceleration α_1 of reel 1.

****

Tape speed = radius x angular velocity = 5x6 cm/s

= 30 cm/s

Tape acceleration = radius x angular acceleration = 5x(-3) cm/s^2

= -15 cm/s^2

Hence

$$r_1\omega_1 = 30 \quad \text{and} \quad r_1\alpha_1 = -15$$

With $r_1 = 15$ cm we have

$$\omega_1 = \frac{30}{15} = \underline{2 \text{ rad/s}} \curvearrowright$$

$$\alpha_1 = -\frac{15}{15} = \underline{-1 \text{ rad/s}^2} \curvearrowright$$

3-8

A uniform and slender bar AB is rotating in a horizontal plane about the fixed hinge at A. At the instant given the angular velocity of the bar is 5 rad/s counterclockwise, and its angular acceleration is 4 rad/s^2 clockwise. Find the total acceleration of its end B.

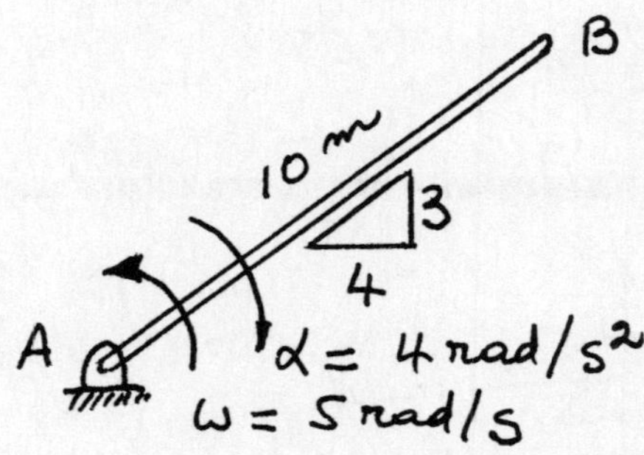

****

The normal acceleration is $a_n = \omega^2 r = (5)^2(10)$ or 250 m/s^2 pointing from B to A. The tangential acceleration is $a_t = \alpha r = (4)(10) = 40$ m/s^2 perpendicular to B in the downward direction.

$a = [(250)^2 + (40)^2]^{1/2} = 253.18\ m/s^2$ as shown in the vector diagram. The angle it makes with the rod is given by $\theta = \arctan\left(\frac{40}{250}\right)$ or $9.1°$ in the indicated direction.

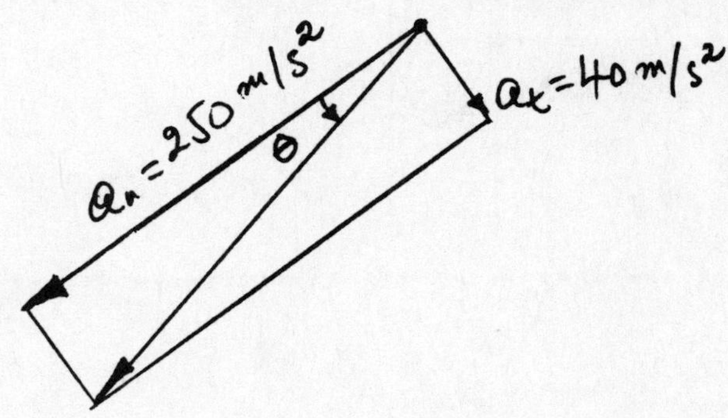

COMBINED TRANSLATION AND ROTATION

3-9

The wheel is rolling to the right without slipping. The relative velocity of point A with respect to point B has the direction:

a) → b) ↗ c) ↑ d) ↖ e) it is zero

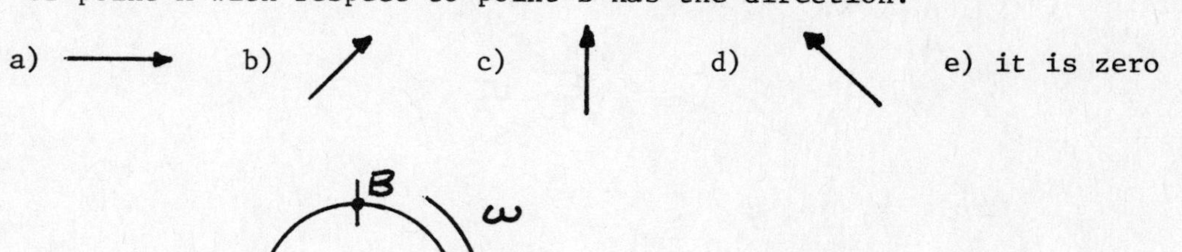

****

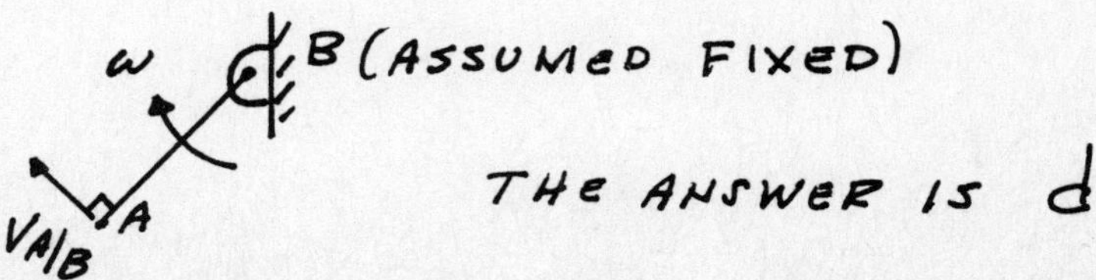

THE ANSWER IS d

3-10

The ends of bar AB are pin-connected to blocks moving in the guides shown. Point B has a velocity of V_B = 6 ft/sec↑ at the instant when AB is horizontal. Find the angular velocity of AB and the velocity of A.

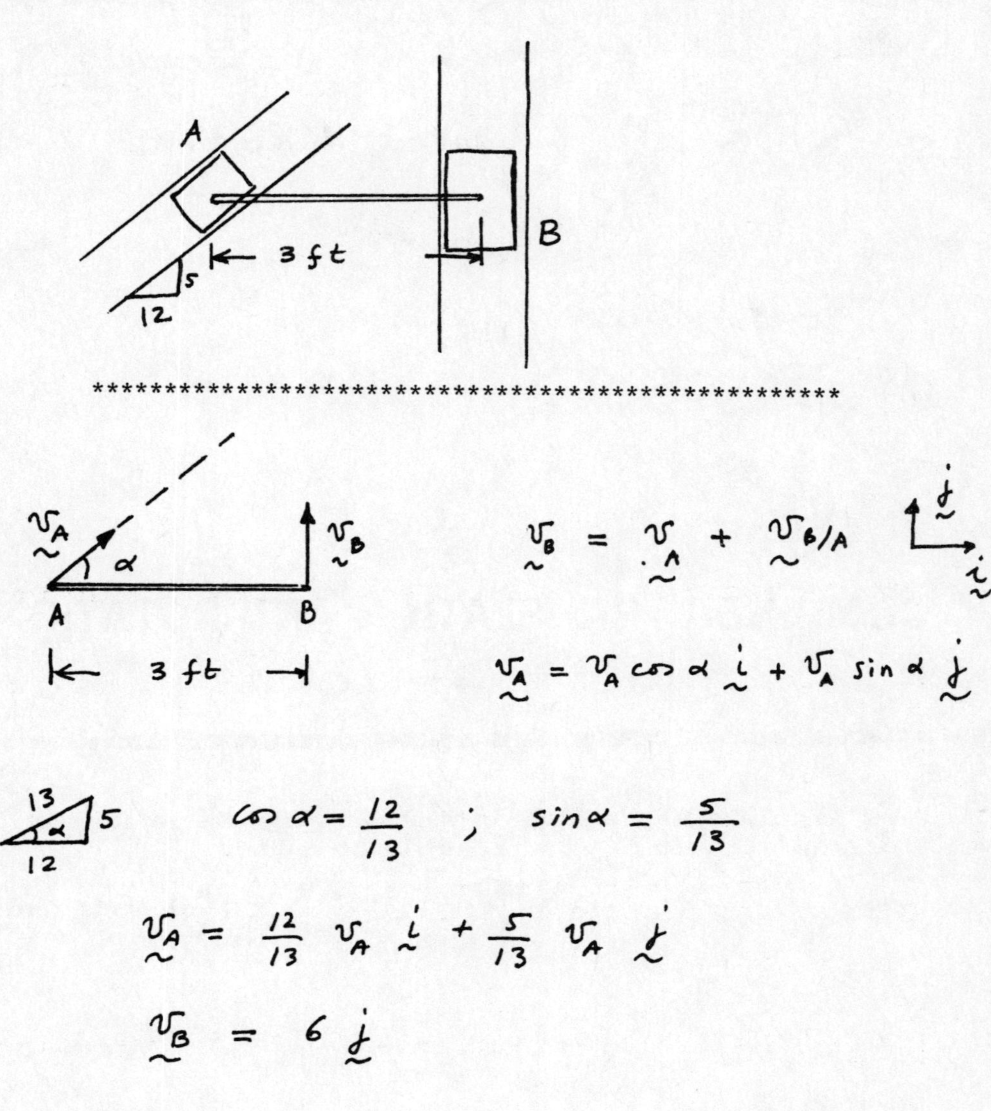

****

$$\underset{\sim}{v}_B = \underset{\sim}{v}_A + \underset{\sim}{v}_{B/A}$$

$$\underset{\sim}{v}_A = v_A \cos\alpha \, \underset{\sim}{i} + v_A \sin\alpha \, \underset{\sim}{j}$$

$$\cos\alpha = \frac{12}{13} \quad ; \quad \sin\alpha = \frac{5}{13}$$

$$\underset{\sim}{v}_A = \frac{12}{13} v_A \underset{\sim}{i} + \frac{5}{13} v_A \underset{\sim}{j}$$

$$\underset{\sim}{v}_B = 6 \underset{\sim}{j}$$

A, B, 3ft, ω, $\underset{\sim}{v}_{B/A}$

$$v_{B/A} = \omega r_{AB} = \omega(3)$$

$$\underset{\sim}{v}_{B/A} = 3\omega \underset{\sim}{j}$$

Substituting :

$$6\underset{\sim}{j} = \frac{12}{13} v_A \underset{\sim}{i} + \frac{15}{13} v_A \underset{\sim}{j} + 3\omega \underset{\sim}{j}$$

$$6\underset{\sim}{j} = \left(\frac{12}{13} v_A\right)\underset{\sim}{i} + \left(\frac{5}{13} v_A + 3\omega\right)\underset{\sim}{j}$$

Comparing both sides of the equation, we obtain :

$$\frac{12}{13} v_A = 0 \quad \text{and}$$

$$6 = \frac{5}{13} v_A + 3\omega$$

or $v_A = 0$ and $\omega = 2$ rad/s ↶ .

3-11

A thin disk is rolling without slip on a conveyor belt that is moving forward at a speed of 5 ft/sec. The disk is rolling with a constant angular velocity and it takes 5 seconds for its center to travel 40 ft. What is the angular velocity of the disk ?

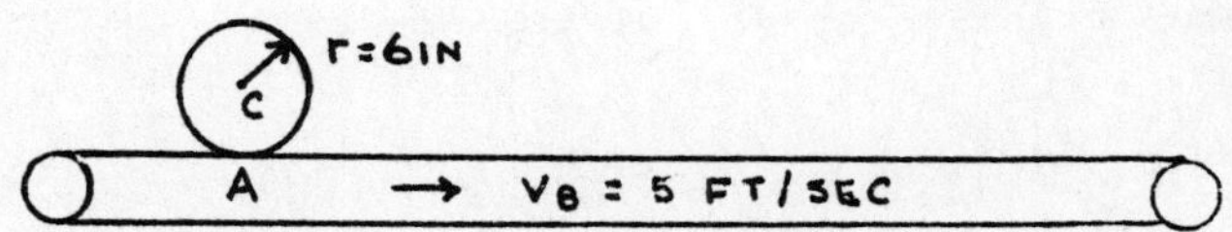

****

THE DISK ROLLS WITHOUT SLIP. THUS THE VELOCITY OF THE POINT ON THE DISK IN CONTACT WITH THE BELT HAS THE SAME VELOCITY AS THE BELT.

$$\vec{V}_A = 5 \text{ FT/SEC } \vec{i}.$$

THE VELOCITY OF THE CENTER OF THE DISK IS GIVEN BY THE RELATIVE VELOCITY EQUATION

$$\vec{V}_C = \vec{V}_A + \vec{\omega} \times \vec{r}_{C/A}$$

$$\vec{V}_C = (5 \text{ FT/SEC})\vec{i} + \omega\vec{K} \times \tfrac{1}{2} \text{ FT } \vec{j}$$

$$\vec{V}_C = (5 - \tfrac{1}{2}\omega)\vec{i} \text{ FT/SEC}$$

HOWEVER ω IS CONSTANT. THUS $\vec{V}_C$ IS CONSTANT AND THE DISTANCE TRAVELED BY THE CENTER OF THE DISK X_C IS

$$x_C = |\vec{V}_C|\, t$$

$$40 \text{ FT} = (5 - \tfrac{1}{2}\omega) \text{ FT/SEC } (5 \text{ SEC})$$

$$\omega = 6 \text{ RAD/SEC } \downarrow$$

3-12

The bar AB of the link as shown has a clockwise angular velocity of 20 rad/s when θ = π/3 rad. Compute the angular velocity of the bar BC and the velocity of the point C at this instant.

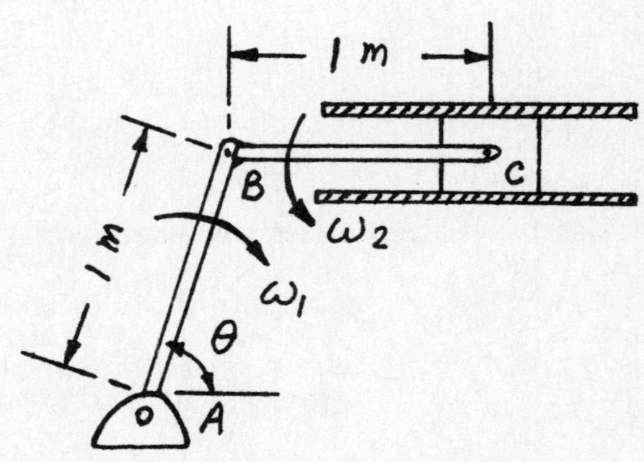

****

$$\vec{V}_B = \vec{V}_A + \vec{\omega}_1 \times \vec{r}_{B/A} = 0 + (-20\,\vec{k}) \times (\cos\tfrac{\pi}{3}\,\vec{i} + \sin\tfrac{\pi}{3}\,\vec{j})$$

$$\vec{V}_B = 17.32\,\vec{i} - 10\,\vec{j} \quad \text{-----------} \quad (i)$$

Also, $\vec{V}_B = \vec{V}_C + \omega_2 \vec{k} \times \vec{r}_{B/C} = V_C \vec{i} + \omega_2 \vec{k} \times (-\vec{i})$

Or, $\vec{V}_B = V_C\,\vec{i} - \omega_2\,\vec{j}$ ----------- (ii)

Comparing coefficients of respective unit vectors for equations (i) & (ii), one obtains

$$\omega_2 = 10 \text{ rad/s}\,; \quad V_C = 17.32 \text{ m/s}$$

3-13

The bar AB rotates with 20 rpm in the clockwise direction. Determine the velocity of the point C and the angular speed of the bar BC .

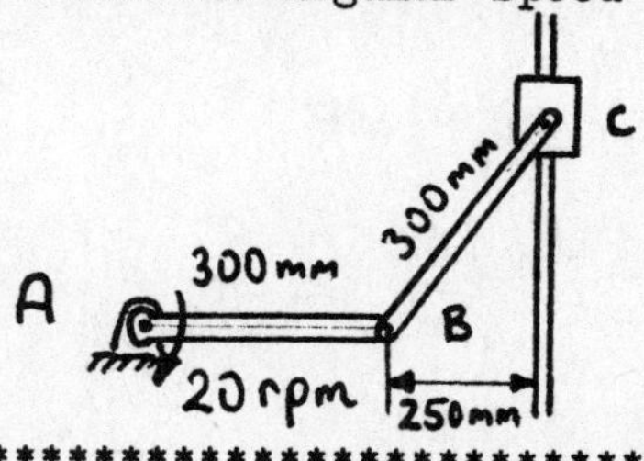

****

AB is rotating about the fixed axis at A ; angular velocity for AB

$$\omega_{AB} = \frac{2\pi n}{60} = \frac{2\pi . 20}{60} = \frac{2\pi}{3} \text{ rad/sec}$$

Velocity of the point B : $V_B = \overline{AB} . \omega_{AB} = 0.3 \times \frac{2\pi}{3} = 0.6283$ m/sec
(see figure below)

According to the chosen coordinate system x, y : $\vec{V}_B = -0.628 \vec{j}$

If we use the relative velocity concept (as an alternative to instantaneous center of rotation concept)

$\vec{V}_C = \vec{V}_B + \vec{V}_{C/B}$ where $|\vec{V}_{C/B}| = \overline{BC} . \omega_{BC}$

$\therefore \quad -V_C \vec{j} = -V_B \vec{j} - \overline{BC} . \omega_{BC} . \cos\phi \, \vec{i} + \overline{BC} . \omega_{BC} . \sin\phi \, \vec{j}$

$\vec{i}$: $\quad 0 = \overline{BC} . \omega_{BC} . \cos\phi \quad , \quad \overline{BC}, \phi \neq 0 \; , \therefore \; \omega_{BC} = 0$

$\vec{j}$: $\quad -V_C = -V_B \rightarrow V_C = V_B$

$\therefore \quad \vec{V}_C = \vec{V}_B = -0.6283 \vec{j}$ (m/sec)

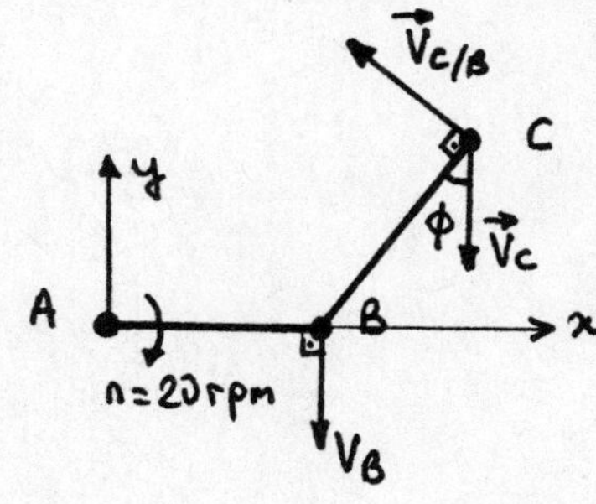

3-14

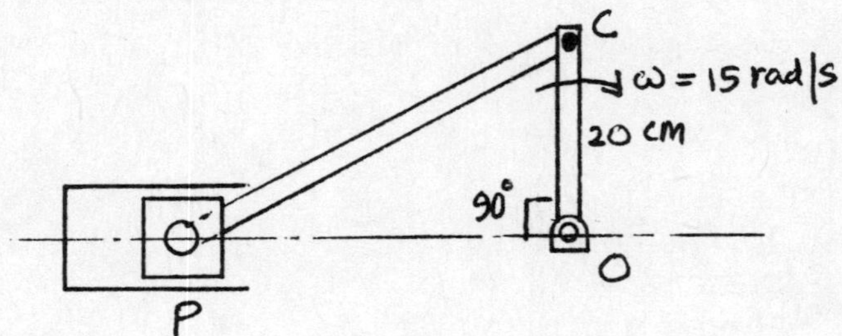

In the shown configuration of the piston-cylinder mechanism, crank OC rotates at angular velocity 15 rad/s. Determine the angular velocity of connecting rod PC and the velocity of piston P.

****

Velocity at C $\quad \underline{V}_C = \omega r = 15 \times 20 \text{ cm/s} \rightarrow$

$\qquad\qquad\qquad\qquad = 3 \text{ m/s} \rightarrow$

Now

$$\underline{V}_P = \underline{V}_C + \underline{V}_{P/C}$$

$\underline{V}_C$, $\underline{V}_{P/C}$, $\underline{V}_P$, α

$\alpha = 0$

velocity of P relative to C, $\underline{V}_{P/C}$ has to be perpendicular to PC because PC is a rigid rod. Since the piston velocity $\underline{V}_P$ is parallel to $\underline{V}_C$ in the given configuration, it follows from the triangle of velocities that

$$\underline{V}_{P/C} = \underline{0} \quad \text{and hence} \quad \underline{V}_P = \underline{V}_C$$

Also

$$V_{P/C} = \omega_{PC} \cdot PC$$

Hence

angular velocity of the connecting rod $\omega_{PC} = \underline{0}$

and

$$\underline{V}_P = \underline{3 \text{ m/s}} \rightarrow$$

3-15

A thin disk 1 m in diameter is connected at its center to an arm A, whose length is equal to the diameter of the disk. At the position shown, the disk is rotating about the y axis at ω_d = 10 rad/s and the arm is rotating about the z axis at ω_a = 15 t rad/s, where t is in seconds. Calculate the velocity of a point P on the periphry of the disk when t = 2 s.

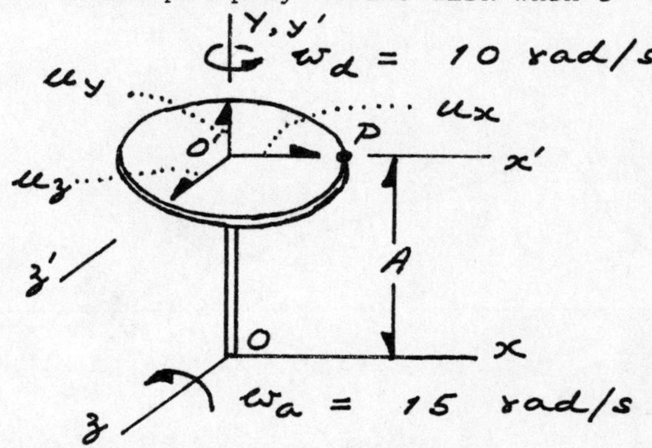

****

Solution

Let u_x, u_y and u_z be the moving unit vectors with O' attached to the center of the disk.

Then, absolute angular velocity ω of the moving unit vectors : $\omega = \omega_d u_y + \omega_a u_z$

Also, position vector R locating O' relative to O :

$$R = A \cdot u_y$$

and position vector $\bar{r}$ locating P relative to O':

$$\bar{r} = r \cdot u_x$$

Hence,

$$\dot{R} = \omega_a u_z \times A u_y = -\omega_a A u_x$$

$$\dot{\bar{r}}_r = 0$$

$$\omega \times \bar{r} = (\omega_d u_y + \omega_a u_z) \times (r\, u_x) = -\omega_d r u_z + \omega_a r u_y$$

$$\text{Velocity } \bar{V} = \dot{R} + \dot{\bar{r}}_r + \omega \times \bar{r} = -\omega_a A u_x + \omega_a r u_y - \omega_d r u_z$$

For the values given in the problem, at t = 2 s, we have

$$\bar{V} = \{(-15\times 2)(1\text{m})\, u_x\} + \{(15\times 2)(\tfrac{1}{2}\text{ m})\, u_y\} - \{(10 \text{ rad/s})(\tfrac{1}{2}\text{ m})\, u_z\}$$

$$= \{-30\, u_x + 15\, u_y - 5\, u_z\} \text{ m/s} \quad \underline{\text{Ans}}.$$

3-16

A piston P in a packaging operation is driven by a crank OC = 6 in via the connecting rod CP as shown in the accompanying illustration. For the position shown, when α = 35° and β = 27.5°, calculate (a) the velocity of the piston P, and (b) the angular velocity of the connecting rod CP if the crank is revolving at a speed of 1575 r.p.m. The connecting rod is 2.5 times as long as the crank.

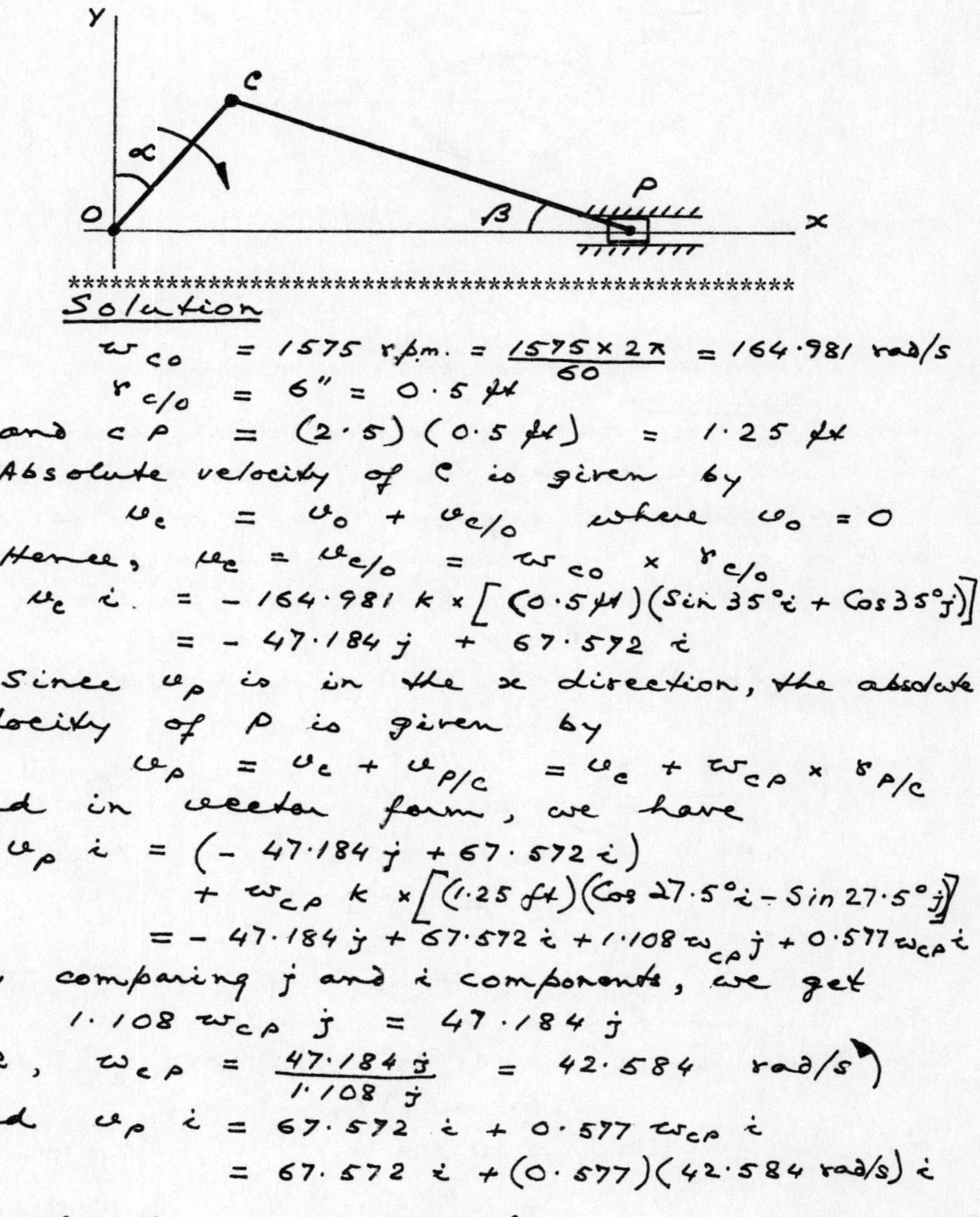

****

<u>Solution</u>

$$\omega_{CO} = 1575 \text{ r.p.m.} = \frac{1575 \times 2\pi}{60} = 164.981 \text{ rad/s}$$

$$r_{C/O} = 6'' = 0.5 \text{ ft}$$

and $$CP = (2.5)(0.5 \text{ ft}) = 1.25 \text{ ft}$$

Absolute velocity of C is given by

$$v_C = v_O + v_{C/O} \quad \text{where } v_O = 0$$

Hence, $$v_C = v_{C/O} = \omega_{CO} \times r_{C/O}$$

OR $$v_C\, i = -164.981\, k \times [(0.5 \text{ ft})(\sin 35°\, i + \cos 35°\, j)]$$

$$= -47.184\, j + 67.572\, i$$

Since v_P is in the x direction, the absolute velocity of P is given by

$$v_P = v_C + v_{P/C} = v_C + \omega_{CP} \times r_{P/C}$$

and in vector form, we have

$$v_P\, i = (-47.184\, j + 67.572\, i)$$
$$+ \omega_{CP}\, k \times [(1.25 \text{ ft})(\cos 27.5°\, i - \sin 27.5°\, j)]$$
$$= -47.184\, j + 67.572\, i + 1.108\, \omega_{CP}\, j + 0.577\, \omega_{CP}\, i$$

By comparing j and i components, we get

$$1.108\, \omega_{CP}\, j = 47.184\, j$$

OR, $$\omega_{CP} = \frac{47.184\, j}{1.108\, j} = 42.584 \text{ rad/s} \circlearrowright$$

and $$v_P\, i = 67.572\, i + 0.577\, \omega_{CP}\, i$$

$$= 67.572\, i + (0.577)(42.584 \text{ rad/s})\, i$$

OR $$v_P = 92.142 \text{ ft/s} \rightarrow$$ <u>Ans.</u>

3-17

Collar A has a velocity of 4 ft/sec to the right in the position shown and bar AB has a clockwise angular velocity of 6 rad/sec. Determine the angular velocity of bar BC and the velocity of collar C.

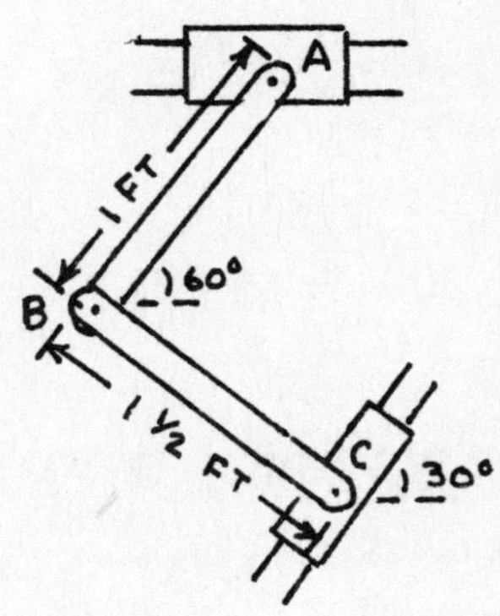

****

$\vec{V}_A = 4\vec{i}$ FT/SEC

$\vec{V}_B = \vec{V}_A + \vec{V}_{B/A} = \vec{V}_A + \vec{\omega}_{AB} \times \vec{r}_{B/A} = 4 \text{ FT/SEC } \vec{i} - 6 \text{ RAD/SEC } \vec{k} \times (-\tfrac{1}{2}\vec{i} - \tfrac{\sqrt{3}}{2}\vec{j}) \text{ FT}$

$= -1.20 \text{ FT/SEC } \vec{i} + 3 \text{ FT/SEC } \vec{j}$

$\vec{V}_C = V_C \tfrac{\sqrt{3}}{2}\vec{i} + \tfrac{V_C}{2}\vec{j} = \vec{V}_B + \vec{\omega}_{BC} \times \vec{r}_{C/B}$

$$\frac{\sqrt{3}}{2} V_C \vec{i} + \frac{1}{2} V_C \vec{j} = -1.20 \text{ FT/SEC } \vec{i} + 3 \text{ FT/SEC } \vec{j} + \omega_{BC} \vec{k} \times 1.5 \text{ FT} (\tfrac{\sqrt{3}}{2}\vec{i} + \tfrac{1}{2}\vec{j})$$

$$\frac{\sqrt{3}}{2} V_C \vec{i} + \frac{1}{2} V_C \vec{j} = (-1.20 \text{ FT/SEC} - .75\,\omega_{BC})\vec{i} + (3 \text{ FT/SEC} + 1.30\,\omega_{BC})\vec{j}$$

EQUATING $\vec{i}$ AND $\vec{j}$ COMPONENTS OF BOTH SIDES OF THE ABOVE EQUATION YIELDS THE FOLLOWING SCALAR EQUATIONS

$$\frac{\sqrt{3}}{2} V_C = -1.20 - 0.75\,\omega_{BC} \qquad (1)$$

$$\frac{1}{2} V_C = 3 + 1.30\,\omega_{BC} \qquad (2)$$

MULTIPLYING (2) BY $\sqrt{3}$ AND SUBTRACTING FROM (1) YIELDS

$$0 = -2.93 - 3.00\,\omega_{BC}$$

$$\omega_{BC} = -0.976 \text{ RAD/SEC}$$

THEN $V_C = 2(3 + 1.30(-0.976)) = 3.46 \text{ FT/SEC}$

3-18

A circular cylinder five feet in diameter is rolling towards the right without slipping on a horizontal floor. At a particular instant the center of the cylinder has a velocity of 4 ft/sec and an acceleration of 2 ft/sec^2. Determine the velocity and acceleration of point A.

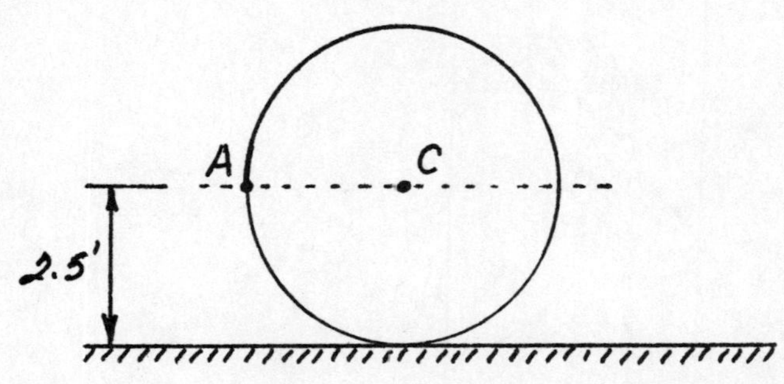

****

$$s = r\theta = a$$

$$\frac{ds}{dt} = \frac{da}{dt} = 4\ \text{ft/sec}$$

$$r\frac{d\theta}{dt} = r\omega = 4\ \text{ft/sec}$$

$$\omega = \frac{4\ \text{ft/sec}}{2.5\ \text{ft}} = 1.6\ \text{rad/sec}$$

$$\frac{d^2s}{dt^2} = \frac{d^2a}{dt^2} = 2\ \text{ft/sec}^2$$

$$r\frac{d^2\theta}{dt^2} = r\alpha = 2\ \text{ft/sec}^2 \qquad \alpha = \frac{2\ \text{ft/sec}^2}{2.5\ \text{ft}} = 0.8\ \text{rad/sec}^2$$

$$V_A = V_c + V_{c/A} = 4\frac{\text{ft}}{\text{sec}} + r\omega = \left(\overrightarrow{4\frac{\text{ft}}{\text{sec}}}\right) + \left(\uparrow 2.5\ \text{ft}\cdot\frac{1.6\ \text{rad}}{\text{sec}}\right)$$

V_A; 4 ft/sec; θ_1; 4 ft/sec

$$\underline{\underline{V_A = 5.66\ \text{ft/sec} \angle 45^\circ}}$$

$$a_A = a_c + a_{c/A} = \overrightarrow{2}\frac{\text{ft}}{\text{sec}^2} + (\overset{\uparrow}{a_t} + \overrightarrow{a_N}) = 2\frac{\text{ft}}{\text{sec}^2} + r\alpha + r\omega^2$$

$$a_A = 2\frac{\text{ft}}{\text{sec}^2} + 2.5\ \text{ft}\cdot\frac{0.8\ \text{rad}}{\text{sec}^2} + 2.5\ \text{ft}\cdot\left(\frac{1.6\ \text{rad}}{\text{sec}}\right)^2$$

$$a_A = \overrightarrow{2}\frac{\text{ft}}{\text{sec}^2} + \overset{\uparrow}{2}\frac{\text{ft}}{\text{sec}^2} + \overrightarrow{6.4}\frac{\text{ft}}{\text{sec}^2}$$

a_A; 2 ft/sec^2; θ_2; 8.4 ft/sec^2

$$\underline{\underline{a_A = 8.63\ \text{ft/sec}^2 \angle 13.4^\circ}}$$

3-19

Bar AC rotates counterclockwise with an angular velocity ω_0 = 5 rad/s. The end C slides on BD and makes the rod BD rotate about B. Find the velocity of the end D.

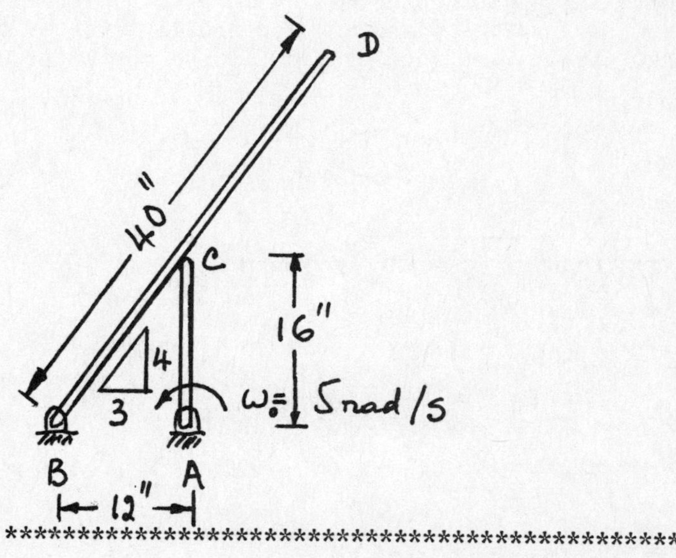

****

The absolute velocity of C is $V_C = \omega_0 \, AC$, or $(5)(16) = 80$ in/s.

V_D, D, 64 in/s, 80 in/s, C′, 48 in/s, B

Resolving the velocity of the point C′ of BD in contact with C at that instant, $(V_{C'})_\perp = 64$ in/s and $(V_{C'})_{\parallel} = 48$ in/s. Therefore

$$\omega_{BD} = \frac{64 \text{ in/s}}{BC'} = \frac{64}{20} = 3.2 \text{ rad/s} \circlearrowleft$$

$V_D = \omega_{BD} \, BD = (3.2)(40) = 128$ in/s $\perp$ to BD at D, and upward as shown. Note that $V_D = 2\,(V_{C'})_\perp$ because $BD = 2\,BC'$

$$\underline{\underline{V_D = 128 \text{ in/s}}} \quad 3\!\!\bigtriangleup\!\! 4$$

3-20

A drum of radius = 4 in. is mounted on a cylinder of 6 in. A cord is wound around the drum and its extremity D is pulled to the left at a constant velocity of 3 in/s causing the cylinder to roll without sliding. Determine: a) the angular velocity and angular acceleration of the cylinder and, b) the velocity and acceleration of the center of cylinder.

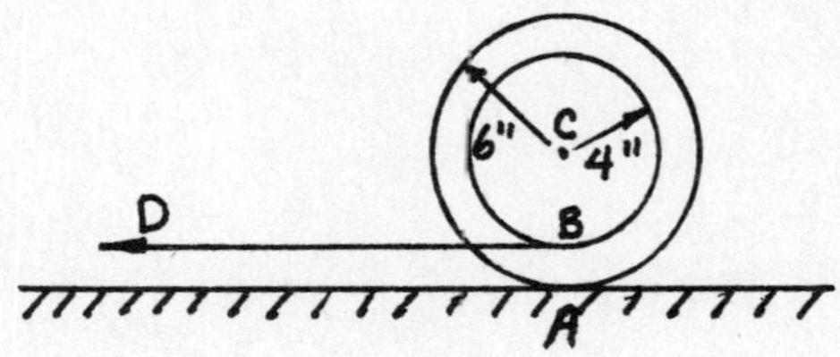

****

THE INSTANTANEOUS CENTER IS AT A

$V_B = 3$ in/s ← $\qquad V_B = r\omega \qquad \omega = \frac{3}{2} = 1.5 \frac{rad}{s}$ ↶

$V_c = r\omega = 1.5(6) = 9$ in/s ←

$a_c = r\alpha = 6\alpha \qquad \vec{a}_c = -6\alpha\vec{i}$

$\vec{a}_c = \vec{a}_B + \vec{a}_{c/B}$ But $a_{B_x}\vec{i} = 0$ SINCE V_B IS CONSTANT

$\vec{a}_c = a_{By}\vec{j} + \vec{a}_{c/B})_t + \vec{a}_{c/B})_n = a_{By}\vec{j} + \alpha\vec{k}\times\vec{r}_{c/B} - \omega^2\vec{r}_{c/B}$

$-6\alpha\vec{i} = a_{By}\vec{j} + \alpha\vec{k}\times(-4\vec{j}) - \omega^2(-4\vec{j})$

$-6\alpha\vec{i} = a_{By}\vec{j} + 4\alpha\vec{i} + 4\omega^2\vec{j}$

$-6\alpha = 4\alpha \quad , \quad -10\alpha = 0 \quad \alpha = 0$

AND $a_c = 6\alpha = 0$

$0 = a_{By} + 4\omega^2$

$a_{By} = -4(1.5)^2 = -9$ in/s² $= a_B$

INSTANTANEOUS CENTER OF ZERO VELOCITY

3-21

Using the Method of Instantaneous Centers, determine the velocities at points B and C for the figure shown if member BD is rotating at 4 radians per second clockwise at the instant shown.

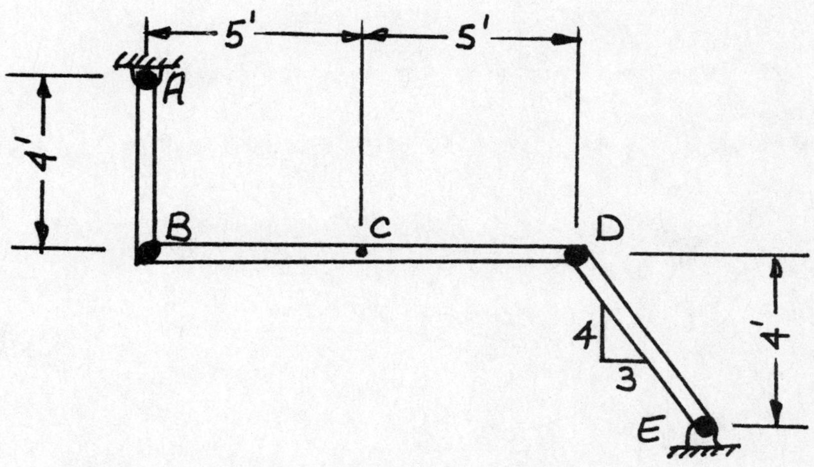

****

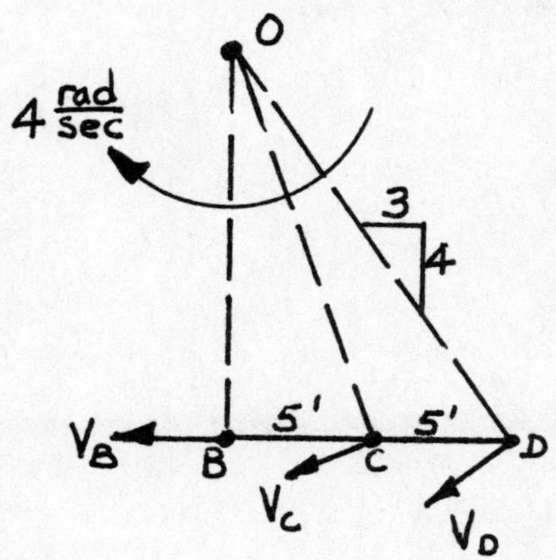

$$r_B = OB = \frac{4}{3}(10') = 13.33 \text{ FT}$$

$$r_C = OC = \sqrt{(13.33')^2 + (5')^2} = 14.24'$$

$$V = r\omega$$

$$V_B = r_B \omega_{BD} = (13.33 \text{ FT})\left(4 \frac{\text{rad}}{\text{sec}}\right) = 53.33 \frac{\text{FT}}{\text{SEC}} \leftarrow$$

$$V_C = r_C \omega_{BD} = (14.24 \text{ FT})\left(4 \frac{\text{rad}}{\text{sec}}\right) = 56.96 \frac{\text{FT}}{\text{SEC}} \quad \text{(slope } 5 : 13.33\text{)}$$

3-22

Determine the velocity of the collar B for the link AB as shown. The velocity of collar A is 20 ft/sec in the downward direction as indicated. The length of the link AB is 20 in.

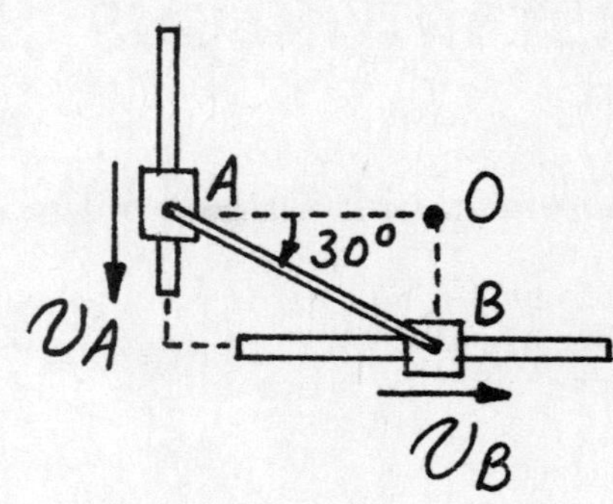

****

The instantaneous center of zero velocity can be located easily at O as shown. Then

$$v_A = r_{OA}\,\omega_{AB}, \qquad 20 = \frac{17.32}{12} \times \omega$$

or, $\omega = 13.86$ rad/sec

Also, $v_B = r_{OB}\,\omega_{AB} = \frac{10}{12} \times 13.86$

or, $v_B = 11.55$ ft/sec

3-23

Consider the rod AB leaning against a wall. If the velocity of B is $12\frac{m}{s}$ to the right, find (a) the velocity of A and (b) the angular velocity $\vec{W}_{AB}$.

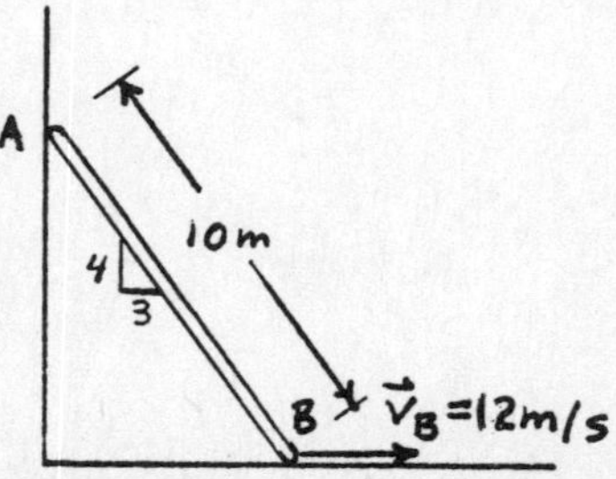

****

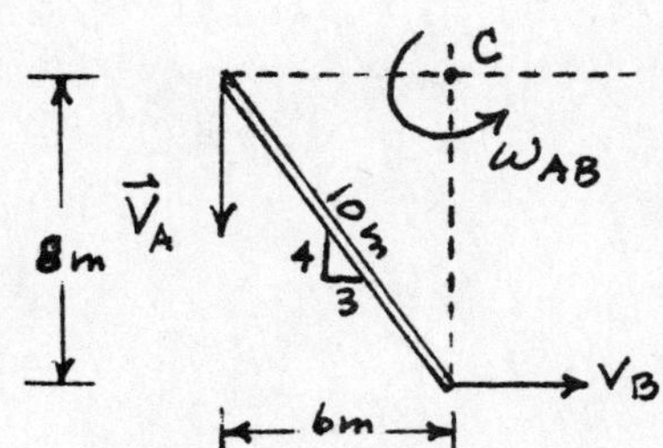

We construct normals to $\vec{V}_A$ and $\vec{V}_B$ and the result is the instantaneous center, labeled C. Now if we consider the rod as, at that instant, rotating about C, we have:

$$V_B = 12 = \omega_{AB}(8)$$

$$\Rightarrow \omega_{AB} = \frac{3}{2} \circlearrowleft \left(\frac{rad}{sec}\right) \quad \left[\text{or } \vec{\omega}_{AB} = \frac{3}{2}k\right]$$

$$\therefore \vec{V}_A = \omega_{AB}(6) = \frac{3}{2}(6) \downarrow$$

$$\therefore \vec{V}_A = 9\left(\frac{m}{sec}\right) \downarrow \quad \left[\text{or } \vec{V}_A = -9j\right]$$

3-24

The motion of the compound gear (two gears that are rigidly attached to each other) is controlled by the two racks which have the velocities shown. The pitch diameters of the gears are 160 and 96 mm. Find the angular velocity of the compound gear.

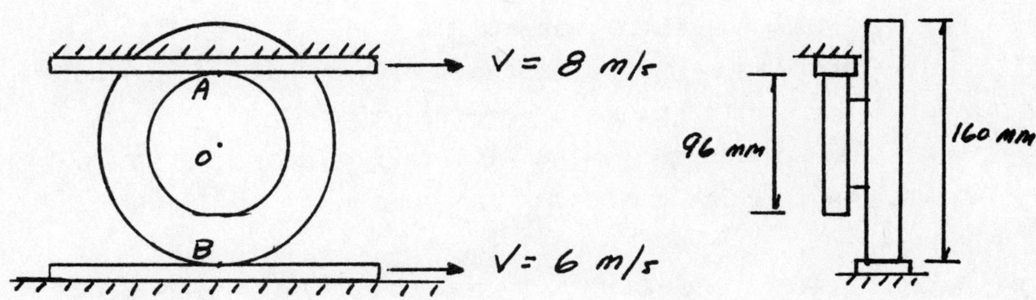

****

FIND C, THE INSTANT CENTER OF THE GEAR.

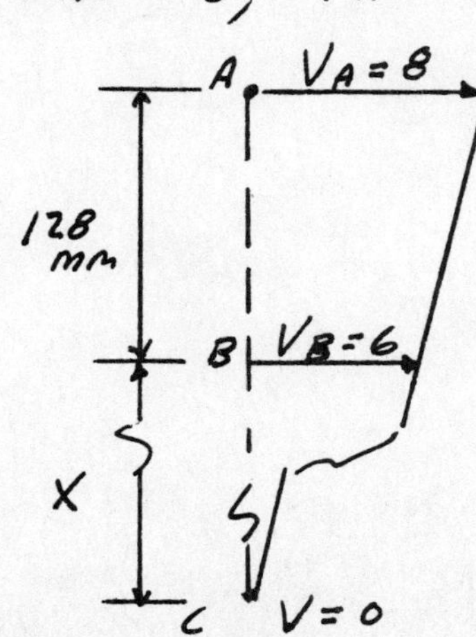

BY SIMILAR TRIANGLES:

$$\frac{X}{6} = \frac{X+128}{8} \Rightarrow X = 384 \text{ mm}$$

$$X = 0.384 \text{ m}$$

$$V_B = r_{CB}\,\omega$$

$$6 = 0.384\,\omega$$

$$\omega = 15.6 \text{ rad/s} \circlearrowright$$

3-25

In the position shown bar BC has an angular velocity of 3 rad/sec clockwise and bar DE has an angular velocity of 1 rad/sec clockwise. Use the instantaneous center method to calculate the angular velocities of bars AB and BC.

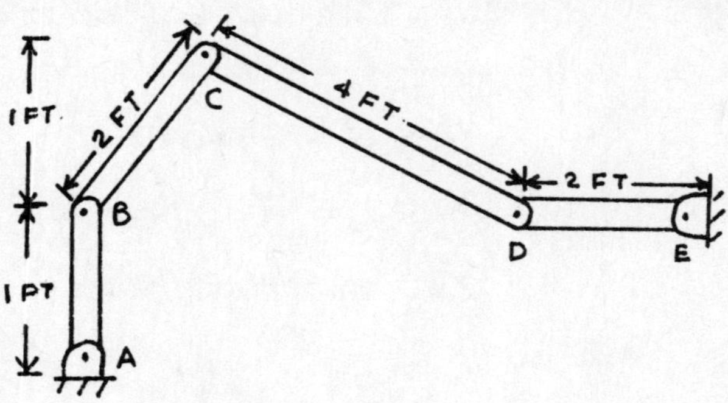

****

RECALL THE FOLLOWING FACTS ABOUT THE INSTANTANEOUS CENTER METHOD

1) A LINE DRAWN PERPENDICULAR TO THE VELOCITY VECTOR OF ANY POINT ON THE RIGID BODY PASSES THROUGH ITS INSTANTANEOUS CENTER.
2) THE DISTANCE BETWEEN A POINT ON A RIGID BODY AND THE BODY'S INSTANTANEOUS CENTER IS THE VELOCITY OF THE POINT DIVIDED BY THE ANGULAR VELOCITY OF THE ROD.

TO SOLVE THIS PROBLEM WE PROCEED AS FOLLOWS:

1) THE VELOCITY OF POINT D IS VERTICAL. THUS THE INSTANTANEOUS CENTER OF BAR CD IS ALONG A HORIZONTAL LINE FROM D.
2) THE VELOCITY OF POINT D IS $V_D = \ell_{DE}\,\omega_{DE}$

$$V_D = 2\text{ FT }(1\text{ RAD/SEC}) = 2\text{ FT/SEC}$$

THUS THE DISTANCE FROM THE INSTANTANEOUS CENTER OF BAR CD TO POINT D IS

$$z_{CD} = V_D/\omega_{CD} = \frac{2\text{ FT/SEC}}{3\text{ RAD/SEC}} = \frac{2}{3}\text{ FT}$$

3) THE DISTANCE FROM THE INSTANTANEOUS CENTER OF CD TO POINT C IS EASILY CALCULATED FROM GEOMETRY

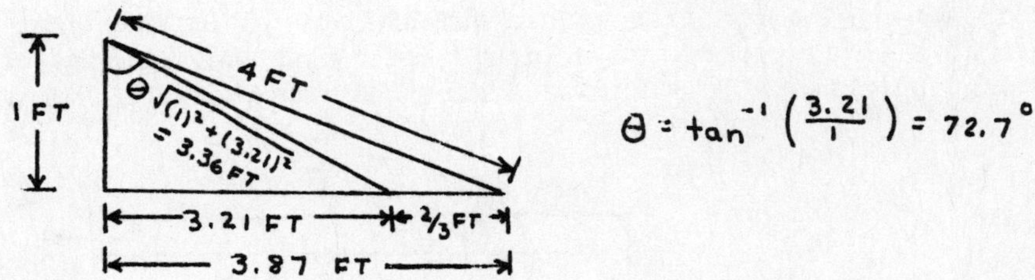

THUS $V_C = (3.36 \text{ FT})(3 \text{ RAD/SEC}) = 10.08 \text{ FT/SEC}$

4) THE VELOCITY OF POINT B IS VERTICAL. THUS THE INSTANTANEOUS CENTER OF BAR BC IS ALONG A VERTICAL LINE DRAWN FROM B.

5) THE VELOCITY OF C IS PERPENDICULAR TO THE LINE DRAWN FROM THE INSTANTANEOUS CENTER OF CD TO C. THUS THIS LINE MUST ALSO GO TO THE INSTANTANEOUS CENTER OF BAR BC. FROM GEOMETRY THIS DISTANCE IS DETERMINED BELOW TO BE 1.81 FT

THUS

$$\omega_{BC} = \frac{V_C}{Z_C} = \frac{10.08 \text{ FT/SEC}}{1.81 \text{ FT}} = 5.53 \text{ RAD/SEC}$$

$1.81 \cos 72.7° = 0.539$ FT

Z_C

72.7°

1.73 FT

C

1 FT

2 FT

B

$$1.73 = Z_C \sin 72.7° \rightarrow Z_C = 1.81 \text{ FT}$$

6) FINALLY $V_B = (1.539 \text{ FT})(5.53 \text{ RAD/SEC}) = 8.51 \text{ FT/SEC}$

THEN

$$\omega_{AB} = \frac{8.51 \text{ FT/SEC}}{1 \text{ FT}} = 8.51 \text{ RAD/SEC}$$

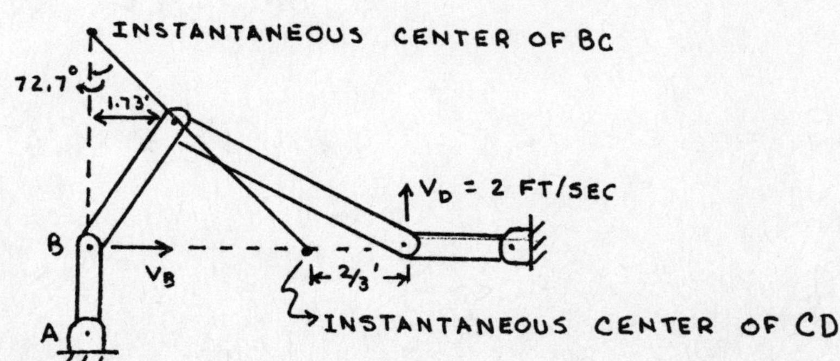

3-26

Bar A moves to the left with a velocity of 10 in/s and bar B moves to the right with a velocity of 6 in/s. Determine the velocity of point E at the position shown. No slipping occurs between the bars and the disk.

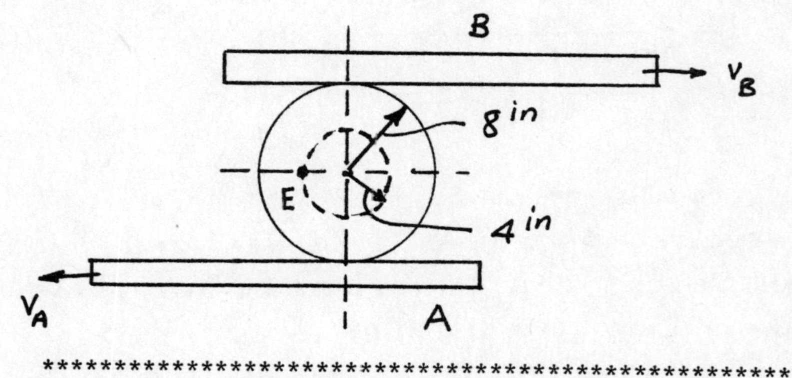

****

B $\quad v_B = 6$ in/s

$v_A = 10$ in/s $\quad$ A

C is the instant center

$$\omega = \frac{v_B}{BC} = \frac{v_A}{AC}$$

$$\frac{6}{e} = \frac{10}{16-e} \quad \text{or} \quad e = 6 \text{ in}.$$

$$\omega = \frac{6}{6} = 1 \text{ rad/s}$$

$$CE = \sqrt{\overline{OE}^2 + \overline{OC}^2} \quad ; \quad OE = 4 \text{ in}, \quad OC = 8 - e = 2 \text{ in}$$

$$CE = \sqrt{4^2 + 2^2} = 4.47 \text{ in}$$

$$v_E = CE(\omega) = 4.47(1) = 4.47 \text{ in/s}.$$

v_E is $\perp$ to CE.

3-27

A crank OC rotating with a clockwise angular velocity ω_0 rad/s drives a rod AB, such that the end A of the rod is constrained to move along the horizontal passing through O. Find the absolute velocity of the end B corresponding to the position shown in terms of ω_0 and a where a has length units.

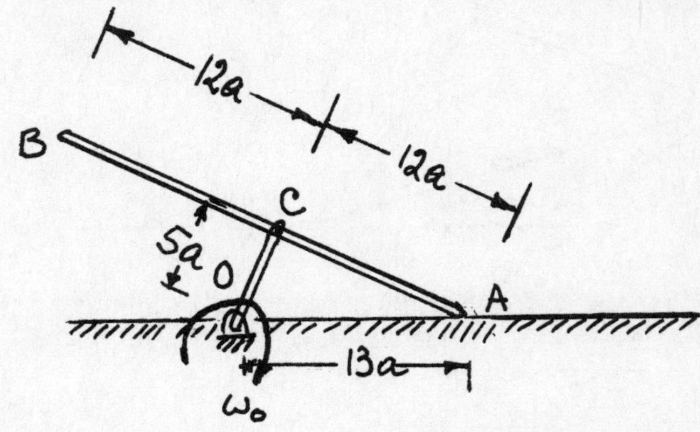

****

The instant center of AB is at the intersection of a ⊥ to OA at A, and the extension of OC.

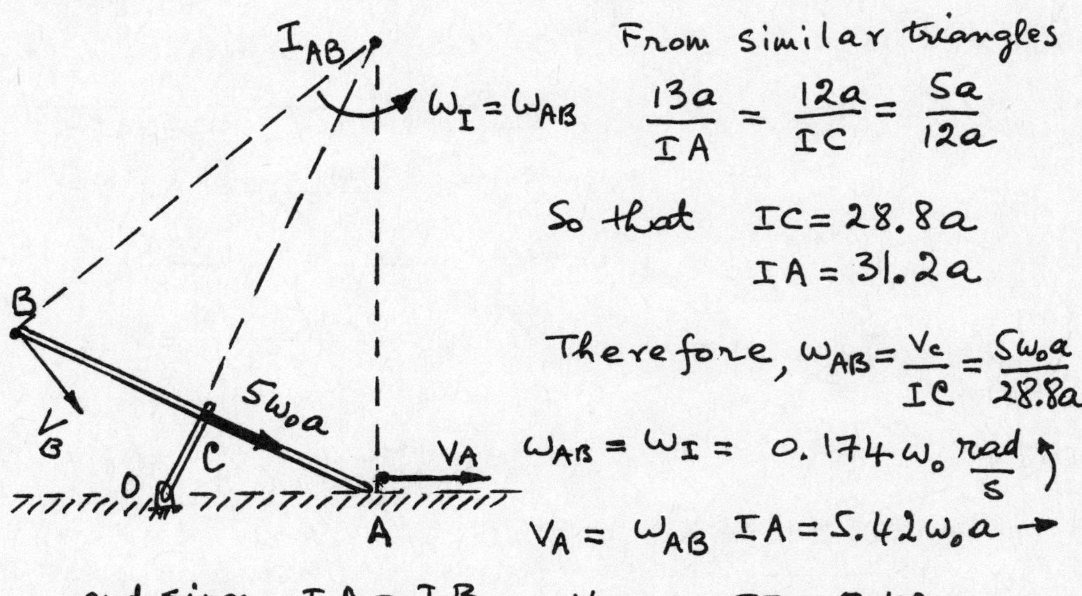

From similar triangles

$$\frac{13a}{IA} = \frac{12a}{IC} = \frac{5a}{12a}$$

So that $IC = 28.8a$

$IA = 31.2a$

Therefore, $\omega_{AB} = \frac{V_c}{IC} = \frac{5\omega_0 a}{28.8a}$

$\omega_{AB} = \omega_I = 0.174\,\omega_0 \frac{\text{rad}}{\text{s}}$ ↻

$V_A = \omega_{AB}\, IA = 5.42\,\omega_0 a$ →

and since $IA = IB$, $V_B = \omega_B\, IB = 5.42\,\omega_0 a$ perpendicular to IB at B as shown.

$$\underline{\underline{V_B = 5.42\,\omega_0 a}} \text{ length/s.}$$

If a is in m, the answer would be in m/s, if in ft it would be in ft/s.

3-28

In the illustration shown, if point A travels at a speed of 15 m/s to the left find the magnitude and velocity of point B by the method of instantaneous center of zero velocity. Neglect all friction along the surfaces.

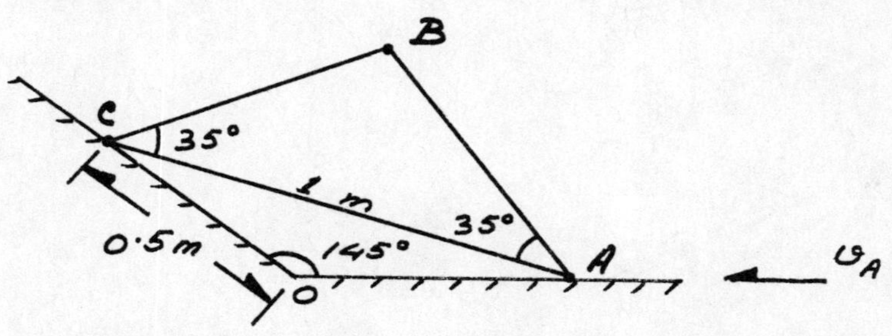

****

<u>Solution</u>

From trignometry, we get

$$\alpha = \sin^{-1}\left(\frac{0.5\,m}{1.0\,m}\sin 145°\right) = 16.665°$$

$$\beta = 90° - 35° - 16.665° = 38.335°$$

$$\theta = 180° - 16.665° - 145° = 18.335°$$

$$\gamma = 16.665° + 18.335° = 35°$$

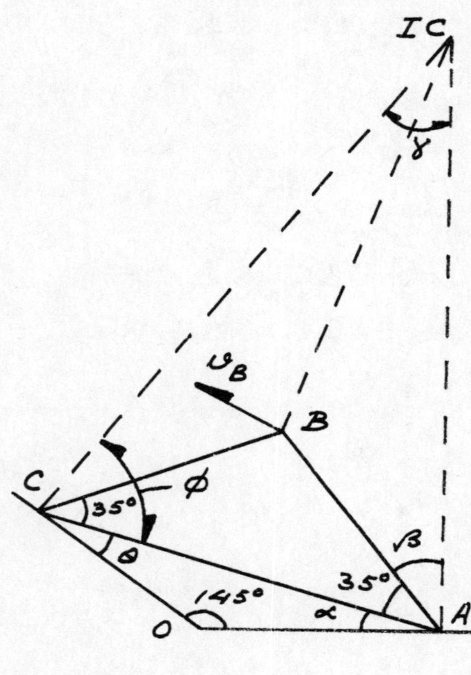

$$\phi = 180° - 35° - 38.335° - 35° = 71.665°$$

By law of Sines, we get

$$\frac{AB}{\sin 35°} = \frac{AC}{\sin(180° - 35° - 35°)}$$

$$\text{OR, } AB = \frac{(1\,m)(0.573)}{(0.939)} = 0.61\,m$$

$$\text{Also, } \frac{r_{IA}}{\sin\phi} = \frac{AC}{\sin\gamma}$$

$$\text{OR, } r_{IA} = \frac{(1\,m)(0.949)}{(0.573)} = 1.656\,m$$

By law of cosines, we get

$$r_{IB}^2 = r_{IA}^2 + AB^2 - 2(r_{IA})(AB)\cos\beta$$

$$\text{OR, } r_{IB} = \sqrt{(1.656\,m)^2 + (0.61\,m)^2 - 2(1.656\,m)(0.61\,m)(0.784)} = 1.236\,m$$

Now, $v_B = \omega_{AB} \times r_{IB}$

and $\omega_{AB} = \dfrac{v_A}{r_{IA}}$

$$\therefore\; v_B = v_A \frac{r_{IB}}{r_{IA}} = \frac{(15\,m/s)(1.236\,m)}{(1.656\,m)} = 11.195\,m/s$$

<u>Ans.</u>

3-29

One end of the rigid link AB is attached to a block which slides on a frictionless surface. The other end is constrained to move along a circle. At the instant shown, Point B is moving downwards with a velocity of 10 m/sec.

(a) What is the velocity of the block ?
(b) What is the angular velocity of the link AB ?

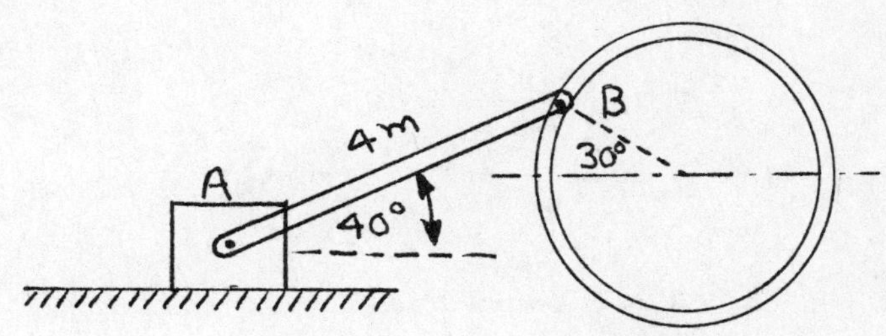

****

(a) Velocity of B $v_B = 10$ m/sec.

$$V_A = v_{Bx} = 10 \sin 30^\circ$$
$$= 5 \text{ m/sec.} \leftarrow$$

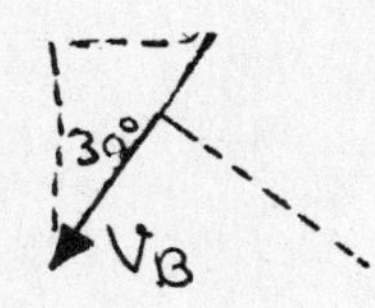

(b) We find the instantaneous center by drawing IC perpendiculars to the direction of velocity.

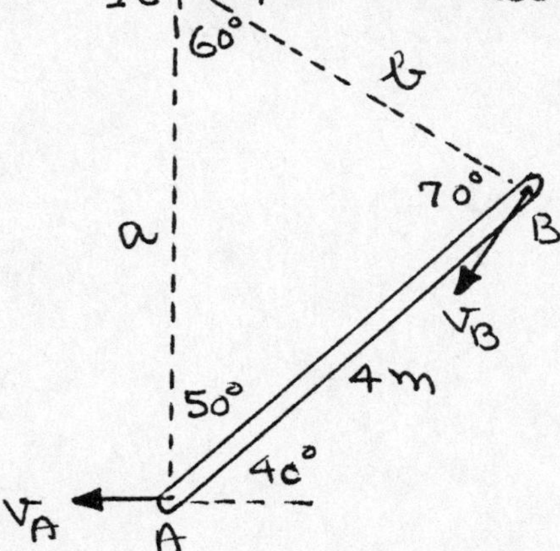

Using Law of Sines

$$\frac{4}{\sin 60^\circ} = \frac{a}{\sin 70^\circ}$$

$$a = 4.34 \text{ m.}$$

$$\omega = \frac{V_A}{a} = \frac{5}{4.34} = 1.15 \text{ rad/sec}$$

3-30

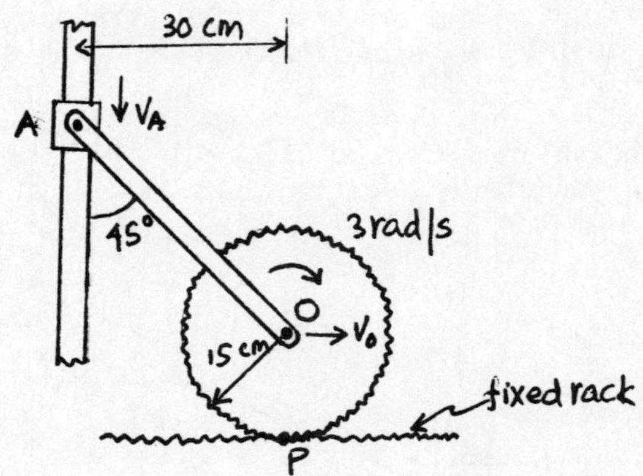

In the mechanism shown, the gear wheel has angular speed 3 rad/s at a certain instant. Determine the angular velocity ω_{OA} of link OA and the velocity v_A of slider A.

****

Since the rack is fixed, instantaneous center for the wheel is its Meshing point P. Hence the velocity of O

$$\underline{V}_O = \omega r = 3 \times 15 \text{ cm/s} \rightarrow \quad = 45 \text{ cm/s} \rightarrow$$

For the rigid link AO, the instantaneous center C is obtained by drawing perpendiculars from velocities $\underline{V}_A$ and $\underline{V}_O$.

$$V_O = \omega_{OA} \times CO \Rightarrow \omega_{OA} = \frac{V_O}{CO} = \frac{45}{30} \text{ rad/s}$$

$$\omega_{OA} = 1.5 \text{ rad/s} \circlearrowleft$$

A 30 cm C ω_{OA} 45° $\underline{V}_A$ 30 cm O $\underline{V}_O$

$$V_A = \omega_{OA} \times CA = 1.5 \times 30 \text{ cm/s} \downarrow$$

$$= \underline{45 \text{ cm/s} \downarrow}$$

3-31

The body shown rolls without slipping on the plane. Find the angular velocity and angular acceleration of the body. Find the acceleration of point B.

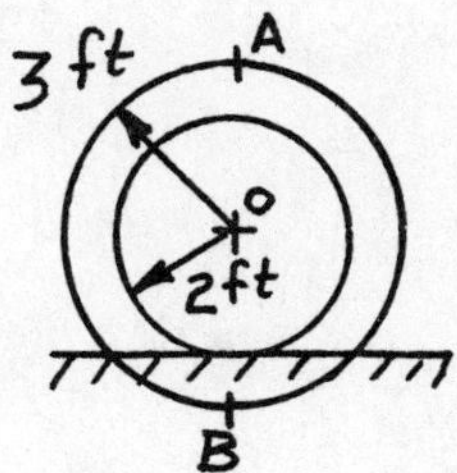

$v_A = 6$ ft/sec to right

$a_0 = 4$ ft/sec^2 to left

****

Defining distances as shown below and using instantaneous center

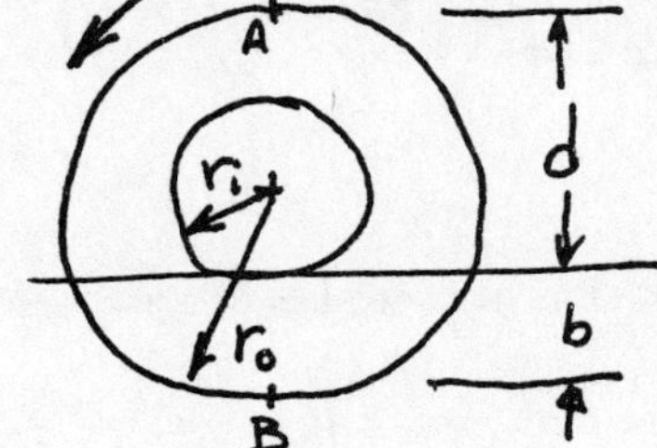

$$V_a = d\omega = 5\omega$$

$$\boxed{\omega = \frac{V_a}{5} = \frac{6}{5} \text{ rdps clockwise}}$$

$$a_o = r_i \alpha = 2\alpha$$

$$\boxed{\alpha = \frac{a_o}{2} = 2 \text{ rdps}^2 \text{ counter clockwise}}$$

$\bar{a}_B$ has a tangential component $a_{BT} = b\alpha$ right

$\bar{a}_B$ has a normal component $a_{BN} = r_o \omega^2$ up

$$\boxed{\bar{a}_B = b\alpha\bar{i} + r_o\omega^2\bar{j} = 2\bar{i} + 4.32\bar{j} \text{ ft/sec}^2}$$

3-32

At the instant shown the rod AB is slipping and A has a velocity of 20 ft/sec. to the left. Find the angular velocity of the rod AB and the velocity of point B.

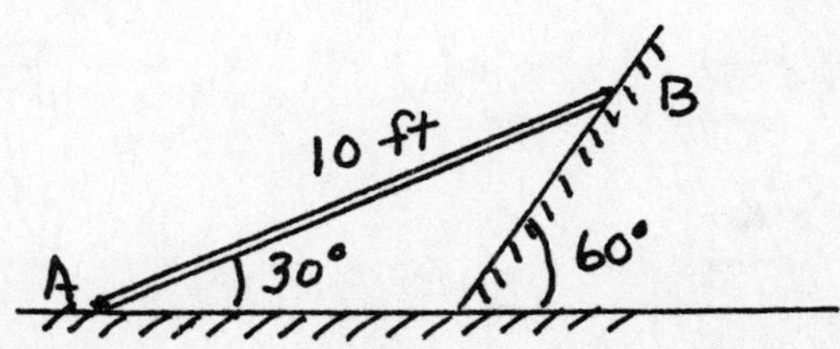

****

FROM THE GEOMETRY OF THE FIGURE:

I.C., 60°, 10', 10'', 60°, B, 60°, 10', A

$$\therefore\ \omega_{AB} = \frac{V_A}{AI} = 2\ \frac{rad}{sec}\ \circlearrowright$$

$$V_B = \omega_{AB}\cdot(BI) = 20\ fps$$

($\perp$ to BI) or 60°

EVALUATION OF ACCELERATIONS

3-33

The wheel is rolling to the right with a constant angular velocity, without slipping. The acceleration of point C has the direction:

a) ← b) ↙ c) ↓ d) ↘ e) it is zero

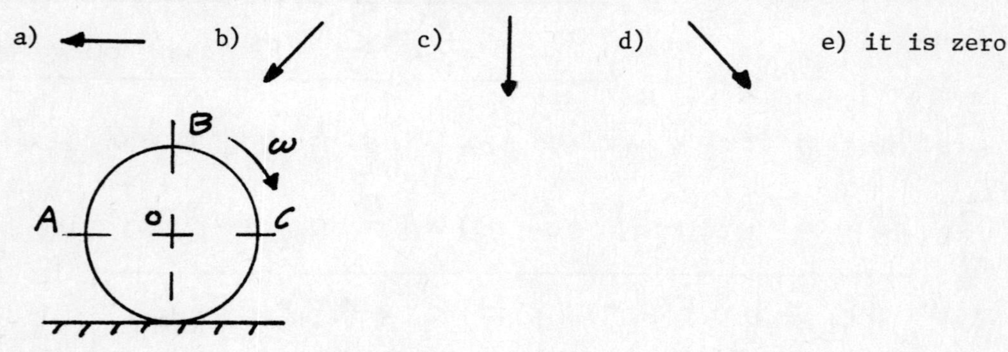

****

WITH $\underline{\omega}$ = CONSTANT, $\underline{\alpha} = 0$ AND $\underline{a}_o = 0$

$$\underline{a}_c = \underline{a}_o + \underline{a}_{c/o}$$

$$a_c = 0 + (\underset{\leftarrow}{r\omega^2} + \underset{\updownarrow}{r\alpha})$$

$$a_c = 0 + \underset{\leftarrow}{r\omega^2} + 0$$

$r\alpha = 0$, $r\omega^2$, O, C, $a_o = 0$

THE ANSWER IS a

3-34

In the link mechanism shown, C is moving down the slot with a constant velocity of 28 inches per second. Find the angular acceleration of member AB.

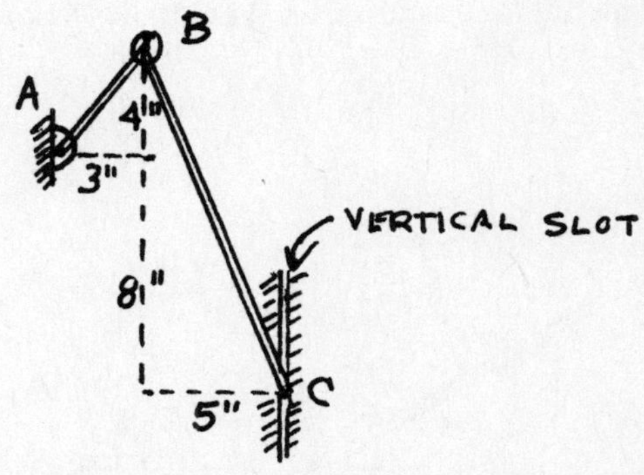

****

I.C OF AB IS A (AB IS ROTATING)

TO LOCATE IC_{BC}:

B, 5", A, 10", 12", C, IC_{BC}, 9", 5"

$$\therefore\ \omega_{BC} = \frac{28}{14} = 2\ rad/sec \ \circlearrowright$$

$$V_B = 15.2 = 30\ ips \ \searrow$$

$$\omega_{AB} = 30/5 = 6\ rad/sec \ \circlearrowright$$

$$a_C = 0$$

$$\bar{a}_B = 5\alpha_{AB}\ (\nwarrow\ 3\text{-}4) + 5.36\ (\swarrow\ 4\text{-}3)$$

Assume α's are $\circlearrowleft$:

$$= -4\alpha_{AB}\bar{i} + 3\alpha_{AB}\bar{j} - 108\bar{i} - 144\bar{j}$$

Then: $\bar{a}_B = \bar{a}_C + \bar{a}_{B/C}$

$$-4\alpha_{AB}\bar{i} + 3\alpha_{AB}\bar{j} - 108\bar{i} - 144\bar{j} =$$

$$13\alpha_{BC}\left(\frac{-12\bar{i} - 5\bar{j}}{13}\right) + 13.\ 2^2\left(\frac{5\bar{i} - 12\bar{j}}{13}\right)$$

Equating $\bar{i}$ and $\bar{j}$ components for the vector equation and solving simultaneously, $\alpha_{AB} = \frac{64}{7}$ or $9.14\ \frac{rad}{sec^2}\ \circlearrowleft$

3-35

For the position shown $\omega_{AB} = 7$ rad/s ↻, $\alpha_{AB} = 5$ rad/s^2 ↺

Find the velocity and acceleration of block C which can freely slide vertically on bar DE.

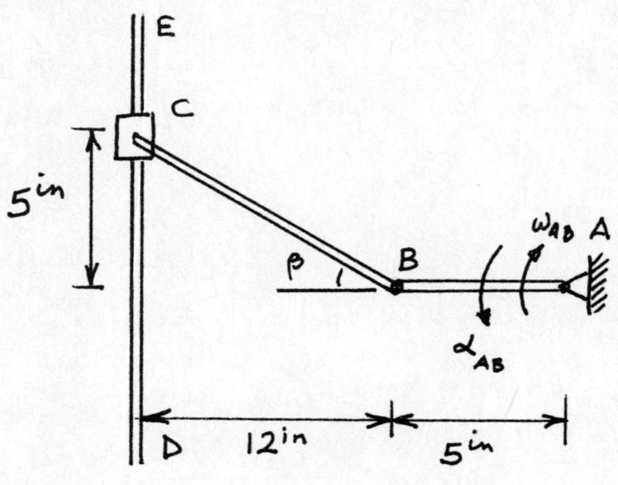

****

The velocities :

$\omega_{AB} = 7$ rad/s

$$v_B = \omega_{AB}\, r_{AB} = 7(5) = 35 \text{ in/s} \uparrow$$

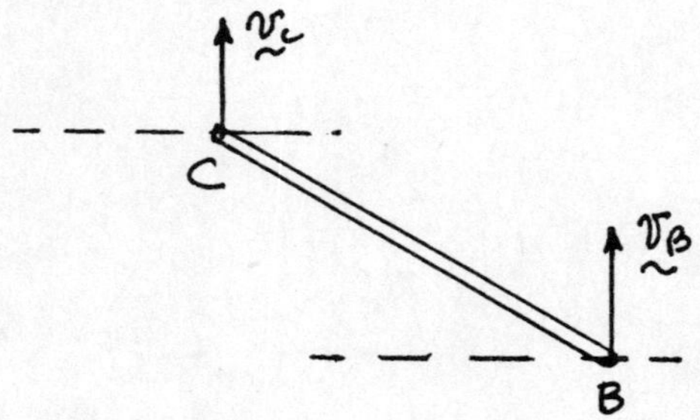

Instant center for BC is at infinity .

$$\omega_{BC} = 0$$

$$v_C = v_B = 35 \text{ in/s} \uparrow$$

The accelerations :

$$\cos\beta = \frac{12}{13} \quad ; \quad \sin\beta = \frac{5}{13}$$

$$(a_B)_t = \alpha_{AB}\, r_{AB} = 5\,(5) =$$
$$= 25\ in/s^2 \downarrow$$
$$(a_B)_n = \omega_{AB}^2\, r_{AB} = (7)^2\,(5) =$$
$$= 245\ in/s^2 \rightarrow$$

$$\underset{\sim}{a}_B = 245\,\underset{\sim}{i} - 25\,\underset{\sim}{j}$$

$$\underset{\sim}{a}_C = \underset{\sim}{a}_B + \underset{\sim}{a}_{C/B} \quad (1)$$

$$(a_{C/B})_n = (\omega_{BC})^2 \cdot r_{BC} = (0)^2\,(13) = 0$$

$$(a_{C/B})_t = \alpha_{BC} \cdot r_{BC} = 13\,\alpha_{BC}$$

$$\underset{\sim}{a}_{C/B} = -13\alpha_{BC}\sin\beta\,\underset{\sim}{i} - 13\alpha_{BC}\cos\beta\,\underset{\sim}{j}$$

$$\underset{\sim}{a}_{C/B} = -13\,\alpha_{BC}\left(\frac{5}{13}\right)\underset{\sim}{i} - 13\,\alpha_{BC}\left(\frac{12}{13}\right)\underset{\sim}{j}$$

$$= -5\,\alpha_{BC}\,\underset{\sim}{i} - 12\,\alpha_{BC}\,\underset{\sim}{j}$$

$$\underset{\sim}{a}_C = -a_C\,\underset{\sim}{j}$$

Substituting in equation (1), we have:

$$-a_C \underset{\sim}{j} = 245 \underset{\sim}{i} - 25 \underset{\sim}{j} - 5\alpha_{BC} \underset{\sim}{i} - 12\alpha_{BC} \underset{\sim}{j}$$

$$-a_C \underset{\sim}{j} = (245 - 5\alpha_{BC}) \underset{\sim}{i} - (25 + 12\alpha_{BC}) \underset{\sim}{j}$$

or

$$245 - 5\alpha_{BC} = 0 \qquad \alpha_{BC} = 49 \text{ rad/s}^2 \circlearrowleft$$

$$a_C = 25 + 5\alpha_{BC} \qquad a_C = 270 \text{ in/s}^2 \downarrow$$

3-36

The 1.6 meter rod AB rotates in a vertical plane. At the instant shown the acceleration of end B is 25 m/s^2 in the direction shown. Find:
a) the angular velocity of the rod.
b) the angular acceleration of the rod.

B L = 1.6 m A
40°
$a_B = 25 \text{ m/s}^2$

****

B, a_{B_N}, 40°, $a_B = 25$, a_{B_T}

$$a_{B_T} = 25 \sin 40° = 16.07 \text{ m/s}^2 \downarrow$$

$$a_{B_N} = 25 \cos 40° = 19.15 \text{ m/s}^2 \rightarrow$$

a) $a_N = r\omega^2 \qquad \omega = \sqrt{\frac{a_N}{r}} = \sqrt{\frac{19.15}{1.6}} = 3.46 \text{ rad/s} \circlearrowright$

b) $a_T = \alpha r \qquad \alpha = \frac{a_T}{r} = \frac{16.07}{1.6} = 10.04 \text{ rad/s}^2 \circlearrowright$

3-37

Gear E rotates with an angular velocity of 4 rad/sec counterclockwise and an angular acceleration of 2 rad/sec^2 clockwise. Determine the instant center of rod CD, and the velocity and acceleration of block D.

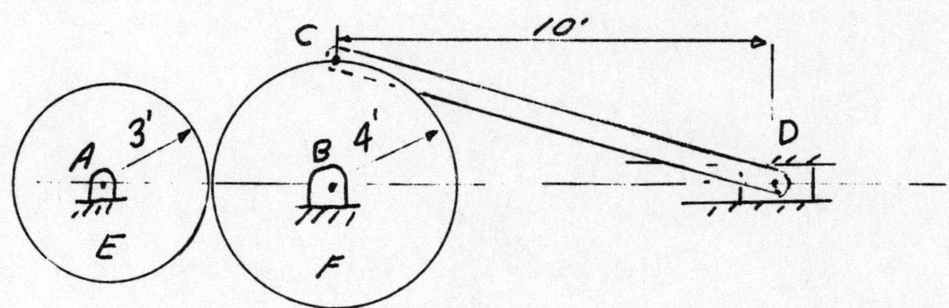

****

Since the velocity of point C and D are horizontal a line ⊥ to each will not meet, the I.C @ ∞ and $\omega_{CD} = 0$

The velocity of a point "P" where they touch is the same for each Gear

$V_P = 3(4)$ or $12\uparrow$ $\quad 12 = 4\omega_F$ or $\omega_F = 3$ ↻

The Velocity of $C = 4 \times 3$ and velocity of $D = V_C$

$\therefore V_D = \underline{\underline{12}}$ ft/sec →

$a_{P_T} = 3(2)$ or $6\downarrow$
$= 4\alpha_F \quad$ so $\alpha_F = 1.5$ ↻

$a_C = a_{C/D} + a_D$

$4(1.5)$ ←, $4(3)^2$ ↓ $= r\alpha_{CD}$, 0 $+ \overrightarrow{a_D}$

$+\uparrow \quad -4(9) = \frac{10}{r} r\alpha_{CD} + 0$

$\alpha_{CD} = -3.6 \quad$ or $\alpha_{CD} = 3.6$ rad/sec^2 ↻

$\overrightarrow{+} \quad -4(1.5) = \frac{4}{r} r(-3.6) + a_D$

$a_D = 14.4 - 6$

$= 8.4 \qquad a_D = \underline{\underline{8.4}}$ ft/sec^2 →

3-38

End A of rod AB is moving to the left with a velocity V_A and an acceleration a_A at the instant shown.

a. What is the angular velocity of the rod at this instant ?
b. What is the acceleration of end B at this instant ?

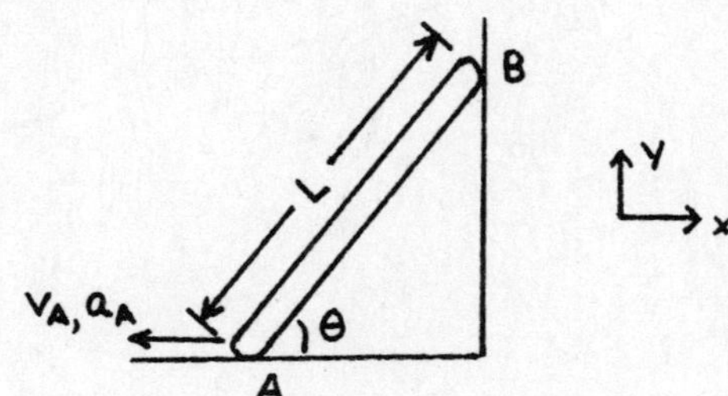

****

a) THE RELATIVE VELOCITY EQUATIONS CAN BE USED TO CALCULATE THE ANGULAR VELOCITY. POINT A HAS ONLY AN $\vec{i}$ COMPONENT OF VELOCITY WHILE POINT B HAS ONLY A $\vec{j}$ COMPONENT. THUS

$$\vec{V}_B = V_B\vec{j} = \vec{V}_A + \vec{\omega} \times \vec{r}_{B/A}$$

WHERE $\vec{\omega} = \omega\vec{k}$ IS THE ANGULAR VELOCITY VECTOR OF THE ROD AND

$$\vec{r}_{B/A} = L\cos\theta\,\vec{i} + L\sin\theta\,\vec{j}$$

IS THE RELATIVE POSITION VECTOR, THE POSITION OF B RELATIVE TO A. THEN

$$V_B\vec{j} = -V_A\vec{i} + \omega\vec{k} \times (L\cos\theta\,\vec{i} + L\sin\theta\,\vec{j})$$

$$= (-V_A - \omega L\sin\theta)\vec{i} + \omega L\cos\theta\,\vec{j}$$

SETTING THE $\vec{i}$ COMPONENT OF THE RIGHT HAND SIDE TO ZERO YIELDS

$$\omega = \frac{-V_A}{L\sin\theta}$$

EQUATING THE $\vec{j}$ COMPONENT OF BOTH SIDES YIELDS

$$V_B = \omega L\cos\theta = \frac{V_A\cos\theta}{\sin\theta} = -V_A\cot\theta$$

b) USING THE RELATIVE ACCELERATION EQUATION

$$\vec{a}_B = \vec{a}_A + \vec{\alpha} \times \vec{r}_{B/A} + \vec{\omega} \times (\vec{\omega} \times \vec{r}_{B/A})$$

WHERE $\vec{\alpha}$ IS THE ANGULAR ACCELERATION VECTOR OF THE BAR AT THIS INSTANT.

$$a_B \vec{j} = -a_A \vec{i} + \alpha \vec{K} \times (L\cos\theta\,\vec{i} + L\sin\theta\,\vec{j})$$
$$-\frac{V_A}{L\sin\theta}\vec{K} \times (V_A\vec{i} - V_A\cot\theta\,\vec{j})$$
$$= \left(-a_A - \alpha L\sin\theta - \frac{V_A^2}{L\sin\theta}\cot\theta\right)\vec{i} + \left(\alpha L\cos\theta - \frac{V_A^2}{L\sin\theta}\right)\vec{j}$$

THE RESULTING SCALAR EQUATIONS ARE

$$0 = -a_A - \alpha L\sin\theta - \frac{V_A^2}{L\sin\theta}\cot\theta$$

$$\alpha = -\frac{1}{L\sin\theta}\left(a_A + \frac{V_A^2\cot\theta}{L\sin\theta}\right)$$

AND THEN

$$a_B = \alpha L\cos\theta - \frac{V_A^2}{L\sin\theta}$$

3-39

At the instant shown, the disk rotates with a constant angular velocity of ω_0 clockwise. Determine the angular velocities and accelerations of rods AB and BC (in terms of ω_0)

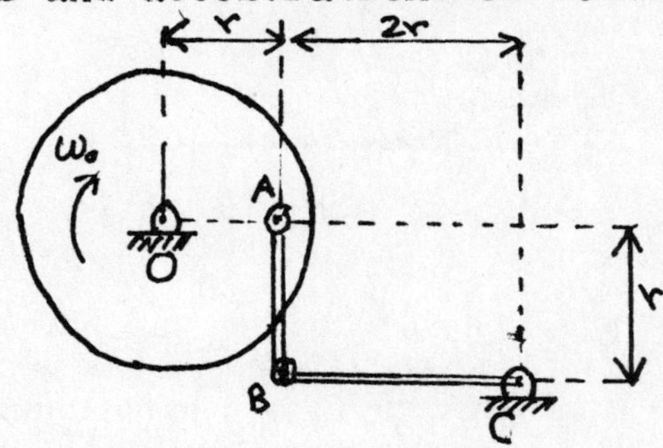

****

(a) $\overline{V}_A = r\omega_0 \downarrow$ But V_B is also $\downarrow$ ∴ AB in translation

∴ $\overline{V}_B = r\omega_0 \downarrow$ and $\omega_{AB} = 0$, $\omega_{BC} = \frac{r\omega_0}{2r} = \frac{\omega_0}{2}$

(b) $\alpha_{WHEEL} = 0$ and $\overline{a}_A = r\omega_0^2 \leftarrow$

Assume α_{BC} and α_{AB} ↺, then $\overline{a}_B = 2r\,\alpha_{BC} \downarrow$

$$+ 2r\left(\frac{\omega_0}{2}\right)^2 \rightarrow$$

Using the relative acceleration equation:

$$\bar{a}_B = \bar{a}_A + \bar{a}_{B/A}$$

$$\frac{r\omega_o^2}{2} \rightarrow + 2r\alpha_{BC} \downarrow = r\omega_o^2 \leftarrow + r\alpha_{AB} \rightarrow$$

Equating components $\alpha_{BC} = 0$

$$\alpha_{AB} = \frac{3}{2}\omega_o^2 \circlearrowleft$$

3-40

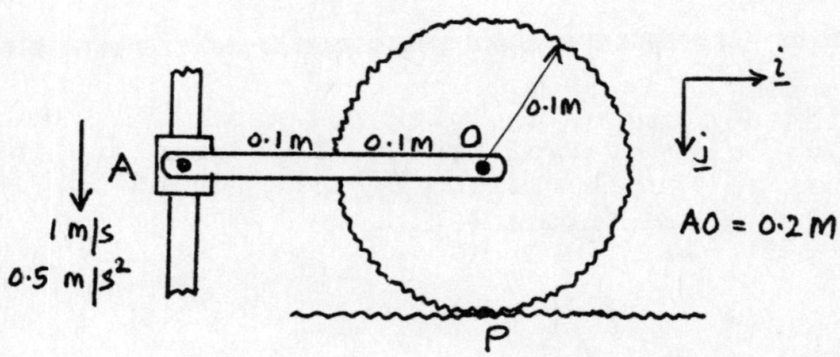

In the configuration shown, the slider A of the mechanism has downward velocity and acceleration 1 m/s and 0.5 m/s^2 respectively. Determine the angular velocity and angular acceleration of the gear wheel, and the angular acceleration of link OA.

****

Since perpendiculars from $\underline{V}_A$ and $\underline{V}_o$ intersect at O, the instantaneous center of link AO is O. Hence:

$$\underline{V}_o = \underline{0}$$

Since P is the instantaneous center of the wheel

$$V_o = \omega_o \times PO = 0$$

Hence

Angular velocity of the wheel $\omega_o = \underline{0}$

Angular velocity of link OA $= \frac{V_A}{OA} = \frac{1}{0.2}$ rad/s ↺ $= \omega_{OA}$

Hence $\omega_{OA} = 5$ rad/s ↺

Chosing the unit vectors $\underline{i}$ and $\underline{j}$ as shown in the figure,

Velocity of A $\underline{V}_A = 1\underline{j}$ m/s

acceleration of A $\underline{a}_A = 0.5\underline{j}$ m/s²

$\underline{V}_o = \underline{0}$

Acceleration at O is always horizontal. Let this be

$\underline{a}_o = a_o \underline{i}$

Acceleration of O relative to A

$$\underline{a}_{O/A} = \omega_{OA}^2 . OA\ \underline{i} + \alpha . OA\ \underline{j} = 5^2 \times 0.2\,\underline{i} + 0.2\alpha\,\underline{j}$$

$$= 5\underline{i} + 0.2\alpha\,\underline{j}$$

α = clockwise angular acceleration of OA.

Using $\underline{a}_o = \underline{a}_A + \underline{a}_{o/A}$ we have,

$$a_o \underline{i} = 0.5\underline{j} + 5\underline{i} + 0.2\alpha\,\underline{j}$$

or $(a_o - 5)\underline{i} - (0.2\alpha + 5)\underline{j} = \underline{0}$

Hence $a_o - 5 = 0$ and $0.2\alpha + 5 = 0$

or, $a_o = \underline{5 \text{ m/s}^2}$ →

$\alpha = -25$ rad/s² ↻ $= \underline{25 \text{ rad/s}^2}$ ↺

3-41

In the apparatus shown, the flywheel turns with a constant angular velocity of 2 rad/sec. counter-clockwise. The piston is constrained to move in the vertical direction and the connecting rod AB is rigid. Find the velocity and acceleration of B when the rod, piston and flywheel are in the position shown. Note that R is one foot long; L is four feet long.

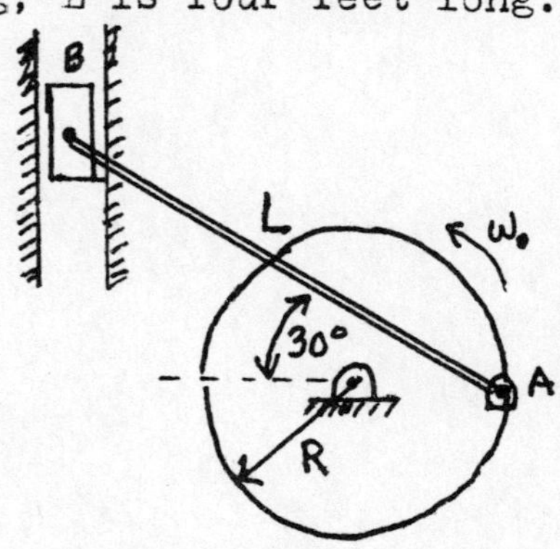

****

(a) $V_A = r \cdot \omega$ or $\overline{V}_A = 2$ ft/sec $\uparrow$

then $\overline{V}_B = \overline{V}_A + \overline{V}_{B/A}$

$$V_B \uparrow = 2 \uparrow + 4 \cdot \omega_{AB} \cdot \cos 30° \uparrow + 4\,\omega_{AB} \sin 30° \rightarrow$$

Equating coefficients, $\omega_{AB} = 0$ & $\overline{V}_B = 2$ ft/sec $\uparrow$

(b) $\overline{a}_A = (a_A)_n \leftarrow = 1 \cdot \omega^2_{WHEEL} \leftarrow = 4$ ft/sec² $\leftarrow$

$\overline{a}_B = \overline{a}_A + \overline{a}_{B/A}$

$$a_B \uparrow = 4 \leftarrow + 4\alpha_{AB} \left(\sin 30° \rightarrow + \cos 30° \uparrow \right)$$

Equating coefficients, $a_B = 4\,\alpha_{AB} \cos 30°$

$$0 = -4 + 4\alpha_{AB} \sin 30°$$

Then $\alpha_{AB} = 2$ rad/sec²

$a_B = 6.93$ ft/sec² OR $\overline{a}_B = 6.93$ ft/sec² $\uparrow$

3-42

In the figure shown, calculate a) the angular velocity and acceleration of the bar BC, b) the velocity and acceleration of the piston C.

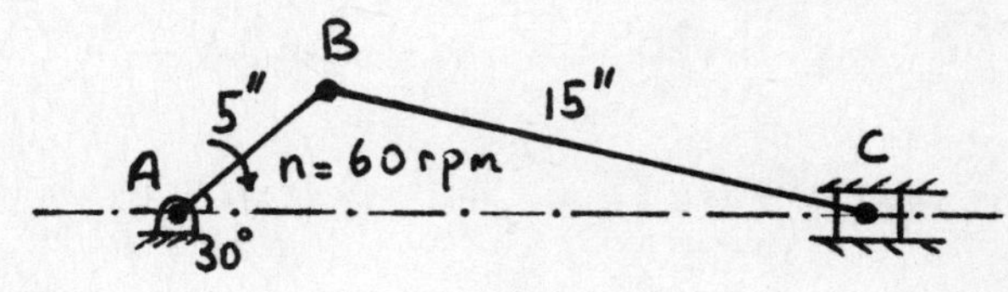

****

Angular speed for AB : $\omega_{AB} = \dfrac{2\pi n}{60} = \dfrac{2\pi \times 60}{60} = 2\pi$

Using sine law, we get $\dfrac{15''}{\sin 30^\circ} = \dfrac{5''}{\sin\beta} \longrightarrow \beta \cong 9.6^\circ$

Velocity of point B :

$$|\vec{V}_B| = \overline{AB}.\omega_{AB} = 5/12 \times 2\pi = 5\pi/6$$

Relative velocity formula :

$$\vec{V}_C = \vec{V}_B + \vec{V}_{C/B} \qquad |\vec{V}_{B/C}| = |\vec{V}_{C/B}| = \overline{BC}.\omega_{BC}$$

$$V_C\vec{i} = \left(\frac{5\pi}{6}\sin 30^\circ\,\vec{i} - \frac{5\pi}{6}\cos 30^\circ\,\vec{j}\right) + \left(\overline{BC}.\omega_{BC}.\cos\beta\,\vec{j} + \overline{BC}.\omega_{BC}.\sin\beta\,\vec{i}\right)$$

$\vec{i}$: $\quad V_C = \dfrac{5\pi}{6}\sin 30^\circ + \overline{BC}.\omega_{BC}.\sin 9.6^\circ$

$\vec{j}$: $\quad 0 = -\dfrac{5\pi}{6}\cos 30^\circ + \overline{BC}.\omega_{BC}.\cos 9.6^\circ$

$\Rightarrow \quad \omega_{BC} \cong 1.84$ rad/sec, $\quad V_C \cong 1.69$ ft/sec

$\therefore \quad \vec{V}_C = 1.69\,\vec{i}$

Acceleration of point C :

$$\vec{a}_C = \vec{a}_B + \vec{a}_{C/B}$$

$$= (\vec{a}_B)_n + (\vec{a}_B)_t + (\vec{a}_{C/B})_n + (\vec{a}_{C/B})_t$$

$(\vec{a}_B)_t = 0$ since $\omega_{AB} = \text{const.}$, $\alpha_{AB} = 0$; $(\vec{a}_B)_n = \overline{AB}.\omega_{AB}^2$

$|\vec{a}_{C/B}|_n = \overline{BC}.\omega_{BC}^2 \qquad |(\vec{a}_{C/B})_t| = \overline{BC}.\alpha_{BC}$

$$a_C\vec{i} = \overline{AB}.\omega_{AB}^2\cos 30^\circ(-\vec{i}) + \overline{AB}.\omega_{AB}^2\sin 30^\circ(-\vec{j}) + \overline{BC}.\omega_{BC}^2.\cos 9.6(-\vec{i}) + \overline{BC}.\omega_{BC}^2\sin 9.6^\circ\vec{j}$$
$$+ \overline{BC}.\alpha_{BC}.\cos 9.6\,\vec{j} + \overline{BC}.\alpha_{BC}.\sin 9.6\,\vec{i}$$

$$\vec{i}: \quad a_c = -\overline{AB}.\,\omega_{AB}^2 \cos 30^\circ - \overline{BC}.\,\omega_{BC}^2.\cos 9.6^\circ + \overline{BC}.\,\alpha_{BC}.\sin 9.6^\circ$$

$$\vec{j}: \quad 0 = -\overline{AB}.\,\omega_{AB}^2.\sin 30^\circ + \overline{BC}.\,\omega_{BC}^2.\sin 9.6^\circ + \overline{BC}.\,\alpha_{BC}.\cos 9.6^\circ$$

After substituting the values, we get

$$a_c = -\frac{5}{12}(2\pi)^2.\cos 30^\circ - \frac{15}{12}(1.8395)^2 \cos 9.6^\circ + \frac{15}{12}\,\alpha_{BC}.\sin 9.6^\circ$$

$$0 = -\frac{5}{12}(2\pi)^2.\sin 30^\circ + \frac{15}{12}(1.8395)^2 \sin 9.6^\circ + \frac{15}{12}.\,\alpha_{BC}.\cos 9.6^\circ$$

from which $\alpha_{BC} = 6.1$ rad/sec^2

and $a_c = -17.14$ ft/sec^2

are obtained. $\therefore$ $\vec{a}_c = -17.14\,\vec{i}$ and the piston is decelerating and going to stop.

3-43

A 1.5 ft diameter flywheel is connected to a piston P by a connecting rod PQ as shown in the accompanying illustration. If the flywheel is rotating with an angular velocity $\omega = 4.5$ rad/s and angular acceleration $\alpha = 9.7$ rad/s^2, calculate for the position shown (a) the angular acceleration of PQ, and (b) the acceleration of P.

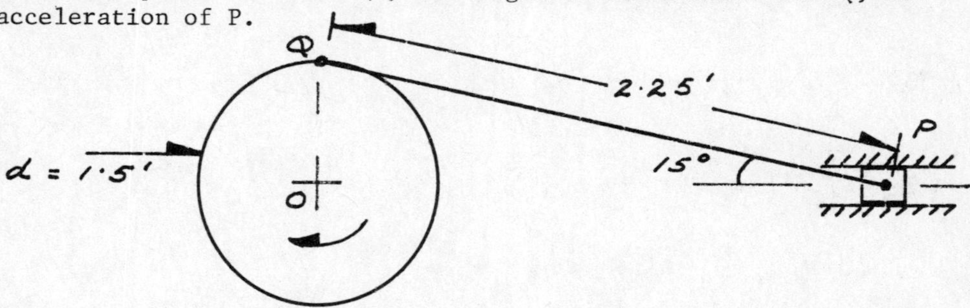

****

The instantaneous center of rotation IC is at ∞,

Hence, $\omega_{PQ} = 0$

$$(a_Q)_t = r\alpha = \left(\frac{1.5'}{2}\right)(9.7\ \text{rad/s}^2) = 7.275\ \text{ft/s}^2 \rightarrow$$

and $(a_Q)_n = (r\omega)\omega = \left(\frac{1.5'}{2}\right)(4.5\ \text{rad/s})^2 = 15.187\ \text{ft/s}^2 \downarrow$

$$\vec{a}_P = \vec{a}_Q + (\vec{a}_{P/Q})_t + (\vec{a}_{P/Q})_n$$

or, $\underset{\rightarrow}{a_P} = [(\underset{\rightarrow}{7.275\ ft/s^2}) + (\underset{\downarrow}{15.187\ ft/s^2})] + (2.25\ ft)\,\alpha\ \underset{15°}{\nearrow} + 0$

By separating the horizontal and vertical components, we get

$+\uparrow \quad 0 = (-15.187\ ft/s^2) + (2.25\ ft)\,\alpha \cos 15°$

OR $\quad \alpha = \dfrac{15.187\ ft/s^2}{(2.25\ ft)(0.965)} = 6.988\ rad/s^2$ **Ans.**

and $\overset{+}{\rightarrow}$ $\quad a_P = (7.275\ ft/s^2) + (2.25\ ft)(6.988\ rad/s^2)\sin 15°$
$= 11.331\ ft/s^2$ **Ans.**

3-44

The slide shown in the figure moves upward with a constant velocity of 10 ft/sec.Calculate a)the angular velocities and accelerations of the bars AB and CB, b)the accelerations of the mass centers of these two bars.

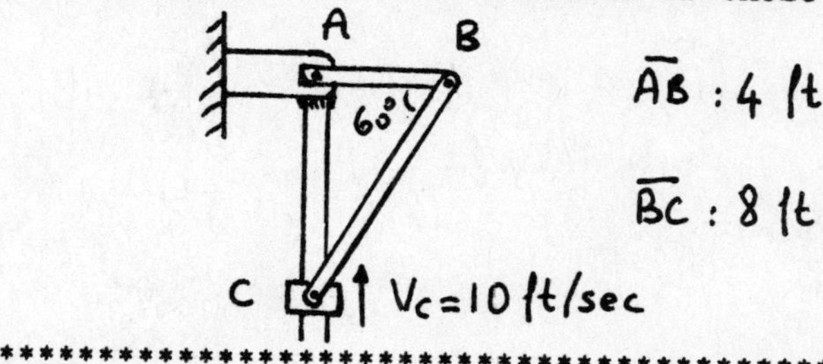

****

Velocities : For the bar $\overline{BC}$, we can write

$\vec{V}_B = \vec{V}_C + \vec{V}_{B/C} \qquad |\vec{V}_{B/C}| = \overline{CB} \times \omega_{BC}$

$V_B\vec{j} = V_C\vec{j} + \overline{CB} \times \omega_{BC} \sin 30°\,\vec{j} + \overline{CB} \times \omega_{BC} \cos 30°(-\vec{i})$

$\vec{i}: \quad 0 = \overline{CB} \times \omega_{BC} \cdot \cos 30 \qquad \overline{CB} \neq 0,\ \cos 30 \neq 0,\ \therefore\ \omega_{BC} = 0$

$\vec{j}: \quad V_B = V_C \quad \therefore \quad \vec{V}_B = V_C\vec{j} = 10\vec{j}$

Bar $\bar{BC}$ is in translational motion (at this moment). $\therefore \vec{V}_{G_1} = \vec{V}_C = 10\vec{j}$ ft/sec

For the bar $\bar{AB}$: $\omega_{AB} = V_B / \bar{AB} = 10/4 = 2.5$ rad/sec

$\vec{\omega}_{AB} = 2.5\vec{k}$

$\therefore$ $V_{G_2} = \omega_{AB} \times \bar{AB}/2 = 2.5 \times 2 = 5$ ft/sec $\vec{V}_{G_2} = 5\vec{j}$ ft/sec

Accelerations : For the bar $\bar{BC}$,

$(a_{B/C})_t$, 30°, B, $(\vec{a}_{G_1})_t$, $(a_{B/C})_n$, G_1, 30°, C, $(\vec{a}_{G_1})_n$

$\vec{a}_B = \vec{a}_C + \vec{a}_{B/C}$ $\qquad$ $\vec{a}_C = 0$ since $\vec{V}_C =$ const.

$(\vec{a}_B)_n + (\vec{a}_B)_t = 0 + (\vec{a}_{B/C})_n + (\vec{a}_{B/C})_t$, $(\vec{a}_{B/C})_n = 0$ since $\omega_{BC} = 0$

$\therefore$ $\bar{AB} \times \omega_{AB}^2(-\vec{i}) + \bar{AB} \times \alpha_{AB}\vec{j} = 0 + 0 + \bar{BC} \times \alpha_{BC} \times \cos 30(-\vec{i}) + \bar{BC} \times \alpha_{BC} \times \sin 30\vec{j}$

$\vec{i}$: $-\bar{AB}\,\omega_{AB}^2 = -\bar{BC}\,\alpha_{BC}\cos 30° \rightarrow 4 \times 2.5^2 = 8 \times \alpha_{BC} \times \cos 30° \rightarrow \alpha_{BC} = 3.608$

$\vec{j}$: $\bar{AB} \cdot \alpha_{AB} = \bar{BC}\,\alpha_{BC}\sin 30° \rightarrow 4 \times \alpha_{AB} = 8 \times 3.608 \times 0.5 \rightarrow \alpha_{AB} = 3.608$ rad/sec^2

Accelerations of the mass centers :

For the bar $\bar{BC}$: $\vec{a}_{G_1} = \vec{a}_C + \vec{a}_{G_1/C}$

$= 0 + (\vec{a}_{G_1/C})_n + (\vec{a}_{G_1/C})_t$ $\qquad$ $\vec{a}_{G_1/C} = 0$ since $\omega_{BC} = 0$

$= \bar{CG}_1\,\alpha_{BC}\cos 30(-\vec{i}) + \bar{CG}_1\,\alpha_{BC}\sin 30\vec{j}$

$= 4 \times 3.608 \times \cos 30(-\vec{i}) + 4 \times 3.608 \times \sin 30\vec{j}$

$= -12.5\vec{i} + 7.217\vec{j}$

For the bar $\bar{AB}$: $\vec{a}_{G_2} = \vec{a}_A + \vec{a}_{G_2/A}$

$= 0 + \bar{AG}_2 \cdot \omega_{AB}^2(-\vec{i}) + \bar{AG}_2 \cdot \alpha_{AB}\vec{j}$

$= 2 \times 2.5^2(-\vec{i}) + 2 \times 3.608\vec{j}$

$= -12.5\vec{i} + 7.217\vec{j}$

3-45

For the position shown the wheel, W, rotates clockwise, and has the acceleration of point P as shown. Member ABC is pinned to the wheel at A and to the block at C.

Determine for this position:

(a) The instant center of ABC.

(b) The velocity of point B.

(c) The acceleration of the block at C.

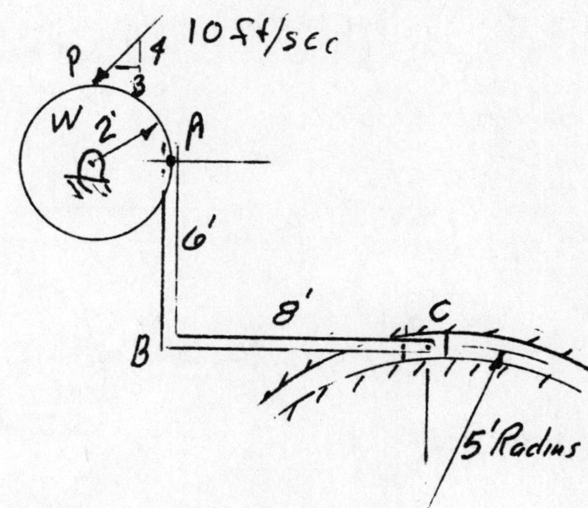

****

The instant center shown on sketch

$a_A = a_N \rightarrow a_t \qquad a_N = V^2/r$ or $r\omega^2$

$= 8 = 2\omega^2$

$\omega_W = 2$

$a_P = 3/5\,10 \rightarrow 4/5\,10$

$a_{P_t} = 6$ since $a_t = 2\alpha \qquad a_{A_T} = 6$

A, I.C

$V_A = 2(2)$, 10'

$\omega_{ABC} = 4/8$

V_C

$V_C = 6(1/2) = 3$

$V_B = 10\,\omega_{ABC} = 10\,1/2$

$V_B = 5$ (3, 4)

$a_C = a_{C/A} \rightarrow a_A$

$\frac{V_C^2}{5} \rightarrow a_{C_T} = 10\,\omega_{ABC}^2 \rightarrow 10\,\alpha_{ABC} \rightarrow 6 \rightarrow 8$

$\frac{3^2}{5} \rightarrow a_{C_T} = 10(1/2)^2 \rightarrow 10\,\alpha_{ABC} \rightarrow 6 \rightarrow 8$

$+\; 1.8 = 4/5\,10\,\alpha_{ABC} - 3/5\,10(1/2)^2 - 6$

$8\alpha_{ABC} = 9.3$ or $\alpha_{ABC} = \frac{9.3}{8}$ or 1.16 rad/sec

$\overset{+}{a_{C_T}} = -4/5\,10\,1/2^2 - 3/5\,10\,(1.16) - 8$

$= -16.96$ or $a_{C_T} = 16.96$

$a_C = 1.8 \rightarrow 16.96$ or $a_C = 17.06$ ft/sec²

(16.96, 1.8)

3-46

In the four bar mechanism shown, AB rotates with a constant angular velocity ω_o of 10 rad/s counterclockwise. Find the angular velocities and accelerations of the coupler BC and the follower DC, and the velocity and acceleration of C.

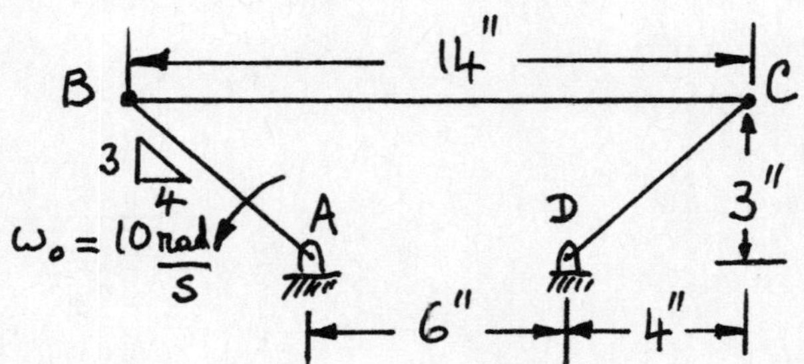

****

1) Velocity Analysis: The rod AB is 5". The velocity of B is 50 in/s as shown. The velocity of C is ⊥ to CD at C. The instant center is therefore at I.

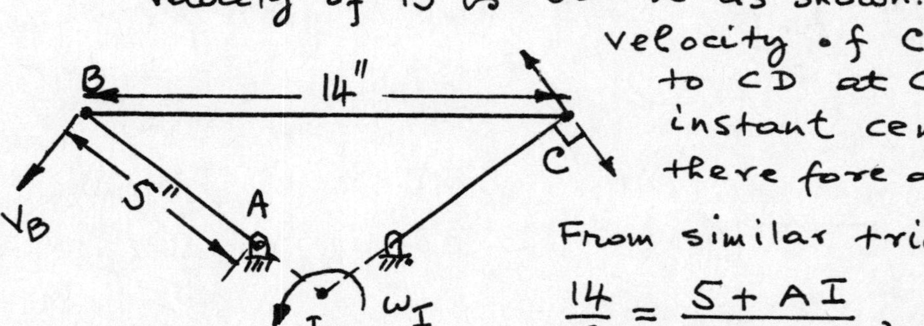

From similar triangles

$$\frac{14}{6} = \frac{5 + AI}{AI}, \quad AI = 3.75",$$

and BI = 8.75". Hence $\omega_{BC} = \frac{50}{8.75} = 5.72\ \text{rad/s}$ ↺

$V_C = \omega_{BC}\ IC = 50\ \text{in/s}$ ↖ and $\omega_{CD} = \frac{V_C}{CD} = 10\ \frac{rad}{s}$ ↻.

2) Acceleration Analysis: The vector equation $\vec{a}_C = \vec{a}_{C/B} + \vec{a}_B$ can be split up into normal and tangential components to yield

$$(\vec{a}_C)_N + (\vec{a}_C)_T = (\vec{a}_{C/B})_N + (\vec{a}_{C/B})_T + (\vec{a}_B)_N + (\vec{a}_B)_T$$

The normal components can be found from from $\omega^2 r$ where the ω's have been determined. The tangential components are $\perp$ to the normal, and the final intersection gives the solution.

$(\vec{a}_B)_N = (10)^2 5 = 500\ \text{in/s}^2 \searrow \quad (\vec{a}_B)_T = 0$

$(\vec{a}_{C/B})_N = (5.72)^2(14) = 458\ \text{in/s}^2 \leftarrow$

$(\vec{a}_C)_N = (10)^2 5 = 500\ \text{in/s}^2 \swarrow$

Projections give

$$\frac{3}{5}(a_C)_T = 2(500)\frac{4}{5}\ ;\ (a_C)_T = 1333.33\ \frac{\text{in}}{\text{s}^2}$$

$$(a_{C/B})_T = \frac{4}{5}(a_C)_T = 1066.67\ \text{in/s}^2.$$

<u>CHECK:</u> $(a_C)_T \frac{3}{5} \stackrel{?}{=} 2(500)\frac{4}{5}$ is satisfied identically.

The angular accelerations now are:

$$\underline{\underline{\alpha_{DC}}} = \frac{(a_C)_T}{5} = \underline{\underline{266.67}}\ \frac{\text{rad}}{\text{s}^2}$$

$$\underline{\underline{\alpha_{BC}}} = \frac{(a_{C/B})_T}{14} = \underline{\underline{76.19}}\ \frac{\text{rad}}{\text{s}^2}$$

The acceleration of C is $a_C^2 = (a_C)_x^2 + (a_C)_y^2$

$(a_C)_x = (OC)_x = -58\ \text{in/s}^2,\ (a_C)_y = -\left[\frac{3}{5}(500) + 1066.67\right]$

$(a_C)_y = (OC)_y = -1366.67\ \text{in/s}^2$, where OC is the resultant acceleration on the diagram.

$$\underline{\underline{a_C = 1367.9\ \text{in/s}^2}} \qquad \tan\theta = \frac{58}{1366.67},\ \underline{\underline{\theta = 2.43^\circ}}$$

3-47

Block A is constrained to move in a slot as shown. Point D has a velocity of 20 in/sec (as shown) and is increasing at a rate of 30 in/sec^2. Determine the instant center of bar AB and the acceleration of block A.

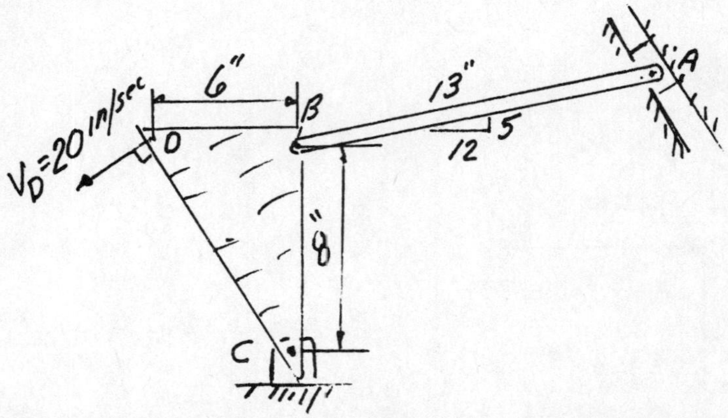

****

To find the instant center draw a line ⊥ to the slot and ⊥ to B.C

Using the instant center and $\omega_{BCD} = 20/10$ or 2

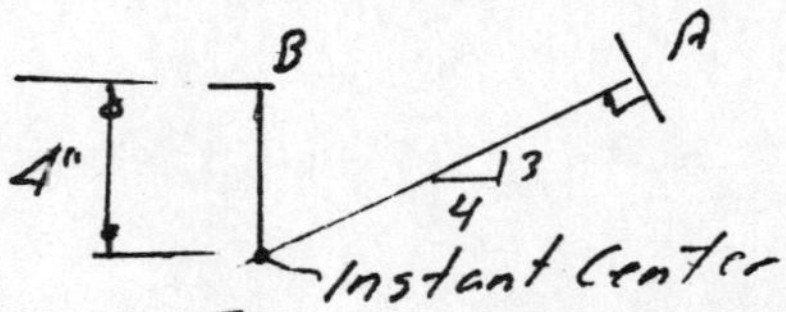

$V_B = 8(2)$ or 16

$V_B = r_{IC}\,\omega_{AB}$ so $\omega_{AB} = 16/4$ or 4 ↻

$a_D = 30 = 10\,\alpha_{ABC}$

$\alpha_{ABC} = \frac{30}{10}$ or 3 ↻

$a_A = a_{A/B} \rightarrow a_B$

$\overset{3}{\nwarrow}4 \quad a_A = 13(4)^2 + 13\alpha_{AB} \rightarrow 8(3) + 8(2)^2$

$\overset{+}{\leftarrow} \quad \frac{3}{5}a_A = \frac{12}{13}(13)16 - \frac{5}{13}13\alpha_{AB} + 24$

$\frac{3}{5}a_A = 216 - 5\alpha_{AB}$

$+\uparrow \quad \frac{4}{5}a_A = -\frac{5}{13}13(16) - \frac{12}{13}(13)\alpha_{AB} - 32$

$\frac{4}{5}a_A = -112 - 12\alpha_{AB}$

Solve the 2 equations $\alpha_{AB} = -75$

$a_A = \frac{5}{3}(216) - 5(-75))$

$= 985$ in/sec^2 $\overset{3}{\nwarrow}4$

3-48

The wheel (W) rolls without slipping on the curved surface, and in the position shown has an angular velocity of 3 rad/sec and an angular acceleration of 4 rad/sec^2 both counterclockwise.

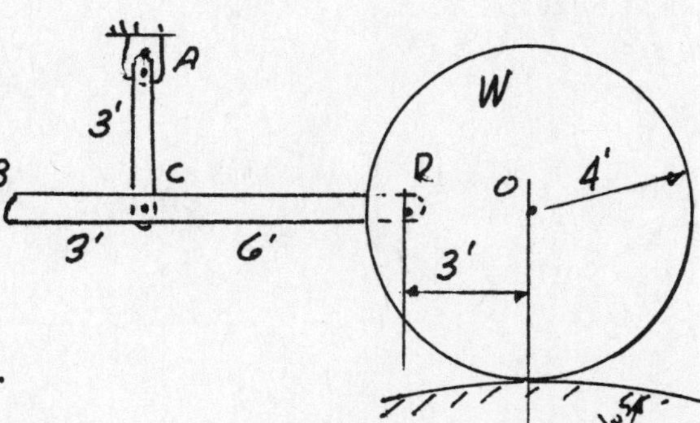

Determine for the position shown:
(a) The instant center of member BCD.
(b) The velocity of point B.
(c) The acceleration of point B.

****

From sketch I.C 8' above point C

$V_D = 5(3)$ or 15

From I.C $\omega_{BCD} = \frac{15}{10}$ or 1.5 (cw)

$\therefore V_B = r_{ic}\,\omega_{BCD}$ or $\sqrt{8^2+3^2}(1.5)$

$= \underline{12.82}$ (direction 8, 3)

I.C, 8, 3, 4, 3, 6, C, V_B, V_D

$a_D = a_{D/O} + a_O$ $\quad$ $a_O = r\alpha + \frac{V_O^2}{R}$ or $4(4) + \frac{12^2}{16}$

$= 3(3)^2 \rightarrow,\ 3(4)\downarrow + 16\leftarrow + 9\downarrow$

$a_D = 21\downarrow + 11\leftarrow$

$a_C = a_{C/D} + a_D$

$4\alpha_{AC} \leftarrow,\ 3(4)^2 \uparrow = 6\alpha_{BCD}\uparrow,\ 6(3/2)^2 \rightarrow + 11\leftarrow + 21\downarrow$

$+\uparrow\ 48 = 6\alpha_{BCD} - 21$ $\quad$ or $\alpha_{BCD} = 11.5$ rad/sec^2 (cw)

$a_B = a_{B/D} + a_D$

$= 9(11.5)\uparrow,\ 9(1.5)^2 \rightarrow + 21\downarrow + 11\leftarrow$

$a_B = 82.5\uparrow + 31.25\rightarrow$ $\quad$ or $a_B = \underline{88.2 \text{ ft/sec}^2}$ (82.5, 31.3)

3-49

Wheel D has an α = 10 rad/sec² and ω = 4 rad/sec when in the position shown. Determine α_{BC}.

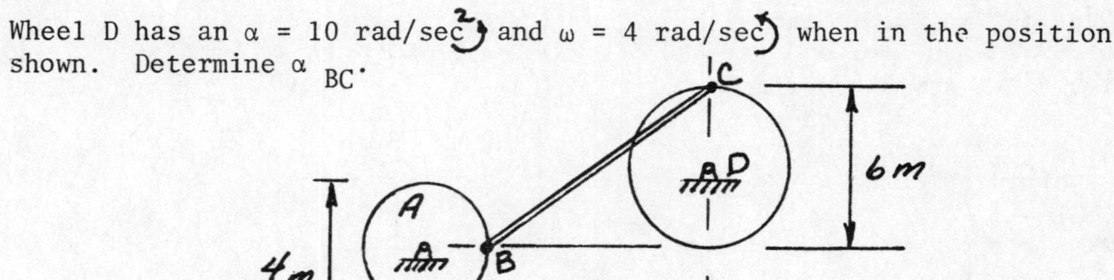

****

FOR ROTATION ABOUT A FIXED POINT D

$$\vec{a}_c = r\alpha\vec{i} + r\omega^2\vec{j} = -(3)(10)\vec{i} - 3(4)^2\vec{j} = (-30\vec{i} - 48\vec{j})\ m/s^2$$

$$\vec{v}_c = r\omega\vec{i} = -(3)(4)\vec{i} = (-12\vec{i})\ m/s$$

$$\vec{a}_B = \vec{a}_c + \vec{a}_{B/c} = -30\vec{i} - 48\vec{j} + \alpha_{BC}\vec{k}\times\vec{r}_{B/c} - \omega_{BC}^2\vec{r}_{B/c}$$

$$= -30\vec{i} - 48\vec{j} + \alpha_{BC}\vec{k}\times(-8\vec{i} - 6\vec{j}) - \omega_{BC}^2(-8\vec{i} - 6\vec{j})$$

SINCE THE VELOCITY OF C IS HORIZONTAL AND B IS VERTICAL, THE INSTANTANEOUS CENTER OF BC WILL BE AT THE BOTTOM OF WHEEL D. THEREFORE

$$\omega_{BC} = \frac{V_c}{r_{I/c}} = \frac{12}{6} = 2\ rad/s$$

$$V_B = r_{I/B}\,\omega_{BC} = 8(2) = 16\ m/s \downarrow \quad \text{AND} \quad \omega_A = \frac{V_B}{r_{AB}} = \frac{16}{2} = 8\ rad/s$$

THEN, $\vec{a}_B = a_{Bx}\vec{i} + a_{By}\vec{j}$ WITH $a_{Bx} = r\omega_A^2 = 2(8)^2 = 128$

$$\vec{a}_{Bx} = -128\vec{i} = -30\vec{i} + 6\alpha_{BC}\vec{i} + (2)^2 8\vec{i}$$

$$-128 + 30 - 32 = 6\alpha_{BC} \qquad \alpha_{BC} = -\frac{130}{6} = -21.7\ rad/s^2$$

$$= 21.7\ rad/s^2$$

3-50

The end A of rod AB moves to the left with a constant velocity of 4 ft/sec. At the instant shown, find (a) the angular acceleration of the rod, and (b) the acceleration of the midpoint G.

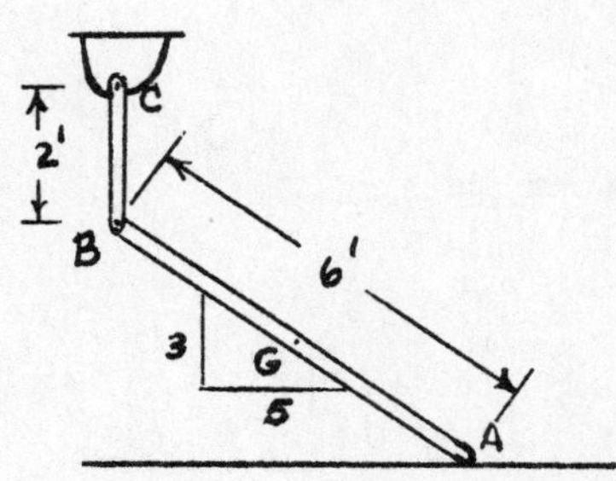

****

Consider Rod AB:

Since $\vec{V}_A$, $\vec{V}_B$ are parallel, the instantaneous center of AB is at infinity $\therefore \omega_{AB} = 0$

$\therefore \vec{V}_B = 4 \text{ ft/sec} \leftarrow$. Since pt. B is also on BC, we observe $V_B = 4 = \omega_{BC} \cdot 2$

$\Rightarrow \omega_{BC} = 2$ ↻

Consider Rod BC:

$\vec{a}_B = \vec{a}_{B,\text{Normal}} + \vec{a}_{B,\text{Tangential}}$

$= r_{CB} \cdot \omega_{BC}^2 \uparrow + r_{BC}\alpha_{BC} \leftrightarrow$ – (at this point we are not sure of the direction of α_{BC})

$\therefore \vec{a}_B = 2 \cdot 2^2 \uparrow + 2\alpha_{BC} \leftrightarrow$ (*)

or

$\vec{a}_B = 8\uparrow + 2\alpha_{BC} \leftrightarrow$ (*)

Consider Rod AB: Since $\vec{V}_A = \text{const.}$, $\Rightarrow \vec{a}_A = 0$

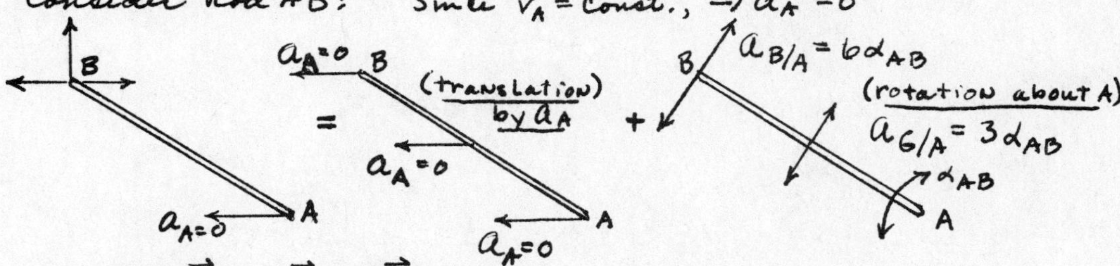

KEY: $\vec{a}_B = \vec{a}_A + \vec{a}_{B/A}$

From (*)

$8\uparrow + 2\alpha_{BC} \leftrightarrow = 0 + 6\alpha_{AB} \swarrow\!\nearrow$ – from this we observe α_{BC} ↺ and α_{AB} ↻:

$\Rightarrow 6\alpha_{AB} \cdot \frac{5}{\sqrt{34}} = 8$

$\Rightarrow \alpha_{AB} = 1.55$ ↻ $\left(\frac{\text{rad}}{\text{s}^2}\right)$

then $\vec{a}_G = \vec{a}_A + \vec{a}_{G/A}$

$= 0 + \nearrow 3\alpha_{AB} = 3(1.55)\nearrow$

$\vec{a}_G = 4.65$ (slope 5/3) (ft/sec^2)

3-51

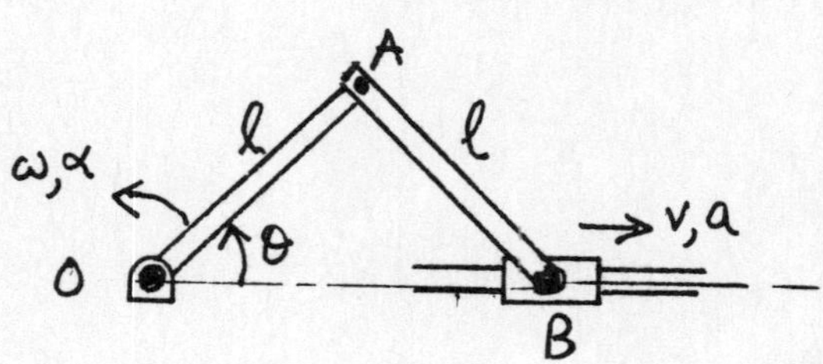

Determine the angular velocity ω and angular acceleration α of link OA in terms of the velocity v and acceleration a of slider B and the angle θ.

****

Let the position of B measured from the "fixed" point O be x.

$$x = l\cos\theta + l\cos\theta = 2l\cos\theta$$

Velocity of B $\quad v = \dfrac{dx}{dt} = -2l\sin\theta\dfrac{d\theta}{dt}$

But $\omega = \dfrac{d\theta}{dt}$. Hence $\quad v = -2l\omega\sin\theta$

or,

$$\omega = -\frac{v}{2l\sin\theta}$$

acceleration of B $\quad a = \dfrac{dv}{dt} = -2l\sin\theta\dfrac{d\omega}{dt} - 2l\omega\cos\theta\dfrac{d\theta}{dt}$

$$= -2l\alpha\sin\theta - 2l\omega^2\cos\theta$$

Substitute for ω:

$$a = -2l\alpha\sin\theta - 2l\frac{v^2}{4l^2\sin^2\theta}\cos\theta$$

Hence

$$\alpha = -\left[\frac{a}{2l\sin\theta} + \frac{v^2\cos\theta}{4l^2\sin^3\theta}\right]$$

4
PARTICLE MOTION RELATIVE TO TRANSLATING AND ROTATING REFERENCE FRAMES

4-1

A pendulum extends from point O and swings in the XY plane with an angular speed of 2 rad/sec. and an angular acceleration of 0.5 rad/sec^2 in the directions shown at the instant shown. Along the circular path in the pendulum bob a particle performs 5 revolutions per second relative to the bob. Consider the instant of time when the pendulum is vertical and the particle is at the bottom of the circle.

a. Calculate the absolute velocity of the particle at this instant.
b. Calculate the absolute acceleration of the particle at this instant.

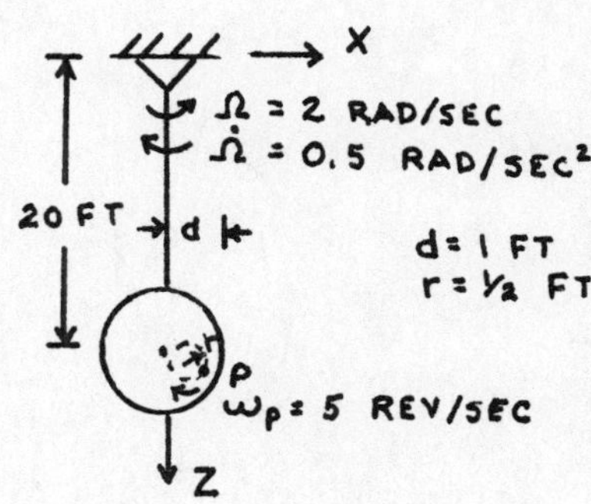

****

a) LET XZ BE A FIXED COORDINATE SYSTEM. LET XZ BE A COORDINATE SYSTEM THAT ROTATES WITH THE PENDULUM WITH ITS ORIGIN AT O.

$$\vec{V}_P = \vec{V}_{P'} + \vec{V}_{P/F}$$

WHERE

$\vec{V}_P$ = ABSOLUTE VELOCITY OF P

$\vec{V}_{P'}$ = VELOCITY OF POINT P' ON THE MOVING FRAME COINCIDING WITH P

$\vec{V}_{P/F}$ = VELOCITY OF P RELATIVE TO MOVING FRAME

$$\vec{V}_{P'} = \vec{\Omega} \times \vec{r}_{P/O} = (2 \text{ RAD/SEC } \vec{j}) \times (1 \text{ FT } \vec{i} + 20.5 \text{ FT } \vec{k})$$
$$= 41 \text{ FT/SEC } \vec{i} - 2 \text{ FT/SEC } \vec{k}$$

$$\vec{V}_{P/F} = -r\,\omega_P \vec{i} = -(\tfrac{1}{2} \text{ FT})(5 \tfrac{\text{REV}}{\text{SEC}})(2\pi \text{ RAD/REV})\vec{i}$$
$$= -15.7 \text{ FT/SEC } \vec{i}$$

$$\vec{V}_P = 25.3 \text{ FT/SEC } \vec{i} - 2 \text{ FT/SEC } \vec{k}$$

(b) $\vec{a}_P = \vec{a}_{P'} + \vec{a}_{P/F} + \vec{a}_c$

WHERE

$\vec{a}_P$ = ABSOLUTE ACCELERATION OF P

$\vec{a}_{P'}$ = ACCELERATION OF POINT P' IN MOVING FRAME COINCIDING WITH P

$\vec{a}_c$ = CORIOLIS ACCELERATION

$$\vec{a}_{P'} = \dot{\vec{\Omega}} \times \vec{r}_{P/O} + \vec{\Omega} \times (\vec{\Omega} \times \vec{r}_{P/O})$$
$$= -0.5 \text{ RAD/SEC}^2 \vec{j} \times (1 \text{ FT } \vec{i} + 20.5 \text{ FT } \vec{k})$$
$$+ 2 \text{ RAD/SEC } \vec{j} \times (41 \text{ FT/SEC } \vec{i} - 2 \text{ FT/SEC } \vec{k})$$
$$= (-81.5 \vec{k} - 14.25 \vec{i}) \text{ FT/SEC}^2$$

$$\vec{a}_{P/F} = -r\,\omega_P^2 \vec{k} = \tfrac{1}{2} \text{ FT } (10\pi \tfrac{\text{RAD}}{\text{SEC}})^2 \vec{k} = -493.5 \tfrac{\text{FT}}{\text{SEC}^2} \vec{k}$$

$$\vec{a}_c = 2\vec{\Omega} \times \vec{V}_{P/F} = 2(2 \text{ RAD/SEC } \vec{j}) \times (-15.7 \text{ FT/SEC})\vec{i}$$
$$= 62.8 \text{ FT/SEC}^2 \vec{k}$$

$$\vec{a}_P = -512.2 \text{ FT/SEC}^2 \vec{k} - 14.25 \text{ FT/SEC}^2 \vec{i}$$

4-2

A pin moves with a velocity of 2 m/sec and an acceleration of 1 m/sec^2 along a slot in a disk which is rotating with an angular velocity of 5 rad/sec and an angular acceleration of 4 rad/sec^2 in the clockwise direction. Calculate the absolute velocity and acceleration of this pin.

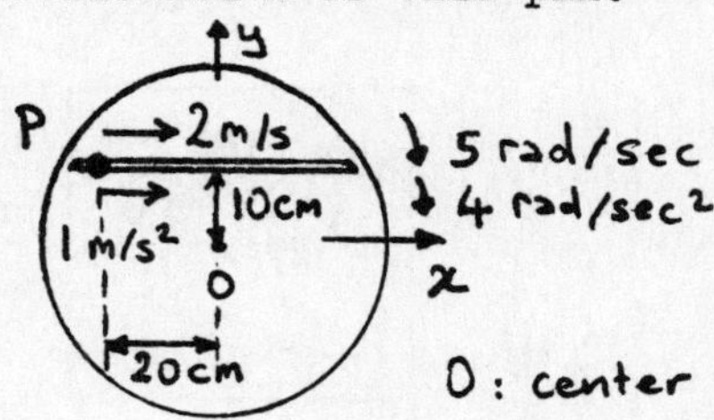

****

$$\overline{OP}^2 = \ell^2 = 10^2 + 20^2 \quad \rightarrow \quad \ell = \overline{OP} = 0.2236 \text{ m}$$

$$|\vec{V}_{p'}| = \overline{OP} \times \omega = 0.2236 \times 5 = 1.118 \text{ m/sec}$$

$$\tan \alpha_1 = \frac{10}{20} \quad \rightarrow \quad \alpha_1 \cong 26.6^\circ$$

Absolute velocity of the pin :

$$\vec{V}_P = \vec{V}_{P/F} + \vec{V}_{P'}$$

$$= (2\vec{i}) + (1.118 \times \sin 26.6^\circ \vec{i} + 1.118 \times \cos 26.6^\circ \vec{j})$$

$$\vec{V}_P = 2.5\vec{i} + 1\vec{j} \qquad |\vec{V}_P| = 2.69 \text{ m/sec}$$

Absolute acceleration of the pin :

$$\vec{a}_P = (\vec{a}_{P/F}) + \vec{a}_{P'} + \vec{a}_c$$

$$= (1\vec{i}) + (\vec{a}_{P'})_n + (\vec{a}_{P'})_t + (2\vec{\omega} \times \vec{V}_{P/F})$$

$$\vec{\omega} = -5\vec{k}, \quad (a_{P'})_t = \overline{OP}.\alpha, \quad (a_{P'})_n = \overline{OP}.\omega^2, \quad \vec{a}_c = 2(-5\vec{k} \times 2\vec{i})$$

$$\therefore \quad \vec{a}_P = (1\vec{i}) + (0.2236 \times 5^2 \cos 26.6^\circ \vec{i} - 0.2236 \times 5^2 \sin 26.6^\circ \vec{j})$$
$$+ (0.2236 \times 4 \sin 26.6^\circ \vec{i} + 0.2236 \times 4 \cos 26.6^\circ \vec{j}) - 20\vec{j}$$

$$\vec{a}_P = 6.4\vec{i} - 21.7\vec{j} \qquad |\vec{a}_P| = 22.62 \text{ m/sec}^2$$

4-3

The plate shown in the figure rotates about point O with 200 rpm in the anti-clockwise direction.The pins move in the slots with a relative velocity of 12 ft/sec in the directions shown.Calculate the absolute velocity and accelerations of the pins 1 and 2.

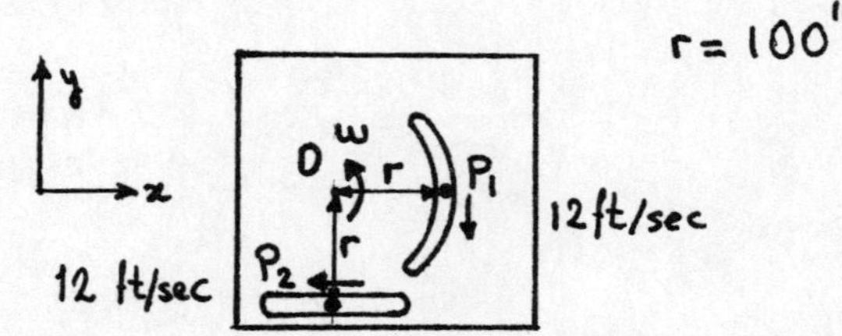

****

Angular speed of the plate :

$$\omega = \frac{2\pi n}{60} = \frac{2\pi \times 200}{60} = 20.94 \text{ rad/sec} \quad \therefore \quad \vec{\omega} = \omega\vec{k} = 20.94\,\vec{k}$$

We choose the coordinate axes as shown in the figure.

Pin. 1. $\vec{V}_P = \vec{V}_{P/F} + \vec{V}_{P'} = -12\vec{j} + \frac{100}{12} \times 20.94\vec{j} = 162.5\vec{j}$

$\vec{a}_P = \vec{a}_{P/F} + \vec{a}_{P'} + \vec{a}_c$

$$\vec{a}_{P/F} = (\vec{a}_{P/F})_n + \cancel{(\vec{a}_{P/F})_t}^{\,0} = \left(\frac{12}{100/12}\right)^2 \times \frac{100}{12}(-\vec{i})$$

$$\vec{a}_{P'} = r\omega^2(-\vec{i}) = \frac{100}{12} \times (20.94)^2(-\vec{i})$$

$$\vec{a}_c = 2\,\vec{\omega} \times \vec{V}_{P/F} = 2 \times (20.94\vec{k}) \times (-12\vec{j})$$

$\vec{a}_P = -3168.75\,\vec{i}$

Pin. 2. $\vec{V}_P = \vec{V}_{P/F} + \vec{V}_{P'} = -12\vec{i} + \frac{100}{12} \times (20.94)\vec{i} = 162.5\vec{i}$

$\vec{a}_P = \vec{a}_{P/F} + \vec{a}_{P'} + \vec{a}_c \qquad \vec{a}_c = 2\,\vec{\omega} \times \vec{V}_{P/F}$

$$= 0 + \frac{100}{12} \times (20.94)^2\vec{j} + 2 \times (20.94\vec{k}) \times (-12\vec{i})$$

$$= 3151.5\,\vec{j}$$

4-4

In an amusement park ride, a person P moves around in a circle represented by radius r_1 in the accompanying illustration at an uniform angular velocity of ω_1 rad/s, while the large wheel of radius r_2 spins at a uniform angular velocity of ω_2 rad/s. Express the velocity of P using the moving coordinate system e_r and e_θ.

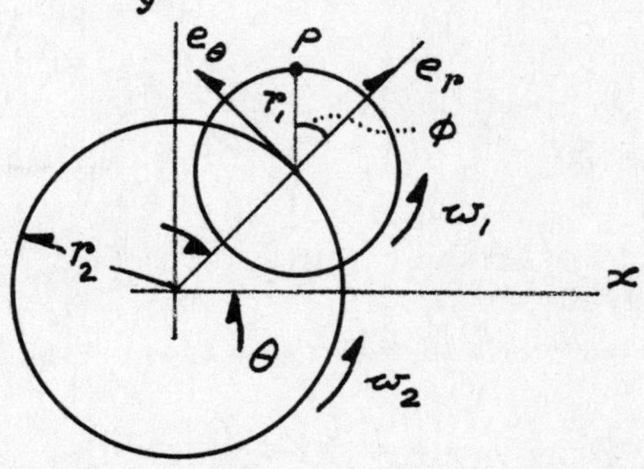

****

<u>Solution</u>

By attaching $\bar{e}_r$ and $\bar{e}_\theta$ to the small wheel representing the rotation of P, we get

$$\bar{\omega} = (\dot{\theta} + \dot{\phi})\,\bar{k}$$

Hence, $$\bar{v} = \dot{\bar{r}}_2 + \dot{\bar{r}}_{r_1} + \bar{\omega} \times \bar{r}_1$$

Now, $$\dot{\bar{r}}_2 = \dot{\theta}\,\bar{k} \times r_2\,\bar{e}_r = r_2\,\dot{\theta}\,\bar{e}_\theta$$

$$\dot{r}_{r_1} = 0$$

$$\bar{\omega} \times \bar{r}_1 = (\dot{\theta} + \dot{\phi})\,\bar{k} \times r_1(\cos\phi\,\bar{e}_r + \sin\phi\,\bar{e}_\theta) = r_1(\dot{\theta} + \dot{\phi})\cos\phi\,\bar{e}_\theta - r_1(\dot{\theta} + \dot{\phi})\sin\phi\,\bar{e}_r$$

$$\therefore \quad \bar{v} = \left[r_2\dot{\theta} + r_1(\dot{\theta} + \dot{\phi})\cos\phi\right]\bar{e}_\theta - \left[r_1(\dot{\theta} + \dot{\phi})\sin\phi\right]\bar{e}_r \qquad \underline{\underline{\text{Ans.}}}$$

4-5

The rigid, bent rod rotates around the vertical at a constant angular speed Ω = 4 rad/s, while the slider S moves along the rod according to

$$r(t) = r_1 + C\cos pt$$

$$r_1 = 0.3 \text{ m}$$

$$C = 0.02 \text{ m}$$

$$p = 20 \text{ s}^{-1}$$

Determine the acceleration of S for the instant t = π/60 seconds.

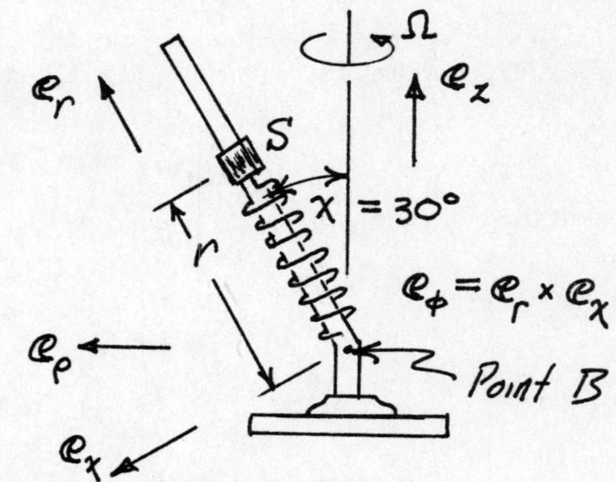

****

Let α be the "fixed" reference and β be the rod.

$$\mathbf{a}_{S/\beta} = \ddot{r}\mathbf{e}_r \; ; \quad 2\boldsymbol{\Omega} \times \mathbf{v}_{S/\beta} = 2\Omega\dot{r}\sin\chi\,\mathbf{e}_\phi$$

$$\mathbf{a}_{B/\alpha} = \dot{\boldsymbol{\Omega}} \times \mathbf{r} = 0$$

$$\boldsymbol{\Omega} \times (\boldsymbol{\Omega} \times \mathbf{r}) = -\Omega^2 r\sin\chi\,\mathbf{e}_\rho$$

$$= -\Omega^2 r\sin\chi\,(\mathbf{e}_\chi\cos\chi + \mathbf{e}_r\sin\chi)$$

$$\mathbf{a}_{S/\alpha} = \mathbf{a}_{S/\beta} + 2\boldsymbol{\Omega} \times \mathbf{v}_{S/\beta} + \mathbf{a}_{B/\alpha} + \dot{\boldsymbol{\Omega}} \times \mathbf{r} + \boldsymbol{\Omega} \times (\boldsymbol{\Omega} \times \mathbf{r})$$

$$= (\ddot{r} - r\Omega^2\sin^2\chi)\mathbf{e}_r - (r\Omega^2\sin\chi\cos\chi)\mathbf{e}_\chi + 2\Omega\dot{r}\sin\chi\,\mathbf{e}_\phi$$

at $t = \frac{\pi}{60}$ seconds $pt = \pi/3$ and

$$r = 0.3\text{ m} + (0.02\text{ m})\cos\pi/3 = 0.3100\text{ m}$$

$$\dot{r} = -(20\text{ s}^{-1})(0.02\text{ m})\sin\pi/3 = -0.3464\text{ m/s}$$

$$\ddot{r} = -(20\text{ s}^{-1})^2(0.02\text{ m})\cos\pi/3 = -4.000\text{ m/s}^2$$

$$\chi = 30°, \quad \Omega = 4\text{ rad/s}$$

$$\mathbf{a}_{S/\alpha} = -(5.240\text{ m/s}^2)\mathbf{e}_r - (2.148\text{ m/s}^2)\mathbf{e}_\chi - (1.386\text{ m/s}^2)\mathbf{e}_\phi$$

4-6

The rod OB rotates counterclockwise at a constant rate of 1 rad/s. The wheel rolls without slipping at the contact point C. At a specific instant, the spoke carrying the slider P is at 90° to OB, P is 0.2 m from B, moving away from B at 2 m/s relative to the spoke, while this relative speed is diminishing at 5 m/s^2.

Evaluate the ground-observed velocity and acceleration of P.

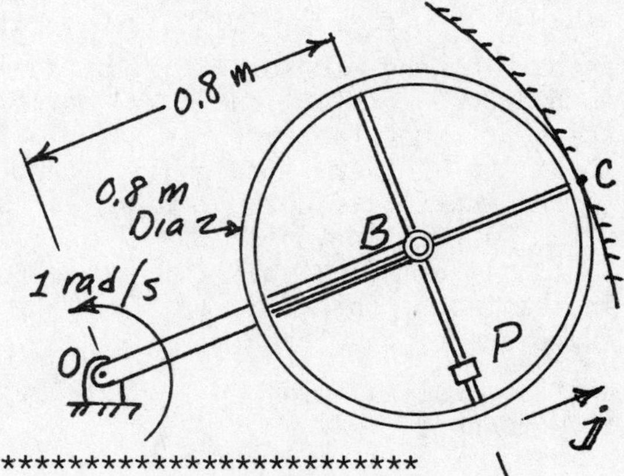

****

Let β denote the wheel. Its angular velocity is

$$\boldsymbol{\Omega} = -\frac{(0.8\,\text{m})(1\,\text{s}^{-1})}{0.4\,\text{m}}\mathbf{k} = -(2\text{ rad/s})\mathbf{k}$$

$$\mathbf{v}_{P/\alpha} = \mathbf{v}_{P/\beta} + \mathbf{v}_{B/\alpha} + \boldsymbol{\Omega}\times\mathbf{r} \qquad (\alpha \text{ denotes ground})$$

$$= [2\mathbf{i} - 0.8\mathbf{i} - (2\,\text{s}^{-1})(0.2\,\text{m})\mathbf{j}]\text{ m/s}$$

$$= (1.2\mathbf{i} - 0.4\mathbf{j})\text{ m/s}$$

$$\mathbf{a}_{P/\beta} = -(5\text{ m/s}^2)\mathbf{i}$$

$$2\boldsymbol{\Omega}\times\mathbf{v}_{P/\beta} = -2(2\,\text{s}^{-1})(2\text{ m/s})\mathbf{j} = -(8\text{ m/s}^2)\mathbf{j}$$

$$\mathbf{a}_{B/\alpha} = (0.8\,\text{m})(1\,\text{s}^{-1})^2\mathbf{j} = -(0.8\text{ m/s}^2)\mathbf{j}$$

$$\dot{\boldsymbol{\Omega}}\times\mathbf{r} = \mathbf{0}$$

$$\boldsymbol{\Omega}\times(\boldsymbol{\Omega}\times\mathbf{r}) = -(2\,\text{s}^{-1})^2(0.2\,\text{m})\mathbf{i} = -(0.8\text{ m/s}^2)\mathbf{i}$$

$\mathbf{a}_{P/\alpha}$ = Sum of above five terms

$$= (-5.8\mathbf{u}_x - 8.8\mathbf{u}_y)\text{ m/s}^2$$

4-7

Bert Barnstormer (nicknamed β) has his aircraft in a vertical tailspin. A point B on the windshield is moving vertically downward at constant speed v_B, while the aircraft spins at a constant angular speed Ω (rad/unit time) about the vertical. Meanwhile, Bert's partner, Paul (P for short), is doing a vertical loop maneuver. At a particular instant, Paul is directly under Bert, moving horizontally at speed v_P, this speed increasing at the rate $\dot{v}_P$. The local radius of curvature of the loop is ρ_C.

What are Paul's velocity and acceleration relative to Bert?

Ω β v_B P v_P

****

$\mathbf{j}$ $\mathbf{i}$ G: Ground

The relative velocity relationship

$$\mathbf{v}_{P/G} = \mathbf{v}_{P/\beta} + \mathbf{v}_{B/G} + \boldsymbol{\Omega} \times \mathbf{r}_{P/B}$$

yields

$$\mathbf{v}_{P/\beta} = \mathbf{v}_{P/G} - \mathbf{v}_{B/G} - \cancelto{0}{\boldsymbol{\Omega} \times \mathbf{r}} = \underline{v_P\,\mathbf{i} + v_B\,\mathbf{j}}$$

The relative acceleration relationship

$$\mathbf{a}_{P/G} = \mathbf{a}_{P/\beta} + 2\boldsymbol{\Omega} \times \mathbf{v}_{P/\beta} + \mathbf{a}_{B/G} + \dot{\boldsymbol{\Omega}} \times \mathbf{r} + \boldsymbol{\Omega} \times (\boldsymbol{\Omega} \times \mathbf{r})$$

yields

$$\mathbf{a}_{P/\beta} = \mathbf{a}_{P/G} - 2\boldsymbol{\Omega} \times \mathbf{v}_{P/\beta} - \cancelto{0}{\mathbf{a}_{B/G}} - \cancelto{0}{\dot{\boldsymbol{\Omega}} \times \mathbf{r}} - \boldsymbol{\Omega} \times \cancelto{0}{(\boldsymbol{\Omega} \times \mathbf{r})}$$

$$= \dot{v}_P\,\mathbf{i} + \frac{v_P^2}{\rho_c}\mathbf{j} - 2(-\Omega\,\mathbf{j}) \times (v_P\,\mathbf{i} + v_B\,\mathbf{j})$$

$$= \underline{\dot{v}_P\,\mathbf{i} + \frac{v_P^2}{\rho_c}\mathbf{j} - 2\Omega v_P\,\mathbf{k}}$$

5 KINETICS OF A PARTICLE

APPLICATIONS OF NEWTON'S SECOND LAW

5-1

Two bars support a body of 20 kg which is rotating with an angular speed of 200 rpm around the axis x-x.Determine the tensions in the bars.

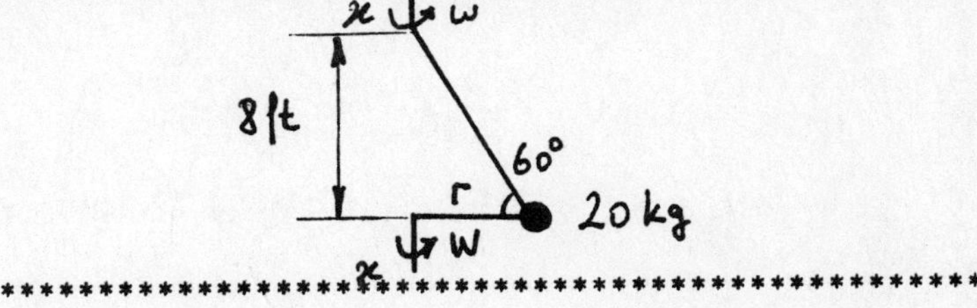

****

Free body diagram for the body :

T_1, $60°$, $20\,kg$, x, T_2, y, W

$\Sigma F_y = 0 \longrightarrow T_1 . \sin 60° - W = 0$ $\qquad T_1 = \dfrac{20 \times 9.8}{\sin 60°} = 226.55\ N$

$\Sigma F_x = m . a_x \longrightarrow T_2 + T_1 . \cos 60° = m \dfrac{V^2}{r}$

Angular speed : $\omega = \dfrac{2 \pi n}{60} = \dfrac{2 \pi \times 200}{60} = 20.94$ rad/sec

$r = \dfrac{8}{\tan 60°} = 4.62\,m$, $V = \omega . r = 20.94 \times 4.62 = 96.72$ m/sec

$$\therefore \quad T_2 = 20 \times \frac{96.72^2}{4.62} - 226.55 \times \cos 60°$$

$$T_2 = 40392.3\ N$$

5-2

Block A supporting another block B is drawn along a rough horizontal plane by a horizontal force P. The coefficient of sliding friction between A and the plane is 0.3; the coefficient of static friction between A and B is 0.4. If the weight of A is 200 lbs and the weight of B is 100 lbs, find the greatest value of P if B does not slip on A.

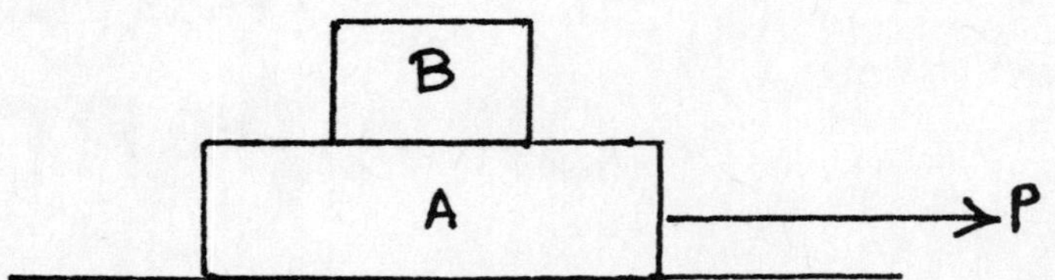

****

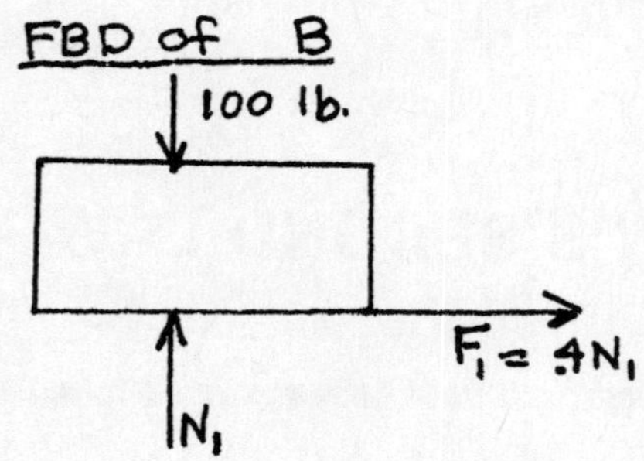

FBD of A

N_1, 200 lb., F_1, P, $F_2 = .3N_2$, N_2

$\Sigma F_y = 0 \qquad N_1 = 100$ lb.

$F_1 = 40$ lb.

$\Sigma F_y = 0 \qquad N_2 = N_1 + 200 = 300$ lb

$F_2 = .3(300) = 90$ lb.

$\Sigma F_x = ma$

$40 = \frac{100}{g} \cdot a$

$a = .4g$

$\Sigma F_x = ma$

$P - 40 - 90 = \frac{200}{g} \cdot a$

$P - 130 = \frac{200}{g} \cdot .4g = 80$

$P = 210$ lb.

5-3

Determine the tension P in the cable which will give the block shown an acceleration of 5 feet per second per second up the incline.

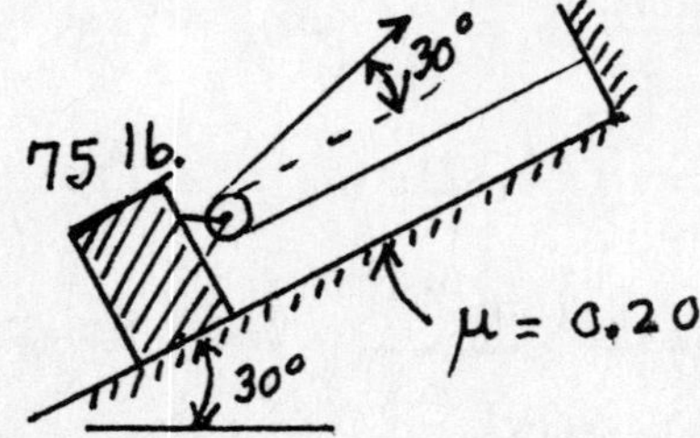

****

FBD OF BLOCK:

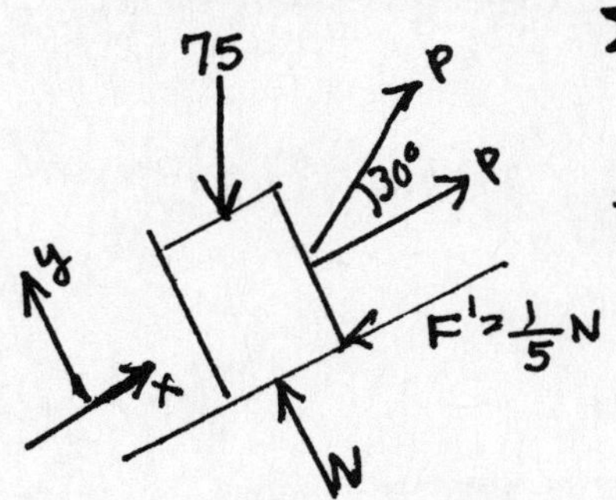

$$\Sigma F_x: -75 \cdot \frac{1}{2} - \frac{1}{5} N + P + P\cos 30^\circ = \frac{75}{g} \cdot 5$$

$$\Sigma F_y: N - 75\cos 30^\circ + P\sin 30^\circ = 0$$

SOLVING SIMULTANEOUSLY

$$P = 31.61 \text{ lb.}$$

5-4

Determine acceleration of the system masses and tension in the cord as indicated. Neglect the masses of the pulley and the cord. Consider the pulley frictionless and the coefficient of kinetic friction between the upper block and horizontal plane to be 0.25.

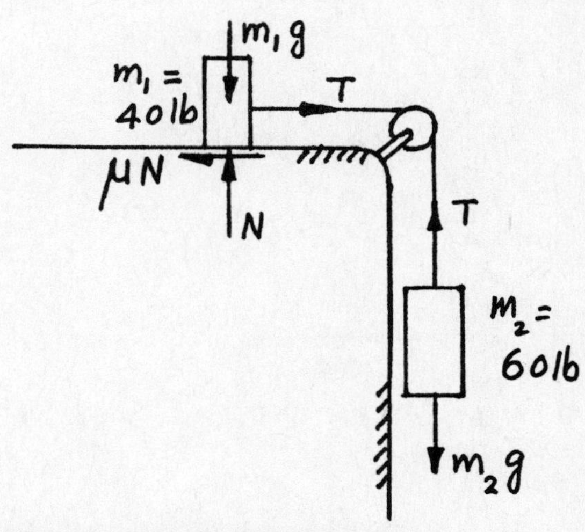

****

Applying Newton's second law of motion, one gets

$$T - \mu N = m_1 a_H \quad \text{---------} \quad (i)$$

$$m_2 g - T = m_2 a_V \quad \text{---------} \quad (ii)$$

Incorporating the information, $a_H = a_V$, and, $N = m_1 g$ in equations (i) & (ii), and adding them,

$$m_2 g - \mu m_1 g = (m_1 + m_2) a_H$$

$$\text{or,}\quad a_H = a_V = \frac{60 - 0.25 \times 40}{(100/32.2)} = 16.1\ ft/sec^2$$

$$\text{Also, from (i)}\quad T = m_1 a_H + \mu N = \frac{40}{32.2} \times 16.1 + 0.25 \times 40$$

$$\text{or,}\quad T = 30\ lb$$

5-5

The system shown consists of two weights suspended from a cord that passes over a pulley. When it is released from rest, the tension in the cord will be:

a) 100 pounds b) 200 pounds c) 240 pounds

d) 300 pounds e) 500 pounds

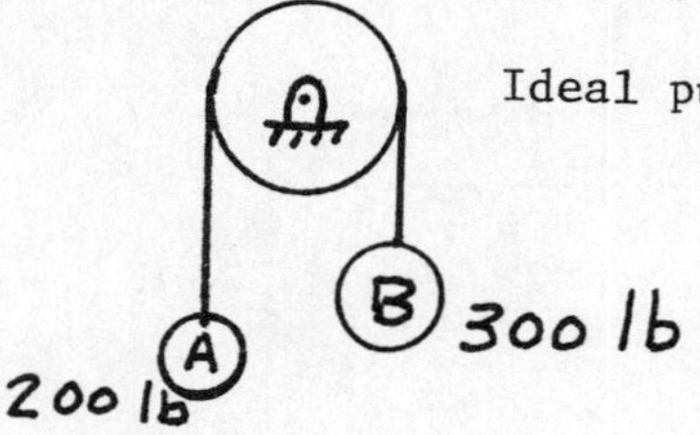

Ideal pulley (weightless and frictionless)

****

A WILL ACCELERATE UPWARDS; B DOWNWARDS

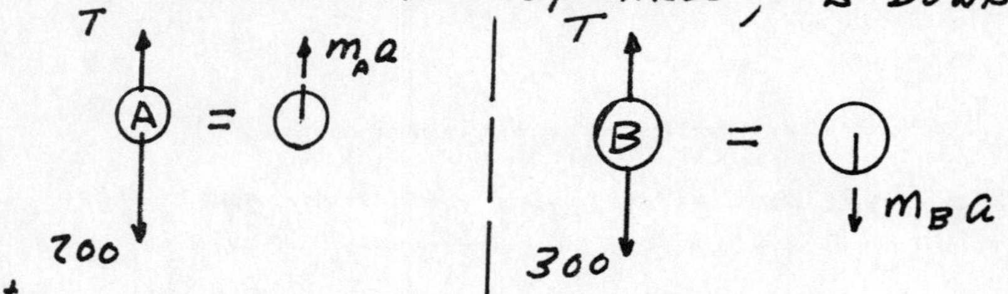

$$+\uparrow \Sigma F = ma = \frac{Wa}{g} \qquad +\downarrow \Sigma F = ma = \frac{Wa}{g}$$

$$\text{a)}\ T - 200 = \frac{200}{32.2} a \qquad \text{b)}\ 300 - T = \frac{300}{32.2} a$$

SOLVE EQUATIONS a) AND b) TO GET $T = 240\ lb$

THE ANSWER IS C.

Note: Since body A accelerates upwards, the free body of A indicates that the tension will have to be greater than 200 pounds. Likewise, the free body of B indicates the tension will have to be less than 300 pounds if it is to accelerate downward. Only answer c is possible, and no calculations are required to solve the problem.

5-6

A stone weighing 5 lbs is fastened to a string and is whirled in a vertical circle of radius 2 feet.

(a) Find the minimum speed of the stone if the string is to stay taut.

(b) If the speed of the stone is 20 ft/sec, what tension must the string be able to withstand?

****

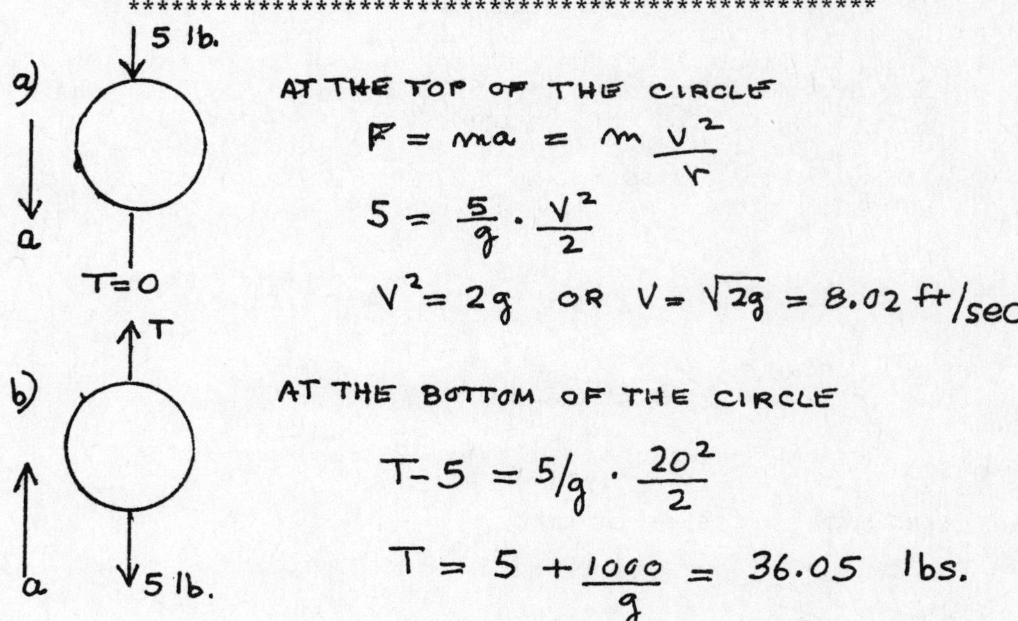

$$F = ma = m\frac{V^2}{r}$$

$$5 = \frac{5}{g}\cdot\frac{V^2}{2}$$

$$V^2 = 2g \quad \text{OR} \quad V = \sqrt{2g} = 8.02 \text{ ft/sec}$$

$$T - 5 = 5/g \cdot \frac{20^2}{2}$$

$$T = 5 + \frac{1000}{g} = 36.05 \text{ lbs.}$$

5-7

The 8 pound collar slides down the smooth circular rod. In the position shown its velocity is 3 ft/s. Find the normal force (the contact force) the rod exerts on the collar.

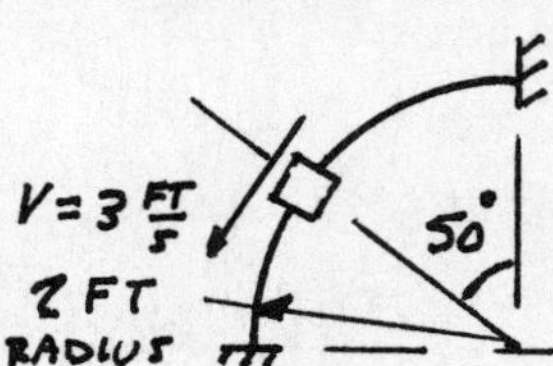

****

FIND NORMAL ACCELERATION: $a_N = \frac{V^2}{\rho} = \frac{3^2}{2} = 4.5 \text{ FT/S}^2$

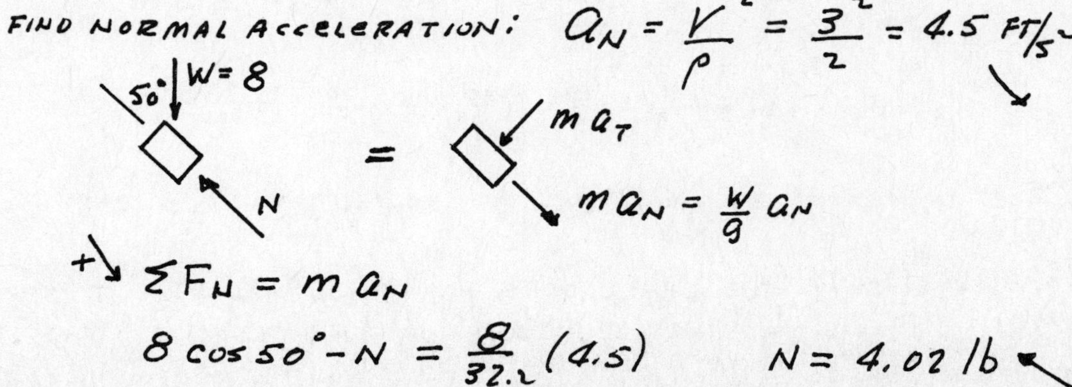

$$+\searrow \Sigma F_N = m a_N$$

$$8\cos 50° - N = \frac{8}{32.2}(4.5) \qquad N = 4.02 \text{ lb}$$

5-8

A two ton pickup truck is traveling down a 10° incline under icy conditions at a speed of 30 mph. The truck driver sees a car stalled at the bottom of the incline and applies his brakes. Due to icy conditions a constant braking force of only 1200 lbs is developed on the truck.

a. If the car is 250 ft. from the truck when the driver applies the brakes does the truck stop before it reaches the car ? How long does it takes the truck to stop ?
b. Now assume that the truck is carrying a 100 lb. crate in the back that is not secured. The values of the friction coefficients between the crate and the truck are $\mu_s = .2$ and $\mu_k = .18$. Does the crate move relative to the truck when the driver applies the brakes?

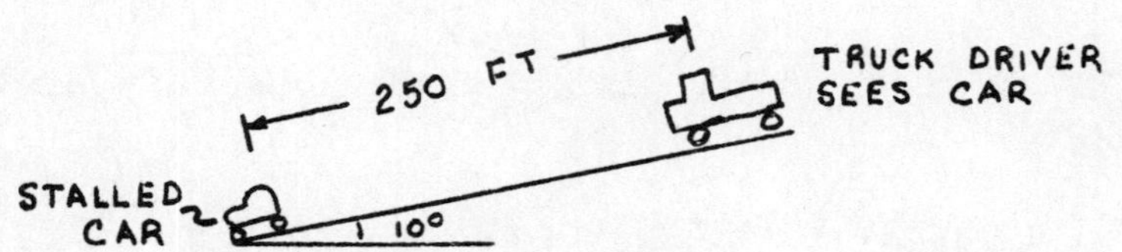

****

(a) FREE BODY DIAGRAM OF TRUCK

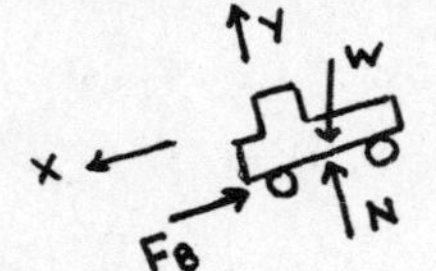

$$\Sigma F_x = m a_x$$

$$-F_B + W \sin 10° = m a_x$$

$$-1200 \text{ LBS} + 4000 \text{ LBS} (\sin 10°) = \frac{2000 \text{ LBS}}{32.2 \text{ FT/sec}^2} a_x$$

$$a_x = -4.068 \text{ FT/sec}^2$$

UNIFORMLY ACCELERATED MOTION RESULTS DUE TO THE BRAKING (a_x = CONSTANT). THUS

$$v^2 = v_0^2 + 2 a_x \Delta x$$

WHERE V_0 = 30 miles per hour , $V = 0$

THEN

$$\Delta x = \frac{-V_0^2}{2a_x} = \frac{-\left[(30\text{mph})(5280 \text{ FT/MILE})(1 \text{ HR}/3600 \text{ SEC})\right]^2}{2(4.068) \text{ FT/sec}^2}$$

$\Delta x = 238$ FT $\longrightarrow$ YES, THE TRUCK STOPS IN TIME.

$$x = a \frac{t^2}{2} + v_0 t$$

$$238 \text{ FT} = -4.068 \text{ FT/sec}^2 \frac{t^2}{2} + \left(\frac{30 \text{mi}}{\text{hr}}\right)\left(\frac{5280 \text{ FT}}{\text{mi}}\right)\left(\frac{1 \text{hr}}{3600 \text{sec}}\right) t$$

THE QUADRATIC EQUATION CAN BE SOLVED TO YIELD $t = 21.63$ sec.

(b) ASSUME THAT THE CRATE DOES NOT SLIDE RELATIVE TO THE TRUCK. THEN THE ACCELERATION OF THE CRATE IS THE

SAME AS THE ACCELERATION OF THE TRUCK

$$-F_B + W \sin 10° = ma_x \qquad (W: 4100 \text{ LBS})$$

$$a_x = -3.83 \text{ FT/SEC}^2$$

FREE BODY DIAGRAM OF CRATE

W_c, N, F

$$\Sigma F_x = 0$$

$$-F + W_c \sin 10° = m_c a_x$$

$$-F + 100 \text{ LBS} \sin 10° = \frac{100 \text{ LBS}}{32.2 \text{ FT/SEC}^2} (-3.83 \tfrac{\text{FT}}{\text{SEC}^2})$$

$$F = 29.3 \text{ LBS}$$

$$\Sigma F_y = 0 \qquad N - W_c \cos 10° = 0$$

$$N = 98.4 \text{ LBS}$$

$$\mu_s N = 19.7 \text{ LBS}$$

YES THE CRATE MOVES RELATIVE TO THE TRUCK AS THE FRICTION FORCE OBTAINED ASSUMING NO RELATIVE MOTION IS GREATER THAN THE ALLOWABLE FRICTION FORCE. THUS THE ASSUMPTION OF NO RELATIVE MOTION IS FALSE.

5-9

An elevator decelerates downward at 2 m/s^2 prior to a stop. Determine the upward reaction exerted by the elevator floor on a man of mass 80 kg.

****

From Newton's second law

Force = Mass x acceleration or $\Sigma F = ma$

$g = 9.81 \text{ m/s}^2$

80 kg, -2 m/s^2, R

$$\downarrow \quad 80 \times 9.81 - R = 80 \times (-2)$$

Hence $R = 80 \times 9.81 + 80 \times 2$

$$= \underline{944.8 \text{ N}}$$

5-10

The system shown is released from rest. Determine the acceleration of the blocks and the tension in the cable.

W_A = 100 lb

W_B = 200 lb

Coefficient of friction μ = 0.25. The pulleys are frictionless.

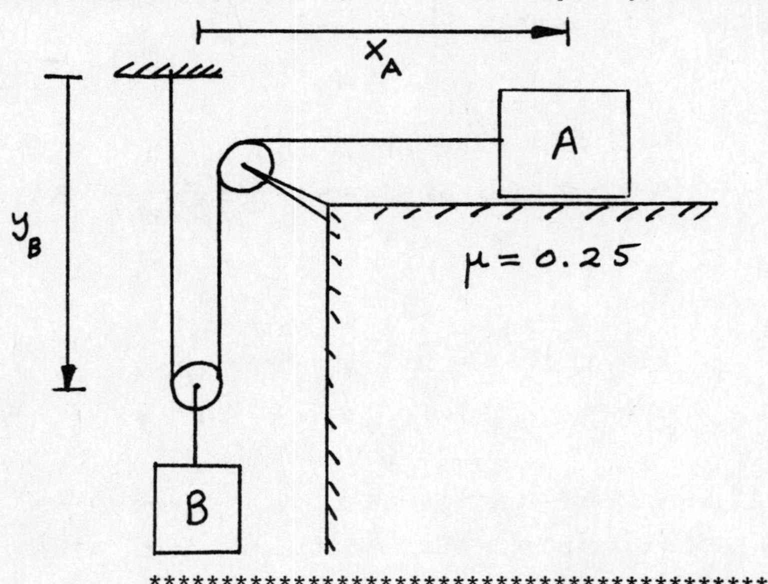

****

Solution :

$$2y_B + x_A = \text{Constant}$$

$$2v_B + v_A = 0$$

$$2a_B + a_A = 0 \qquad (1)$$

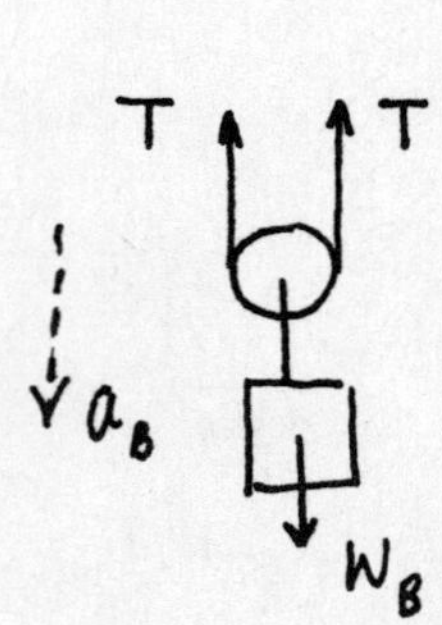

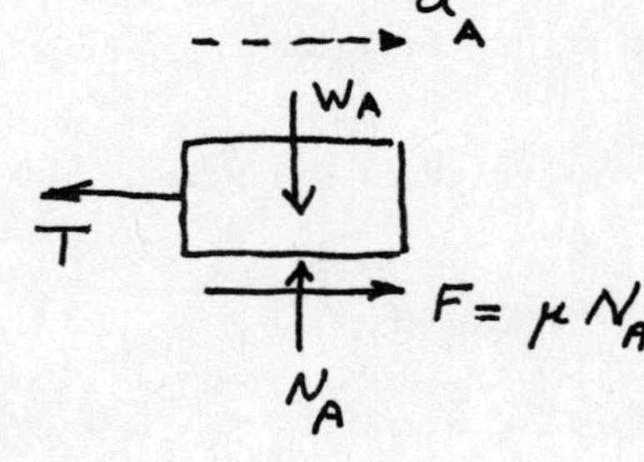

$$N_A = W_A \qquad (3)$$

$$W_B - 2T = \frac{W_B}{g} a_B \qquad (2)$$

$$\mu N_A - T = \frac{W_A}{g} a_A \qquad (4)$$

Solving (1), (2), (3) and (4) we obtain:

$a_B = 0.25g = 8.05 \text{ ft/s}^2 \downarrow$

$a_A = -2a_B = -0.5g \rightarrow$ or $a_A = 16.1 \text{ ft/s}^2 \leftarrow$

$T = 75 \text{ lb}$.

5-11

The member OA rotates about a horizontal axis through O with a constant counterclockwise velocity of 3 rad/sec.. As it passes the position where θ is zero, a small mass m is placed upon it at a radial distance r of 18 inches. If the mass is observed to slip at θ of 45 degrees, determine the coefficient of friction between the mass and the member OA.

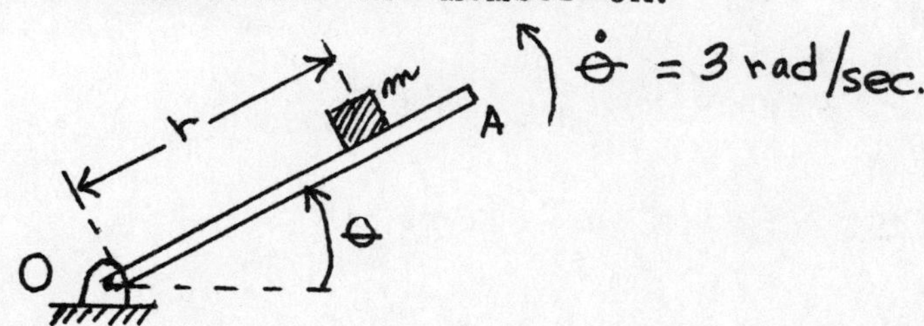

****

FBD OF BLOCK: (AT PT. OF SLIPPING)

θ, r, mg, $F' = \mu N$, N

$$\Sigma F_r: -\frac{mg}{\sqrt{2}} + \mu N = m a_r$$

$$= m\left(0 - \tfrac{3}{2} \cdot 9\right)$$

$$\Sigma F_\theta: N - \frac{mg}{\sqrt{2}} = 0$$

NOTE:

$$\bar{a} = (\ddot{r} - r\dot{\theta}^2)\bar{e}_r + (r\ddot{\theta} + 2\dot{r}\dot{\theta})\bar{e}_\theta$$

HERE $\ddot{r} = 0$ (up to slipping pt.)

$\dot{r} = 0$ $\dot{\theta} = 3$

$r = 1.5 \text{ ft}$ $\ddot{\theta} = 0$

Substituting for N & solving:

$\mu = 0.407$

5-12

Using the Dynamic Equilibrium Method, determine the distance that Block A will travel if A has a mass of 50Kg and is traveling at 6 $\frac{m}{s}$ up the slope at the instant shown.

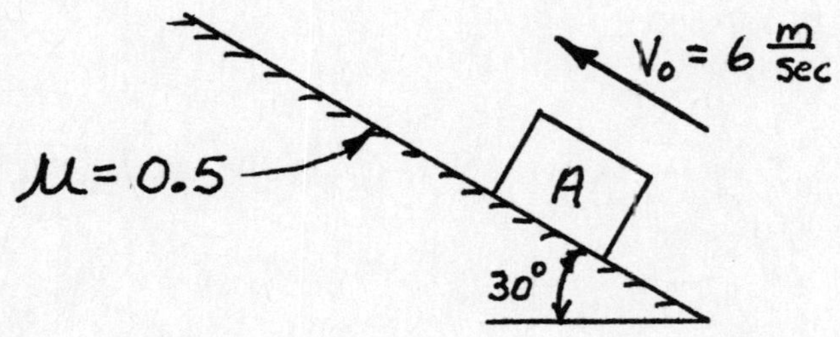

****

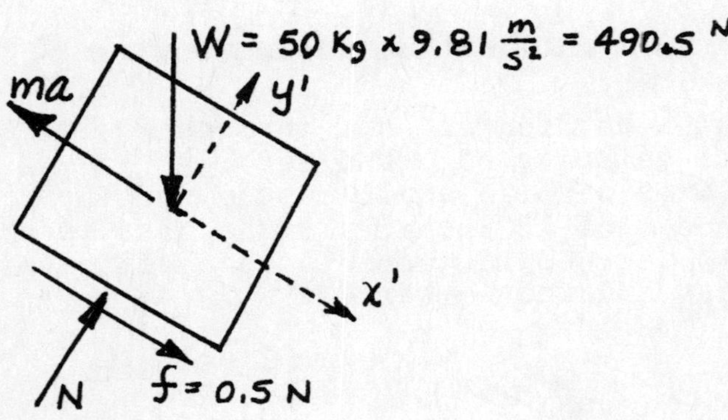

FBD BLOCK A

$\Sigma F_{y'} = 0 \qquad N - W \cos 30° = 0$

$N = (490.5^N)(\cos 30°) = 424.79^N$

$f = 0.5 N = 212.39^N$

$\Sigma F_{x'} = 0 \quad -ma - f - W(\sin 30°) = 0$

$-(50 \text{ Kg})\, a = 212.39^N + (490.5^N)\sin 30° = 457.64^N$

$a = \frac{-457.64^N}{50 \text{ Kg}} = -9.15 \frac{m}{sec^2}$

$V^2 = V_o^2 + 2as$

$S = \frac{V^2 - V_o^2}{2a} = \frac{0^2 - (6\frac{m}{s})^2}{2(-9.15 \frac{m}{sec^2})} = 1.97m$

5-13

Body A weighs 64.4 lb, body B weighs 32.2 lb, and body C weighs 16.1 lb. The coefficient of friction between A and the plane is 0.40 and the coefficient of friction between B and the plane is 0.30. Draw a free body diagram showing the magnitude of all the forces acting on body B as the whole system is released from rest.

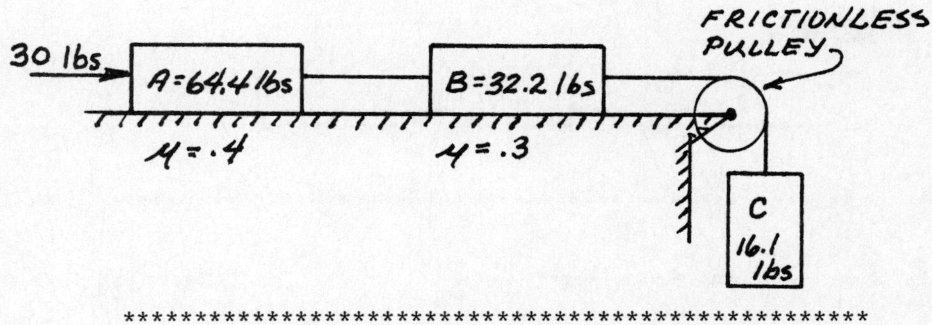

****

ASSUME THEY MOVE TOGETHER, $a_A = a_B = a_C = a$

$\Sigma F_y = 0$ $N_A = 64.4$ FRICTION FORCE $= .4(64.4) = 25.8$ lbs

$N_B = 32.2$ " " $= .3(32.2) = 9.66$ lbs

30, T, 25.8   T, T_1, 9.66 T_1, 16.1

$\Sigma F = ma$

$T + 30 - 25.8 = 2(a)$ $\quad T_1 - T - 9.66 = 1(a)$ $\quad 16.1 - T_1 = \frac{1}{2}(a)$

$T + 4.2 = 2a$

$T_1 - (2a - 4.2) - 9.66 = a, \; T_1 = 3a + 5.46$

$16.1 - (3a + 5.46) = \frac{1}{2}a$

$10.64 = 3.5a, \; a = 3.04 \text{ ft/s}^2$

$T = 2(3.04) - 4.2 = 1.88$ lbs $\quad T_1 = 16.1 - \frac{3.04}{2} = 14.58$ lbs

THEREFORE ASSUMPTION IS CORRECT

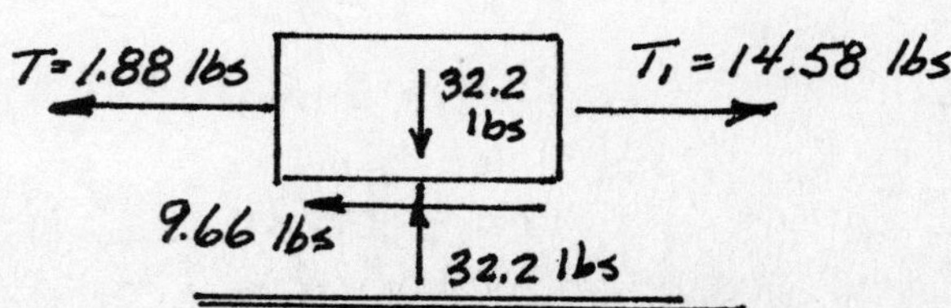

5-14

A locomative A pulls a freight car B when they are travelling at a speed of 40 mph. Suddenly the engineer notices that a person stands on the railway and applies the brakes. The brake forces are 5000 and 3000 lb on A and B, respectively. Determine the distance required to stop the train. What is the force in the coupling between the locomative A and the car B ?

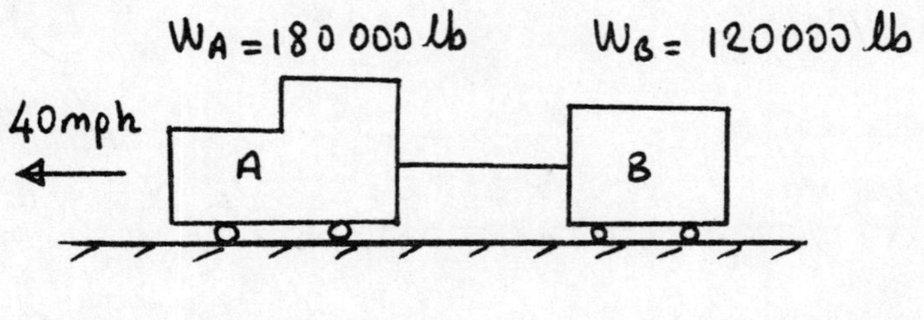

****

A and B are moving together. They can be considered as a single system : Newton's Second Law $\sum\vec{F}=m\vec{a}$ gives

$$-F_A - F_B = (m_A + m_B)\, a_x \qquad (\overset{+}{\leftarrow} \text{ positive direction})$$

$$-5000 - 3000 = \frac{(1.8+1.2)\times 10^5}{32.2}\cdot a_x \longrightarrow a_x = -0.858 \text{ ft/sec}^2$$

This is uniformly decelerating motion. ∴ we can write

$$V^2 = V_{x_0}^2 + 2\, a_x \cdot \Delta x$$

$$V_{x_0} = 40 \text{ mph} = \frac{40\times 5280}{3600} = 58.67 \text{ ft/sec}.$$

$$0 = 58.67^2 + 2\times(-0.858)\times \Delta x \longrightarrow \Delta x = 2004.36 \text{ ft}$$

stopping distance.

Now, let us consider the locomative A ; Newton second law

$$-F_A - T = m_A \cdot a_x$$

$$-5000 - T = \frac{1.8\times 10^5}{32.2}(-0.858) \longrightarrow T = -200 \text{ lb}_f$$

force in the coupling.

CHECK: For B

$$T - F_B = m_B\, a_x$$

$$T - 3000 = \frac{1.2\times 10^5}{32.2}\cdot(-0.858) \longrightarrow T \cong 200 \text{ lb}_f.$$

5-15

A crate weighing 500 lb is to be pulled up a 29.5° slope concrete ramp by a force P, as shown in the accompanying illustration. If the coefficient of friction $\mu = 0.27$ between the concrete and the crate, calculate (a) the constant value of P (= P_1) needed to bring the crate initially at rest to a velocity of 15 ft/s in 10 s, and (b) the new value of P (= P_2) required to maintain the same velocity.

y, W, P, a, v, x, $\mu = 0.27$, 29.5°

****

Solution

W, W_n, W_t, θ, $\theta = 29.5°$

Resolving W into the components W_t and W_n, we get

$$W_t = W \sin\theta = (500 \text{ lb})(0.492) = 246.211 \text{ lb}$$

$$W_n = W \cos\theta = (500 \text{ lb})(0.870) = 435.177 \text{ lb}$$

For acceleration, we have

$$v = v_0 + at \; ; \quad a = \frac{v - v_0}{t}$$

Hence $a = \dfrac{(15 \text{ ft/s}) - (0)}{(10 \text{ s})} = 1.5 \text{ ft/s}^2$

W_n, P_1, F, W_t, N, F_i

Force F_i is given by

$$F = ma \; ; \quad F_i = \frac{W}{g} a$$

Hence $F_i = \dfrac{(500 \text{ lb})(1.5 \text{ ft/s}^2)}{(32.2 \text{ ft/s}^2)} = 23.29 \text{ lb}$

(a) Force P_1 :

$\Sigma F_y = 0 \; ; \quad W_n = N = 435.177 \text{ lb}$

$F = \mu N ; \quad F = (0.27)(435.177 \text{ lb}) = 117.497 \text{ lb}$

$\Sigma F_x = 0 \; ; \quad P_1 = W_t + F_i + F$

$= (246.211 \text{ lb}) + (23.29 \text{ lb}) + (117.497 \text{ lb})$

$= 386.998 \text{ lb.}$ **Ans.**

(b) Force P_2 :

After reaching a velocity of 15 ft/s, acceleration $a = 0$

$\therefore \; F = ma \; ; \quad F_i = \dfrac{W}{g} a = 0$

Hence, $P_2 = W_t + F_i + F$

$= (246.211 \text{ lb}) + 0 + 117.497 \text{ lb}) = 363.708 \text{ lb.}$

Ans.

5-16

A particle having a mass of 50 kg travels along a track whose equation is given by $y = kx^2$. Initially, the particle starts from rest at A (2,8). Find the tangential and normal components of the acceleration as well as the normal force exerted by the track on the particle when the particle at B (1,2), then at C (0,0). all distances are given in m (meters).

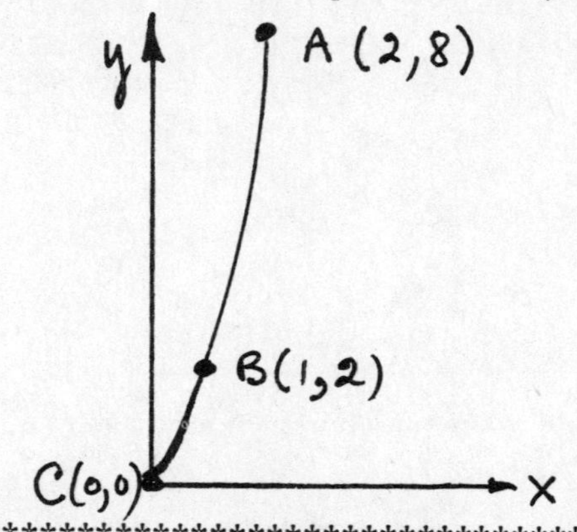

****

Substitute coordinates of A into the eq. of the curve. $8 = k(2)^2$. Hence $k = 2$

The radius of curvature is $\rho = (1+y'^2)^{3/2}/(y'')$

or with $y' = 4x$, $y'' = 4$; $\rho = (1+16x^2)^{3/2}/4$

At B and C $\rho_B = 17.52$ m. and $\rho_C = 0.25$ m.

The speed at any point is $(2gh)^{1/2}$

$$V_B = [2(9.81)(6)]^{1/2} = 10.85 \text{ m/s}$$

$$V_C = [2(9.81)\,8]^{1/2} = 12.53 \text{ m/s}$$

The normal accelerations are

$$(a_n)_B = \frac{V_B^2}{\rho_B} = \frac{(10.85)^2}{17.52} = \underline{\underline{6.72 \text{ m/s}^2}}$$

$$(a_n)_C = \frac{V_C^2}{\rho_C} = \frac{(12.53)^2}{0.25} = \underline{\underline{628 \text{ m/s}^2}}$$

The slope at B is $y'_B = 4(1) = 4$

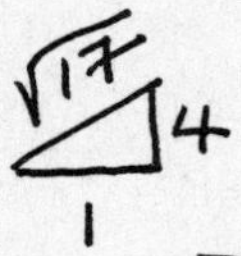

A freebody diagram showing $\vec{F} = m\vec{a}$ gives at B

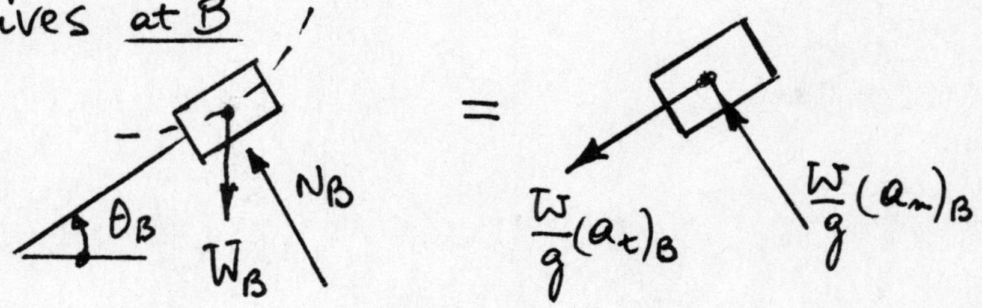

Summation along the normal gives

$N_B - W_B \cos\theta_B = \frac{W}{g}(a_n)_B$. Solving for N_B,

$N_B = \left[\frac{1}{\sqrt{17}} + 6.72\right] W = (6.96)(50) = \underline{\underline{348\ N}}$

along the tangent

$W \sin\theta_B = \frac{W}{g}(a_t)_B$ $\qquad$ $\underline{\underline{(a_t)_B = 9.52\ m/s^2}}$

At C

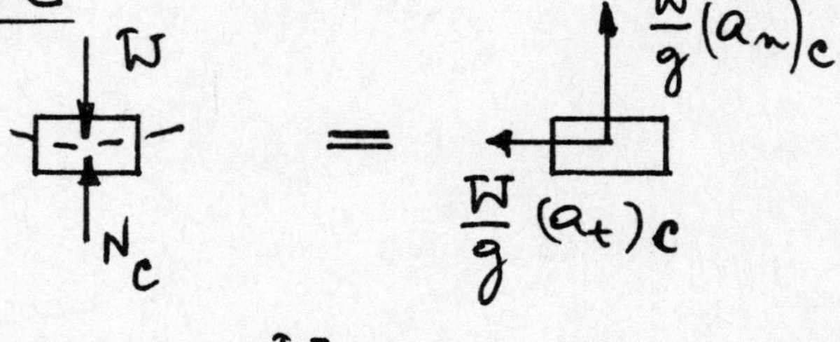

$N_c - W = \frac{W}{g}(a_n)_c \; ; \; N_c = W\left[1 + \frac{(a_n)_c}{g}\right]$

$\underline{\underline{N_c = 3250.8\ N \uparrow}}$

$\frac{W}{g}(a_t)_c = 0$ $\qquad$ $\underline{\underline{(a_t)_c = 0}}$

This is expected since C is the position of maximum velocity.

5-17

A 20 lb. parcel is pushed down an incline with velocity of 15 ft/sec. The coefficient of friction is μ = .4. (a) Find the velocity when the parcel hits the end of the incline. (b) If, inadvertently, the parcel is pushed up the incline, how far would it travel before coming to rest?

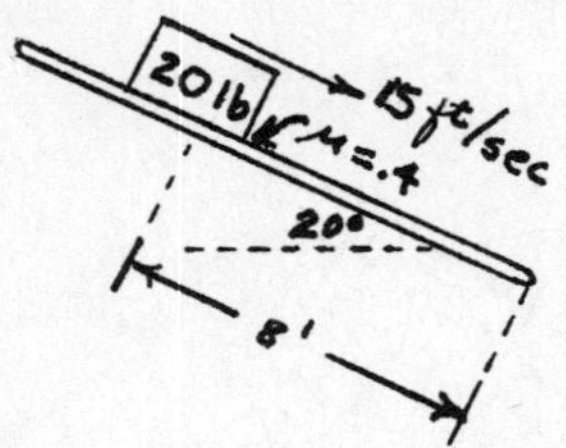

****

(a) F.B.D:

$+\nearrow \Sigma F_y = 0: \quad N - W\cos 20° = 0$

$\therefore N = W\cos 20°$

$N = .9397W$

$\therefore F = \mu N = .4(.9397)W$

$F = .3759W$

$\nwarrow^{+} \Sigma F_x = ma_x$

$F - W\sin 20° = \frac{W}{g}a_x$

$.3759W - .3420W = \frac{W}{g}a_x$

$.0339 = \frac{a_x}{g}$

$\therefore a_x = .0339g = (.0339)(32.2)$

$a = a_x = 1.091 \text{ ft/sec}^2$ (20°)

Since we have uniformly accelerated motion:

$v_f^2 = v_0^2 + 2as$

but $v_0 = 15$, $a = 1.091 \text{ ft/sec}^2$, $s = -8$

$\therefore v_f^2 = 225 + 2(1.091)(-8)$

$= 225 - 17.456$

$v_f^2 = 207.54$

$\Rightarrow v_f = 14.4 \text{ ft/sec}$

(b)

as in (a) $N = .9397W$

$F = \mu N = .3759W$

$\searrow^{+} \Sigma F_x = ma_x$

$.3759W + W\sin 20° = \frac{W}{g}a_x$

$.7179 = .3759 + .3420 = \frac{a_x}{g}$

$a_x = .7179g$

$= .7179(32.2)$

$a = a_x = 23.116 \text{ ft/sec}^2$

if $\searrow^{+}$ $\therefore v_f^2 = v_0^2 + 2as$

$0 = 15^2 + 2(23.116)s$

$\therefore s = -\frac{225}{46.232} = -4.87$

$\therefore s = 4.87$ feet up the incline.

5-18

Block A weighs 100 pounds and rests on block B which weighs 200 pounds as shown below. The coefficient of static friction between the blocks is 0.25. The coefficient of kinetic friction between block B and the inclined plane is 0.20. Determine the maximum force, P, which can be applied to the cable so that block A will not slide off block B.

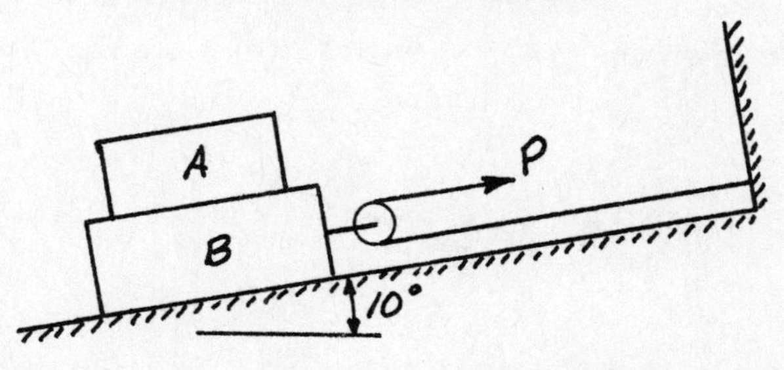

****

FBD of Pulley:

T, P, P

$$\overset{+}{\rightarrow} \Sigma F_x = m a_x = 0$$

$$2P - T = 0 \qquad T = 2P$$

FBD of Block A:

100 lb, .25N₁, N₁

$$+\!\uparrow \Sigma F_y = m a_y = 0$$

$$N_1 - 100 \cos 10 = 0 \qquad N_1 = 98.48 \text{ lb}$$

$$\overset{+}{\rightarrow} \Sigma F_x = m a_x$$

$$.25(98.48) - 100 \sin 10 = \frac{100}{32.2} \cdot a_x$$

$$a_x = 2.34 \text{ ft/sec}^2$$

FBD of Block B:

98.48 lb, 24.62 lb, 200 lb, 2P, .20N₂, N₂

$$+\!\uparrow \Sigma F_y = m a_y = 0$$

$$N_2 - 98.48 - 200 \cos 10 = 0$$

$$N_2 = 295.44 \text{ lb}$$

$$\overset{+}{\rightarrow} \Sigma F_x = m a_x$$

$$-24.62 - .2(295.44) - 200 \sin 10 + 2P = \frac{200}{32.2}(2.34)$$

$$P = 66.51 \text{ lb}$$

5-19

A small sphere of weight W is held as shown by two wires AB and CD. Wire AB is then cut. Determine:
(a) the tension in wire CD before AB was cut.
(b) the tension in wire CD just after AB is cut.
(c) the acceleration of the sphere just after AB is cut.

****

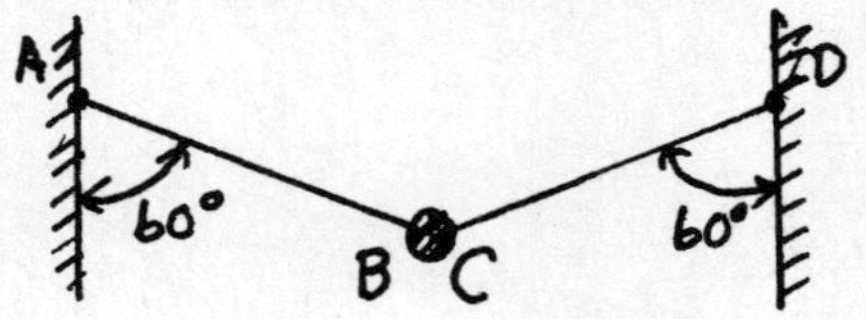

A D 60° B C 60°

(a) FBD OF SPHERE

T 30° T 30° W

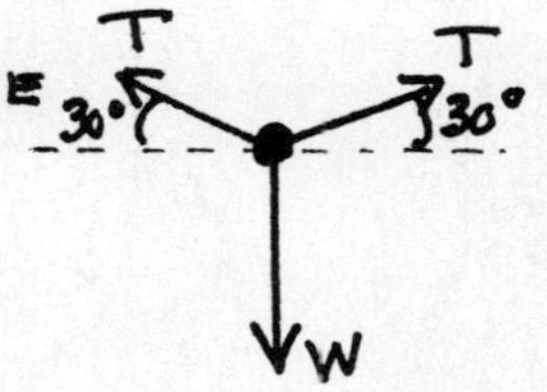

$T_{AB} = T_{BC} = T$

$\Sigma F_y = 0$ (equilibrium)

$\frac{T}{2} + \frac{T}{2} - W = 0$

$T = W$

(b) FBD OF SPHERE

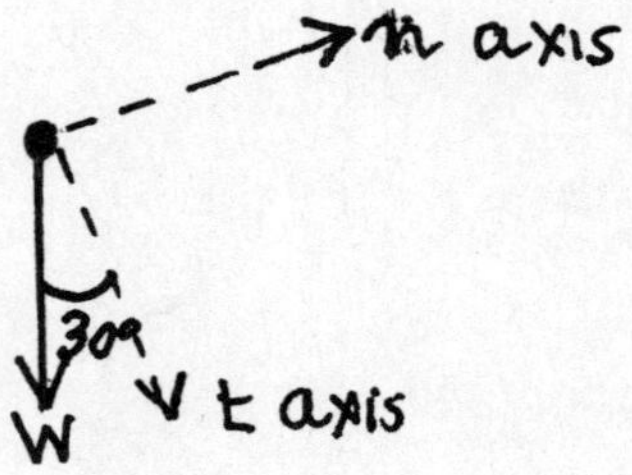

n axis, 30°, W, t axis

$V = 0; \therefore a_n = 0$

$\Sigma F_n = 0$

$T - W \sin 30° = 0 \qquad T = W/2$

(c) $\Sigma F_t = m a_t$

$W \cos 30° = \frac{W}{g} \cdot a_t \qquad \therefore a_t = a = g \cos 30°$

$= 0.866\, g$

5-20

A car enters a curve, radius = 200 ft., at a speed of 40 ft/sec. The car increases its speed at a constant rate of 5 ft/sec^2. $\mu = .8$
A) Draw a clear free body diagram of the car.
B) Define your origin and the positive directions you assume.
C) At what speed will the car begin to skid?

****

TOTAL $\vec{a} = \vec{a}_T + \vec{a}_N$

$\sum F = ma$

$\mu W = \frac{W}{g} a$

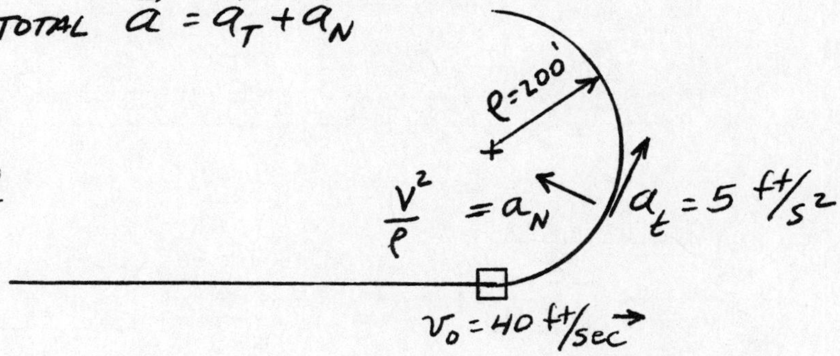

$a = .8g$

$= 25.76 \text{ ft/s}^2$

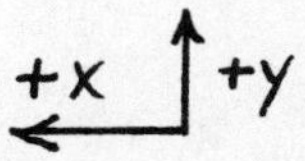

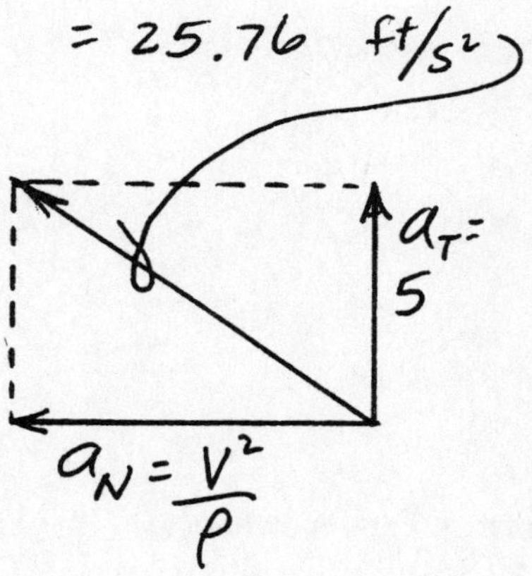

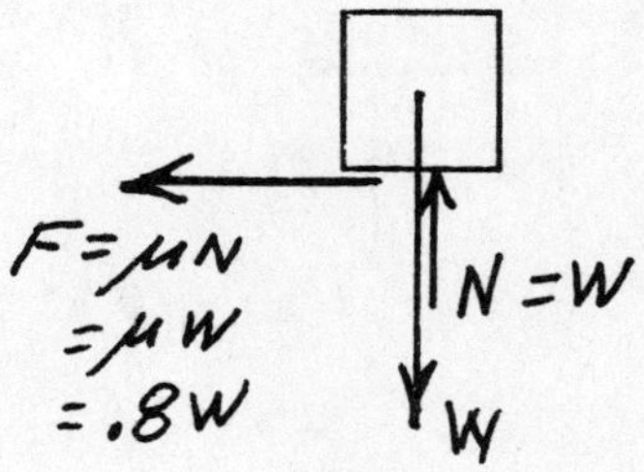

$\rho = 200'$

$\frac{V^2}{\rho} = a_N$

$a_t = 5 \text{ ft/s}^2$

$v_0 = 40 \text{ ft/sec}$

+x, +y

$a_T = 5$

$a_N = \frac{V^2}{\rho}$

$F = \mu N = \mu W = .8W$

$N = W$

W

F.B.D. OF CAR (ELEV).

$5^2 + a_N^2 = (25.76)^2$, $a_N = 25.27 \text{ ft/s}^2 = \frac{V^2}{\rho}$

$V = \sqrt{(25.27)(200)} = \boxed{71.1 \text{ ft/s}}$ ANS.

5-21

If a two-mass system shown in the accompanying illustration, mounted on frictionless wheels, is subjected to mechanical vibrations, determine their equations of motion.

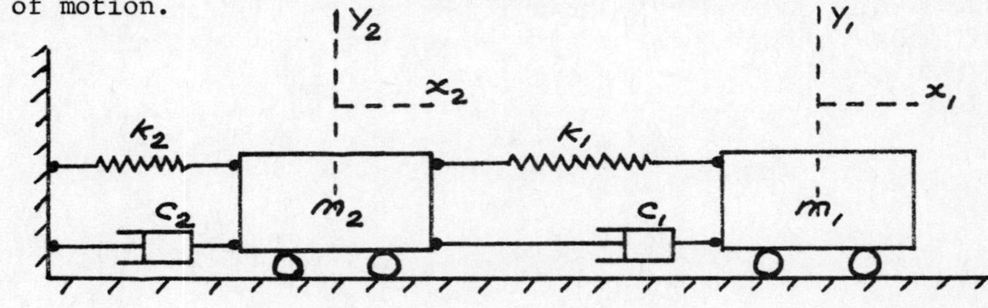

****

<u>Solution</u>

$k_2 x_2$; $c_2 \dot{x}_2$ ← m_2 ⇄ $k_1(x_1 - x_2)$; $c_1(\dot{x}_1 - \dot{x}_2)$ ← m_1

$$m_2 \ddot{x}_2 = -k_2 x_2 + k_1(x_1 - x_2) - c_2 \dot{x}_2 + c_1(\dot{x}_1 - \dot{x}_2)$$

and $m_1 \ddot{x}_1 = -k_1(x_1 - x_2) - c_1(\dot{x}_1 - \dot{x}_2)$

By rearranging the terms, we have

$$m_2 \ddot{x}_2 + k_2 x_2 + c_2 \dot{x}_2 - k_1(x_1 - x_2) - c_1(\dot{x}_1 - \dot{x}_2) = 0$$

<u>Ans.</u>

and $m_1 \ddot{x}_1 + k_1(x_1 - x_2) + c_1(\dot{x}_1 - \dot{x}_2) = 0$ <u>Ans.</u>

5-22

The force exerted by a fixed magnet on a steel block varies inversely as the square of the distance between the block and the magnet. When the distance between the two objects is L_o, the force of attraction is measured to be F_o. If the block starts from rest when $x = L_o$, find its velocity when it is at a distance of $L_o/3$ from the magnet.

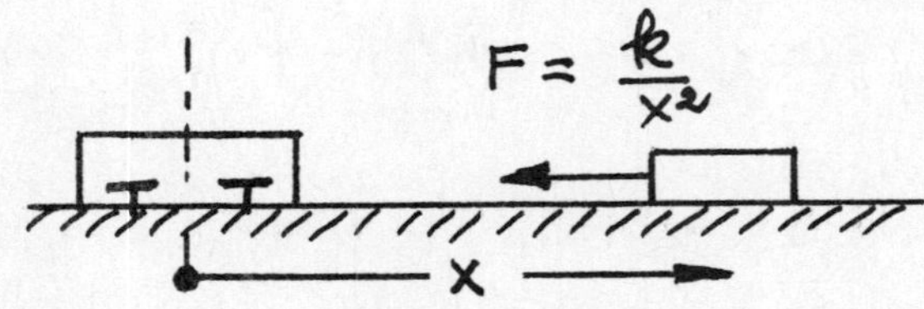

****

Substituting, $F_o = \dfrac{k}{L_o^2}$, or $k = F_o L_o^2$.

which gives $F = \dfrac{F_o L_o^2}{x^2}$ ← The equation

of motion is $m a_x = -\frac{F_o L_o^2}{x^2}$. Replacing

a_x by $v\frac{dv}{dx}$, and integrating leads to

$$\int_0^V v\,dv = F_o L_o^2 \int_{L_o}^{\frac{L_o}{3}} -\frac{dx}{x^2}$$

or

$$\left[\frac{v^2}{2}\right]_0^V = F_o L_o^2 \left[\frac{1}{x}\right]_{L_o}^{\frac{L_o}{3}}$$

finally

$$V = 2(F_o L_o)^{1/2}$$

IMPULSE AND MOMENTUM

5-23

Force P as shown acts on block A which weighs 100 lbs. and is initially at rest. If the coefficient of friction is 0.1 between the block and plane, find the velocity of A when (a) time t is 4 seconds (b) time t is 6 seconds.

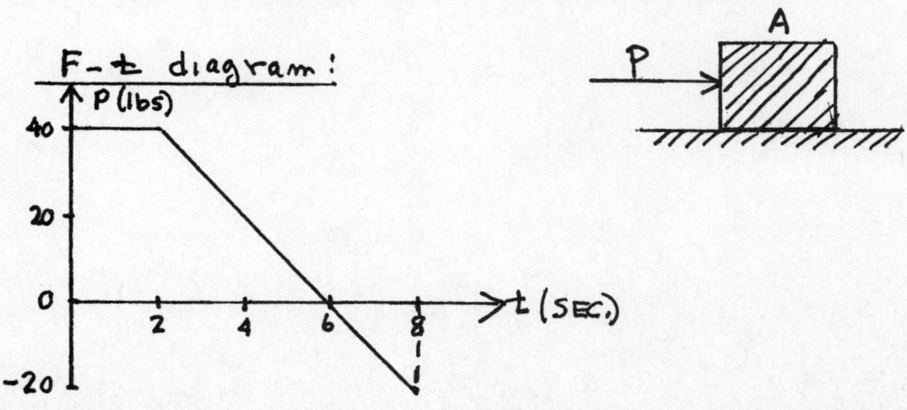

****

FBD OF BLOCK

100 lb

P

$F' = \frac{1}{10} N$

$= 10$ lb.

N = 100 lb

SINCE $P_0 = 40 > F'$, block moves immediately.

BASIC EQUATION:

$$m_1 V_1 + Imp_{1-2} = m_1 V_2$$

(a) FOR $t = 4$, $0 + 80 + 60 - 10 \cdot 4 = \frac{100}{g} \cdot V_2$

$V_2 = 32.2$ ft/sec

(b) FOR $t = 6$, $0 + 80 + 80 - 10 \cdot 6 = \frac{100}{g} \cdot V_3$

$V_3 = 32.2$ ft/sec

5-24

A golf ball of mass 50 gm is hit with a club. If the initial velocity of departure of the ball is 20 m/s, what is impulse applied to the ball?

If the period of contact between the ball and the club was 0.05s, what was the average force on the ball during the strike?

****

Use Impulse = Change of momentum

or, $F. \Delta t = p_2 - p_1$

Initial momentum $p_1 = 0$

Final Momentum $p_2 = m v_2 = 0.05 \times 20 = 1.0$ N.s

Time interval of impulse $\Delta t = 0.05$ seconds

Hence

$$F \times .05 = 1.0 - 0$$

or,

Impulsive force $F = \frac{1}{.05} = \underline{20 \text{ N}}$

5-25

A particle weighing 2 N attached to a string 1 m long rotates in the XY plane at a constant angular velocity of 0.4 rad/s, as shown in the accompanying illustration. Calculate (a) the kinetic energy, (b) the linear momentum and (c) the angular momentum of the particle about the fixed point of rotation A when θ = 35°.

y, j, g, A, i, x, $\theta = 35°$, $\omega = 0.4$ rad/s, $L = 1$ m, W = 2 kg

****

Solution

$$m = \frac{W}{g} = \frac{2\text{ N}}{9.81\text{ m/s}^2} = 0.204\text{ kg}$$

(a) Kinetic energy:

$$\bar{\omega} = \omega\bar{k} \; ; \quad \bar{r} = l(\cos\theta\,\bar{i} - \sin\theta\,\bar{j})$$

$$\bar{v} = \bar{\omega}\times\bar{r} = (\omega l\sin\theta\,\bar{i} + \omega l\cos\theta\,\bar{j})$$

$$\therefore\; T = \frac{1}{2}mv^2 = \frac{1}{2}m\bar{v}\,\bar{v}$$

$$= \frac{1}{2}m(\omega l\cos\theta\,\bar{j} + \omega l\sin\theta\,\bar{i})(\omega l\cos\theta\,\bar{j} + \omega l\sin\theta\,\bar{i})$$

$$= \frac{1}{2}m\,\omega^2 l^2$$

$$= \frac{1}{2}(0.204\text{ kg})(0.4\text{ rad/s})^2(1\text{ m})^2 = 0.016\text{ N.m.}$$

<u>Ans.</u>

(b) Linear momentum:

$$L = m\bar{v}$$

$$= m\omega l(\sin\theta\,\bar{i} + \cos\theta\,\bar{j})$$

$$= (0.204\text{ kg})(0.4\text{ rad/s})(1\text{ m})(0.573\,\bar{i} + 0.819\,\bar{j})$$

$$= (0.046\,\bar{i} + 0.066\,\bar{j})\text{ kg.m/s}$$

<u>Ans.</u>

(c) Angular momentum:

$$H = \bar{r}\times m\bar{v}$$

$$= l(\cos\theta\,\bar{i} - \sin\theta\,\bar{j})\times m(\omega l\sin\theta\,\bar{i} + \omega l\cos\theta\,\bar{j})$$

$$= ml^2\omega\,\bar{k}$$

$$= (0.204\text{ kg})(1\text{ m})^2(0.4\text{ rad/s})\,\bar{k}$$

$$= 0.082\,\bar{k}\text{ N.m.}$$

<u>Ans.</u>

5-26

A particle having a mass of 2 lb is acted upon by a force $\vec{F} = \{3\vec{i} + t\vec{j}\}$ lb for t = 10 sec. Determine the final velocity of the particle if it had an initial velocity $\vec{V} = \{2\vec{i} + \vec{j}\}$ ft/sec. The particle is moving on a smooth horizontal plane.

****

Applying the impulse momentum principle,

$$m\vec{V_1} + \int \vec{F}\,dt = m\vec{V_2}$$

$$m\vec{V_1} = \frac{2}{32.2}(2\vec{i} + \vec{j}) = 0.124\vec{i} + 0.062\vec{j}$$

$$\int \vec{F}\,dt = \int_0^{10} (3\vec{i} + t\vec{j})\,dt = \left[3t\vec{i} + \frac{t^2}{2}\vec{j}\right]_0^{10}$$

$$= 30\vec{i} + 50\vec{j}$$

Therefore, $0.124\vec{i} + 0.062\vec{j} + 30\vec{i} + 50\vec{j} = \frac{2}{32.2}\vec{V_2} = \frac{1}{16.1}\vec{V_2}$

or, $\vec{V_2} = 485\vec{i} + 806\vec{j}$

And $V_2 = 940.67$ ft/sec, $\alpha = 59°$,

$\vec{V_2}$ 59°

5-27

A 200-lb block is moving to the right with a velocity of 5 ft/sec at the instant shown. Suddenly a force F = 50 lb is applied to the block for an 8-sec time interval. Determine the velocity of the block at the end of the given time period. The coefficient of friction between the block and the plane is 0.1.

$V_1 = 5$ ft/Sec

F = 50 lb

μN

****

Initial momentum of the particle + Impulse imparted to the particle by the external force system = Final momentum of the particle

$$mV_1 + \int (F - \mu N)\,dt = mV_2$$

$$\frac{200}{32.2} \times 5 + (50 - 0.1 \times 200)\,8 = \frac{200}{32.2} \times V_2$$

Solving, $V_2 = 43.64$ ft/sec

5-28

A 0.322-lb particle moving along a straight line is acted upon by a force which varies with time as follows:

F (t) = $0.3\,t^2 - 0.6t$ [t] : sec [F] : lb

If the initial velocity of the particle is 20 ft/s, when will the particle first have zero velocity?

****

Principle of impulse and momentum:

$$mv_1 + \int_{t_1}^{t_2} F(t)\,dt = mv_2$$

F → m

$v_1 = 20$ ft/s , $v_2 = 0$, $m = \frac{0.322}{32.2} = 0.01$ slugs

Substituting :

$$(0.01)\,20 + \int_0^t (0.3t^2 - 0.6t)\,dt = 0$$

or

$$0.1\,t^3 - 0.3\,t^2 + 0.2 = 0$$

This equation has three roots. The smallest acceptable root gives : $t = 1$ sec.

5-29

The 200 pound block is at rest on the plane when the force P is applied. P varies with the relation P = 200t pounds, where t is in seconds. The coefficient of friction, both static and kinetic, is 0.50. Find the velocity of the block 2 seconds after P is applied.

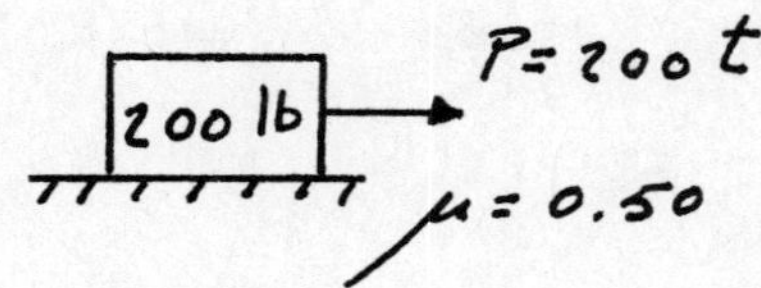

****

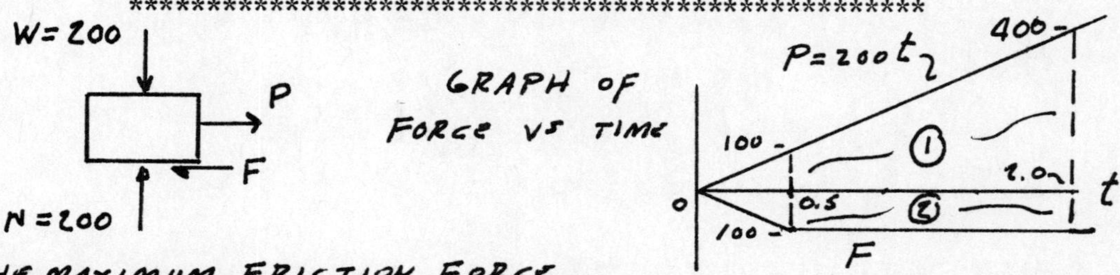

THE MAXIMUM FRICTION FORCE F THAT CAN BE DEVELOPED IS $F_{MAX} = \mu N = 0.5(200) = 100$ lb.
THE BLOCK WILL REMAIN IN EQUILIBRIUM ($P = F$) UNTIL $P > 100$ lb. AT $P = 200t$, $100 = 200t$, $t = 0.5$ SECONDS.
AT THAT TIME, 0.5 s, F WILL REMAIN CONSTANT AT 100 lb.

$$\int \Sigma F_x \, dt = m \, \Delta V_x \quad ; \quad \text{THEN FROM GRAPH:}$$

$$\text{AREA ①} - \text{AREA ②} = m V_x = \frac{W}{g} V_2$$

$$\left(\frac{100 + 400}{2}\right)(2 - 0.5) - 100(2 - 0.5) = \frac{200}{32.2} V_2$$

$$V_2 = 36.2 \text{ FT/s} \rightarrow$$

Note: The block begins to move at t=0.5 seconds. From t=0 to t=0.5, the area under the force vs time graph will equal zero. That is, the positive area due to force P will be balanced by the negative area from the friction force F.

5-30

A toy rocket of mass 1 kg is placed on a frictionless horizontal surface and the rocket engine is ignited. Knowing that the engine delivers a force equal to (0.25 + 0.5t)N, determine a) the velocity of the rocket 0.5 sec. after ignition and, b) how much velocity is gained between 0.5 seconds and burnout (seven seconds after ignition). Assume constant mass for the rocket. What would be the value for a) and b) if the coeficient of friction between the rocket and the surface was .01

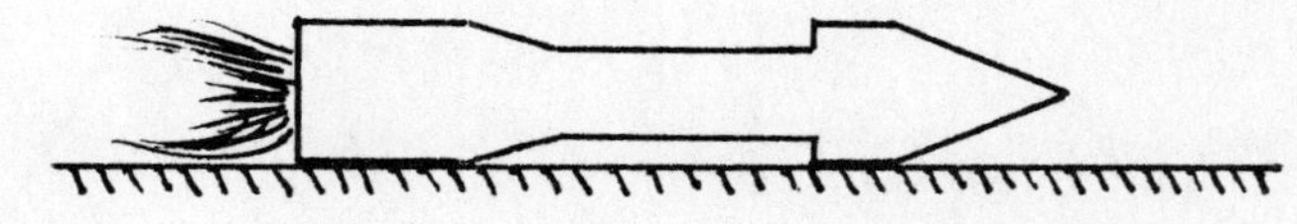

****

a) $$mV_1^{\,0} + \int_{t_1}^{t_2} F\,dt = mV_2$$

$$\int_0^{.5} (.25 + .5t)\,dt = 1V_2 = .25t + \frac{.5t^2}{2}\Big|_0^{1/2} = V_2$$

$$.25(1/2) + .25(1/2)^2 = V_2 = .1875 \text{ m/s}$$

b) $$m^1(V_2 - V_1) = .25t + .25t^2\Big|_{.5}^{7} = 14 - .1875 = 13.81 \text{ m/s}$$

ASSUMEING CONSTANT CONDITIONS, THE FRICTION FORCE $= \mu N =$ CONST. $= .01(1)(9.81) = .0981\ N$

$$\int F\,dt = \int_0^{.5} -.0981\,dt + \int_0^{.5} (.25 + .5t)\,dt = mV_2$$

$$-.0981(.5) + .1875 = .1385 \text{ m/s} = V_2$$

AND $$m(V_2 - V_1) = 13.81 - .0981(6.5)$$

$$V_2 - V_1 = 13.17 \text{ m/s}$$

5-31

A 0.1 kg ball is flying horizontally at 120 m/s. It is struck by a force at 30 degrees as shown. The impulse graph is given. What is the final velocity of the ball?

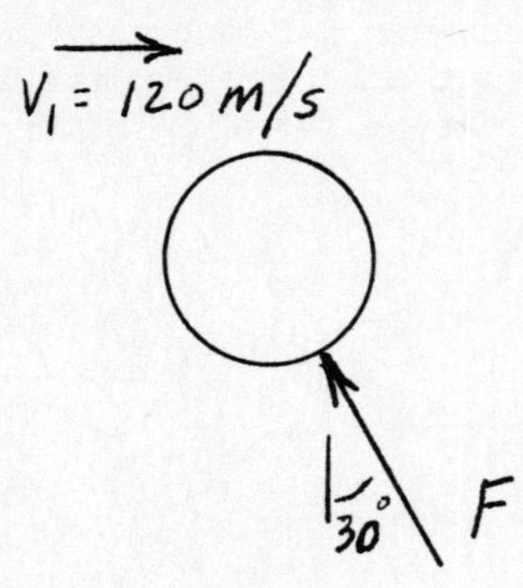

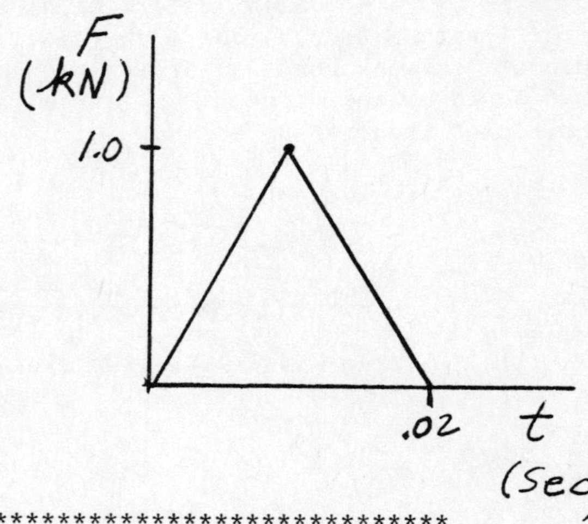

****

Impulse = Area under Curve

$$= \frac{1}{2}bh = \frac{1}{2}(.02)(1000\text{ N}) = \underline{\underline{10\text{ N}\cdot\text{s}}}$$

X COMPONENT: $\vec{M_1 V_{1x}} + \vec{IMP_x} = \vec{M_1 V_{2x}}$

$$(.1)\overrightarrow{(120)} + \overleftarrow{5} = .1(V_{2x})$$

$$12 - 5 = 7 = .1V_2 \quad , \quad V_{2x} = 70\text{ M/s} \rightarrow$$

Y COMPONENT: $\vec{MV_{1y}} + \vec{IMP_y} = \vec{MV_{2y}}$

$$0 + (10\text{ N}\cdot\text{s})\cos 30 = (.1)V_{2y}$$

$$V_{2y} = 86.6\text{ M/s} \uparrow$$

86.6, 70, θ

$$V = \sqrt{86.6^2 + 70^2} = 111\text{ M/s}$$

$$\theta = \tan^{-1}\frac{86.6}{70} = 51°$$

ANS

5-32

A steel ball is dropped vertically on an inclined metal plate. Its velocity just before it strikes the plate is v. It rebounds horizontally with a velocity v'. Find the coefficient of restitution between the ball and the plate.

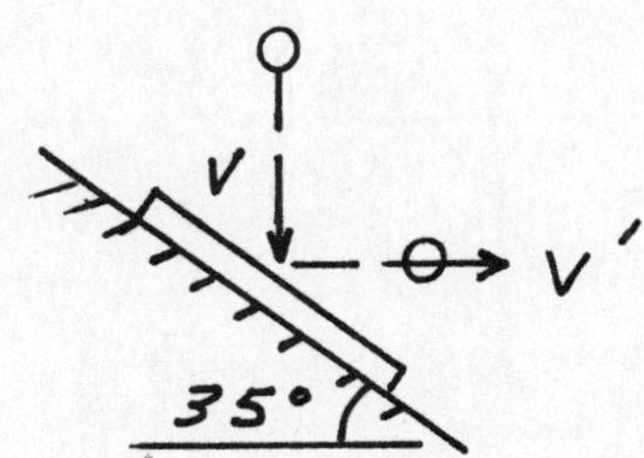

****

BEFORE IMPACT — AT IMPACT ($\int F \cdot dt$) — AFTER IMPACT

$V_x = V \sin 35°$ ↘
$V_y = V \cos 35°$ ↙

$V_x' = V' \cos 35°$ ↘
$V_y' = V' \sin 35°$ ↗

SINCE THE IMPACT IS ONLY IN THE y-DIRECTION, THE X COMPONENT OF V IS NOT CHANGED.

$$V_x = V_x' \Rightarrow V \sin 35° = V' \cos 35°$$

$$V' = V \tan 35°$$

FOR THE IMPACT IN THE y DIRECTION:

$$e = \frac{\text{RELATIVE VELOCITY AFTER IMPACT}}{\text{" \quad " BEFORE IMP}} = \frac{V_y'}{V_y}$$

$$e = \frac{(V')\sin 35°}{V \cos 35°} = \frac{(V \tan 35°)\sin 35°}{V \cos 35°} = \tan^2 35°$$

$$e = 0.49$$

5-33

Using the Impulse-Momentum Method, determine the velocity of the 500# weight at t = 5 sec, if it is subjected to P = 100# and starts from rest.

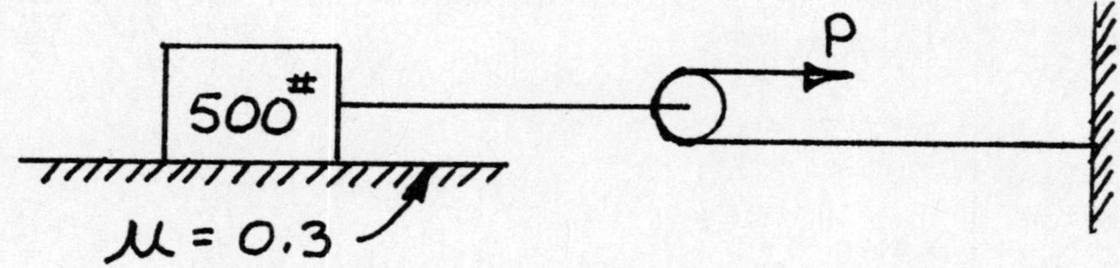

****

500#

2P = 200#

f

N

FBD BLOCK

$N = 500^{\#} \qquad f = 0.3\,N = 150^{\#}$

$\therefore$ Unbalanced Force $= 200^{\#} - 150^{\#} = 50^{\#}$

$$F \cdot t = m\,v$$

$$(50^{\#})(5\,\text{sec}) = \left(\frac{500^{\#}}{32.2\,\frac{FT}{Sec^2}}\right) v$$

$$v = \frac{250\ \text{lb-sec}}{500\ \text{lb}/32\ \frac{FT}{Sec^2}} = 16.1\ \frac{FT}{sec} \rightarrow$$

5-34

A 2000 kg pickup truck is traveling down a 10° incline at 100 km/hr when the driver suddenly notices that his engine is overheating. He applies the brakes, but due to wet conditions a braking force of only 4000 N is developed. How long does it take the truck to stop ?

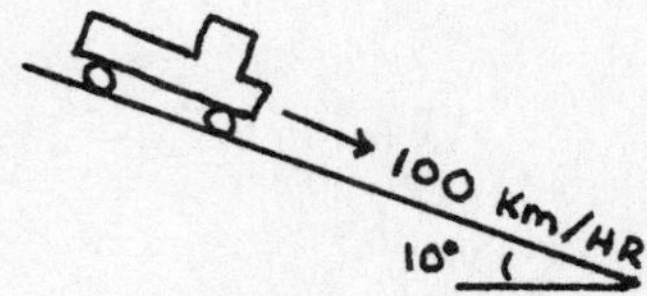

****

A SUBSCRIPT 1 REFERS TO QUANTITIES AS THE DRIVER APPLIES HIS BRAKES. A SUBSCRIPT 2 REFERS TO QUANTITIES AS THE TRUCK STOPS.

PRINCIPLE OF IMPULSE-MOMENTUM:

$$m\vec{V}_1 + \vec{I}mp_{1\to 2} = m\vec{V}_2$$

$$|\vec{v}_1| = 100 \text{ Km/hr}\left(\frac{1000\text{ m}}{\text{Km}}\right)\left(\frac{1\text{ HR}}{3600\text{ SEC}}\right) = 27.77 \text{ m/SEC}$$

$$|\vec{v}_2| = 0 \text{ m/SEC}$$

W, N, F

IN THE DIRECTION TANGENT TO THE ROAD'S SURFACE

$$Imp_{1\to 2} = \int_0^t (W\sin 10° - F)\,dt$$

$$= (W\sin 10° - F)\,t$$

$$= [(2000\text{ Kg})(9.81\text{ m/SEC}^2)\sin 10° - 4000\text{ N}]\,t$$

$$= -593\,t \text{ N-SEC}$$

THUS

$$(2000\text{ Kg})(27.77\text{ m/SEC}) - 593\,t \text{ N-SEC} = 0$$

$$t = 93.66 \text{ SEC}$$

5-35

A golf ball weighing 1.62 ounces is driven off the tee at an angle of 20 degrees with the horizontal with a velocity of 150 mph. Assuming that the golf club stays in contact with the ball for 0.03 seconds, determine the average force on the ball.

****

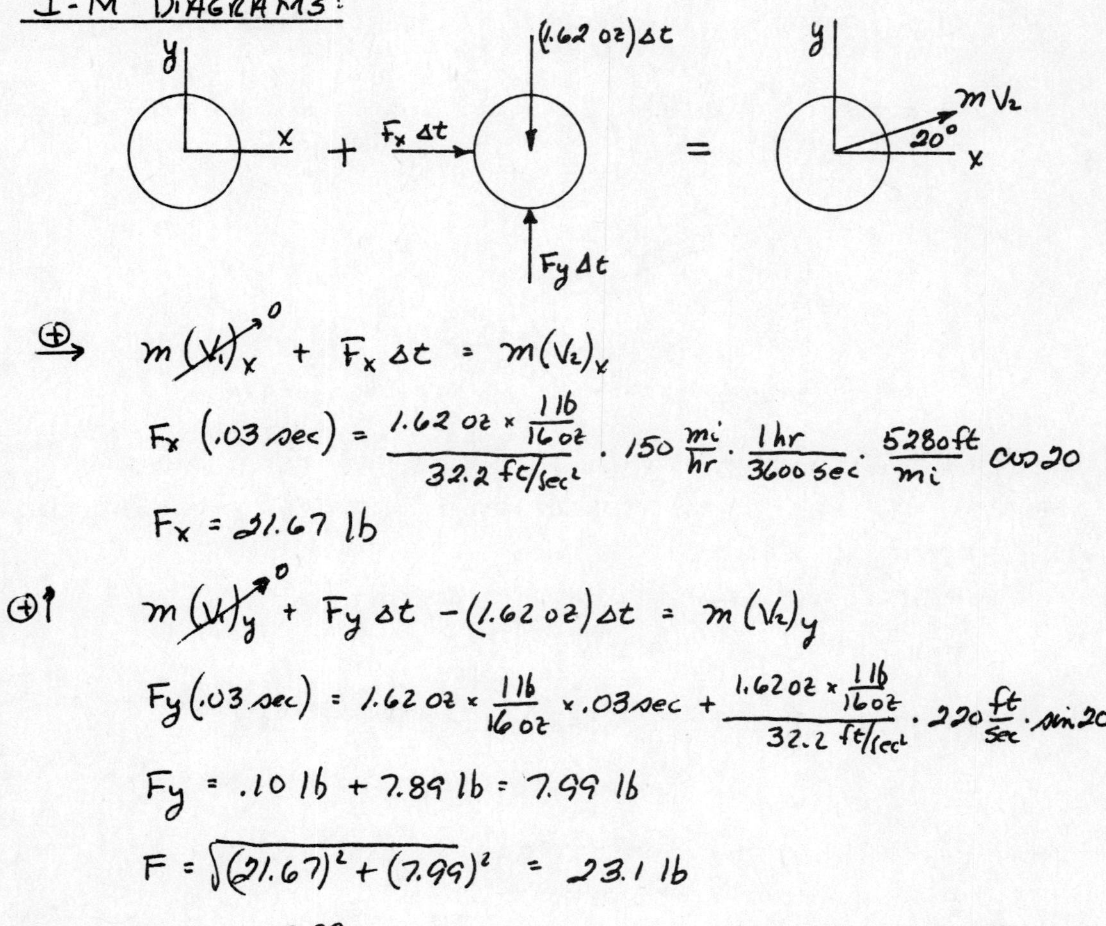

$\tan\theta = \frac{7.99}{21.67} \qquad \theta = 20.24°$ $\qquad$ **F = 23.1 lb ∠20.24°**

5-36

A particle, m = 1 kg, has a velocity of $2m/_s$ in the x-direction when t = 0. Forces, F_1 and F_2, act on the particle in the x and y directions as indicated. The <u>magnitudes</u> change with time as indicated on the graph. Find the velocity of the particle after 2 seconds.

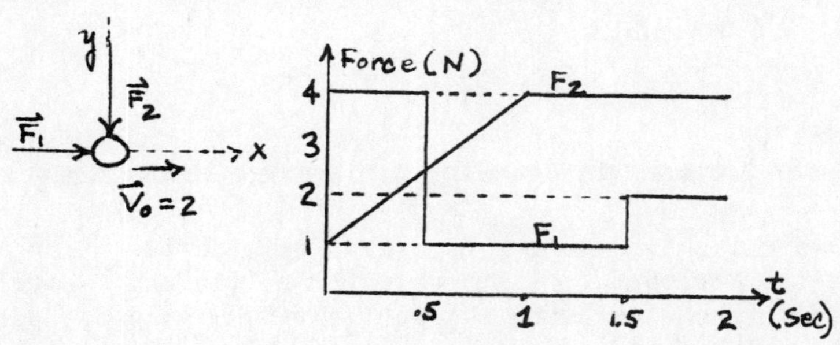

****

$$m V_x)_1 + \int_{t_1}^{t_2} F_x\,dt = m V_x)_2$$

State 1 is at $t=0$
State 2 is at $t=2$

$$\therefore\ 1\cdot(2) + \int_0^2 F_1\,dt = 1\cdot V_x)_2$$

$$2 + 4 = V_x)_2$$

$$\therefore\ V_x)_2 = 6\ \frac{m}{sec} \rightarrow$$

$$m V_y)_1 + \int_{t_1}^{t_2} F_y\,dt = m V_y)_2$$

$$1\cdot 0 - \int_0^2 F_2\,dt = 1\cdot V_y)_2 \qquad (\vec{F_2} = -F_2 j)$$

$$-6.5 = V_y)_2 \qquad \therefore\ V_y)_2 = 6.5\ \frac{m}{sec} \downarrow$$

$\therefore\ \vec{V}_{FINAL}$: (6, 6.5, V_f)

$$V_f = \sqrt{6^2 + 6.5^2} = \sqrt{78.25}$$

$$V_f \cong 8.85\ m/sec$$

$$\therefore\ V_f = 8.85 \quad (6,\ 6.5)$$

ANGULAR MOMENTUM

5-37

A 5-lb particle is moving with a velocity of 8 ft/sec parallel to the direction of the positive z-axis. Determine its angular momentum about the origin when it is located at the point P as shown.

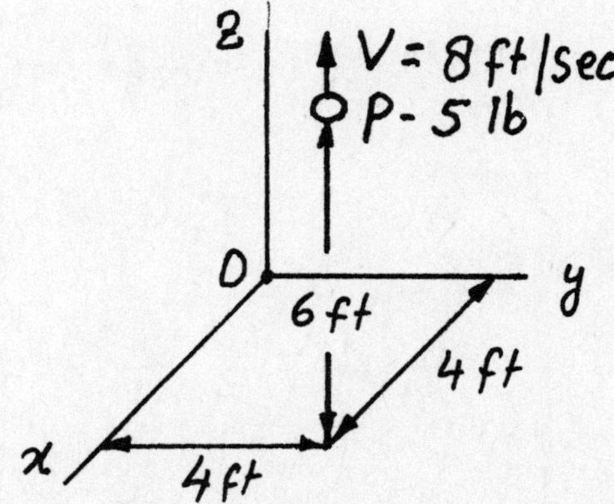

****

Angular momentum of particle = $H_o = \vec{r} \times (m\vec{V})$

or, $H_o = (m)(4\vec{i} + 4\vec{j} + 6\vec{k}) \times (8\vec{k})$

or, $H_o = (m)(32\,\vec{i} \times \vec{k} + 32\,\vec{j} \times \vec{k} + 0)$

or, $H_o = \frac{5}{32.2}(-32\,\vec{j} + 32\,\vec{i}) = \{4.97\vec{i} - 4.97\vec{j}\}$ lb-ft-sec

5-38

A particle of mass 2 kg is moving along a circular path with a velocity time relationship given by v=10t+3

(a) What is the angular momentum of the particle at time t=1 Sec.?

(b) What is the rate of change of Angular momentum ?

The radius of the circle is 1.5 meters.

****

(a) Angular momentum $H = I\omega$

$$I = mr^2 = 2(1.5)^2 = 4.5 \text{ kg-m}^2$$

$$\omega\Big|_{t=1} = \frac{v|_{t=1}}{r} = \frac{10(1)+3}{1.5} = 8.67 \text{ rad/sec.}$$

$$H\Big|_{t=1} = I\omega = 4.5\,(8.67) = 39.0 \text{ kg}\frac{m^2}{sec.}$$

(b) Rate of change of angular momentum.

$$\frac{dH}{dt} = \frac{d}{dt}(I\omega) = I\frac{d\omega}{dt} = \frac{I}{r}\frac{dv}{dt}$$

$$\frac{dH}{dt} = \frac{I}{r}\frac{d}{dt}(10t+3) = \frac{I}{r}10 = \frac{4.5}{1.5}(10)$$

$$= 30 \text{ kg}\frac{m^2}{Sec^2}$$

5-39

A satellite describes a circular orbit around the earth of radius r_o, when its speed at A is suddenly increased in order to change its orbit into an elliptical one whose maximum distance from the center of the earth is 5 r_o as shown. Find a) the speed while the satellite was describing a circular orbit, b) the increase in kinetic energy at A in order to change its orbit into an elliptical one, and c) the speed at B while it travels on the elliptical orbit.

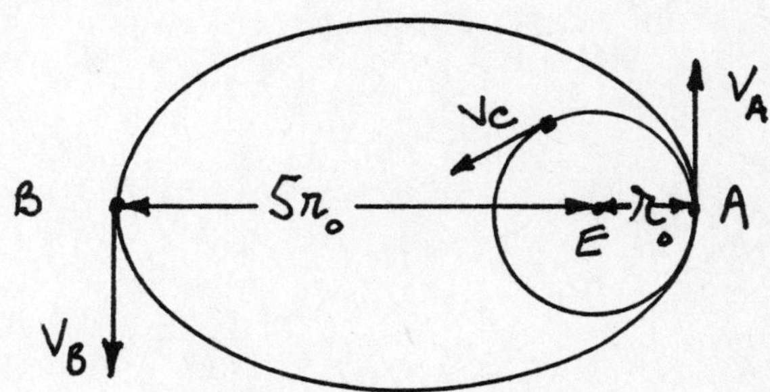

****

a) While on circular orbit around the earth centered at E (earth not shown), $\vec{F} = m\vec{a}$ along the radius gives $\frac{mV_c^2}{r_o} = \frac{GMm}{r_o^2}$. With $GM = gR_e^2$, we have (Re = radius of the earth) $\underline{\underline{V_c^2 = \frac{gR_e^2}{r_o}}}$

b) Conservation of angular momentum between A and B gives

$$mV_A r_o = mV_B(5r_o) \quad ; \quad V_A = 5V_B$$

Conservation of energy between A and B gives

$$\frac{1}{2}mV_A^2 - \frac{mgR_e^2}{r_o} = \frac{1}{2}mV_B^2 - \frac{mgR_e^2}{5r_o}$$

Eliminating V_B gives $V_A^2 = \frac{5}{3}\frac{gR_e^2}{r_o}$

The increase in kinetic energy is $\underline{\underline{T_A = 1.67\,T_c}}$

c) $\underline{\underline{V_B}} = \frac{1}{5}V_A = \underline{\underline{\left[gR_e^2/15r_o\right]^{1/2}}}$ using part (b).

5-40

A thin rod is attached to a particle which is rotating about the center of a frictionless table. The length of the portion of the rod between the particle and center of rotation is controlled to vary with time. The angular speed of the particle at t=0 is 4 rad/sec.

a. Calculate the angular speed of the particle as a function of time.
b. Express the acceleration as a function of time, in radial and transverse coordinates.
c. Calculate the maximum tension in the rod.

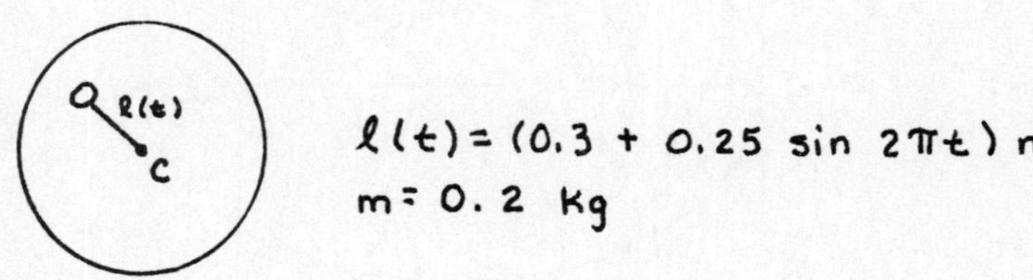

****

a) THE ONLY NET FORCE ACTING ON THE PARTICLE IS THE TENSION IN THE ROD DIRECTED TOWARD THE CENTER OF ROTATION. THUS ANGULAR MOMENTUM ABOUT C IS CONSERVED. THE VELOCITY OF THE PARTICLE CAN BE EXPRESSED AS

$$\vec{v} = \ell \dot{\theta} \vec{i}_\theta + \dot{r} \vec{i}_r$$

WHERE $\vec{i}_r$ AND $\vec{i}_\theta$ ARE UNIT VECTORS IN THE RADIAL AND TRANSVERSE DIRECTIONS RESPECTIVELY. THE ANGULAR MOMENTUM OF THE PARTICLE IS

$$\begin{aligned}\vec{H} &= \ell(t)\vec{i}_r \times m\vec{v} \\ &= \ell(t)\vec{i}_r \times [(m\ell\dot{\theta})\vec{i}_\theta + m\dot{r}i_r] \\ &= m\ell^2(t)\dot{\theta}\vec{i}_z\end{aligned}$$

WHERE $\vec{i}_z$ IS A UNIT VECTOR PERPENDICULAR TO $\vec{i}_\theta$ AND $\vec{i}_r$. SINCE ANGULAR MOMENTUM IS CONSERVED

$$m\ell^2(t)\dot{\theta}(t) = h$$

WHERE h IS A CONSTANT WHICH CAN BE EVALUATED FROM THE INITIAL CONDITION

$$h = (0.2 \text{ Kg})(0.3 \text{ m})^2(4 \text{ RAD/SEC}) = 0.072 \ \frac{\text{Kg-m}^2}{\text{sec}}$$

THEN

$$\dot{\theta}(t) = \frac{h}{m\ell^2(t)} = \frac{0.072 \text{ Kg-m}^2/\text{sec}}{0.2 \text{ Kg}(0.3 + 0.2 \sin 2\pi t)^2} = \frac{0.36}{(0.3 + 0.2 \sin 2\pi t)} \frac{\text{RAD}}{\text{SEC}}.$$

b) IN RADIAL AND TRANSVERSE COORDINATES THE ACCELERATION IS GIVEN BY

$$\vec{a} = (\ddot{\ell} - \ell\dot{\theta}^2)\vec{\iota}_r + (\ell\ddot{\theta} + 2\dot{\ell}\dot{\theta})\vec{\iota}_\theta$$

NOTE THAT THE ANGULAR MOMENTUM IS CONSTANT

$$m\ell^2\dot{\theta} = h$$

BUT $\frac{d}{dt}(m\ell^2\dot{\theta}) = 0 = m[2\ell\dot{\ell}\dot{\theta} + \ell^2\ddot{\theta}] = \ell m[\ell\ddot{\theta} + 2\dot{\ell}\dot{\theta}]$

HENCE THERE IS NO TRANSVERSE COMPONENT OF ACCELERATION FOR CENTRAL FORCE MOTION.

$$\ell(t) = (0.3 + 0.2 \sin 2\pi t) \text{ m}$$

$$\dot{\ell}(t) = (0.4\pi \cos 2\pi t) \text{ m/sec}$$

$$\ddot{\ell}(t) = (-0.8\pi^2 \sin 2\pi t) \text{ m/sec}^2$$

$$\vec{a} = \left[-0.8\pi^2 \sin 2\pi t - \frac{(0.3 + 0.2 \sin 2\pi t)(.36)^2}{(0.3 + 0.2 \sin 2\pi t)^2}\right]\vec{\iota}_r \text{ m/sec}^2$$

$$= [-0.0389 - 7.92 \sin 2\pi t] \frac{\text{m}}{\text{sec}^2} \vec{\iota}_r$$

c) FREE BODY DIAGRAM OF PARTICLE

O ↘ T

$$\Sigma F_r = m a_r$$

$$-T = m(-0.0389 - 7.92 \sin 2\pi t) \frac{\text{m}}{\text{sec}^2}$$

THE MAXIMUM TENSION OCCURS WHEN $\sin 2\pi t = 1$

$$T_{MAX} = (0.2 \text{ Kg})(0.0389 + 7.92) \text{ m/sec}^2$$

$$= 1.59 \text{ N}$$

5-41

One end of a light inextensible string is attached to a particle of mass 100 gm resting on a frictionless table. The other end of the string passes through a small hole O on the table. Initially the particle is swung with angular velocity 2 rad/s at radius r = 50 cm, keeping the string taut. If the string is slowly pulled through the hole until the rotating radius of the particle becomes 10 cm, what is the resulting angular velocity? Determine the corresponding tension T in the string.

****

Since the table top is frictionless there is no transverse force on the particle; the moment on m about the axis of rotation is zero. Hence the angular momentum about this axis is conserved.

m, r, T, ω, O, hole

Initial transverse velocity of m $= \omega_1 r$

Initial angular momentum $= r_1 (m\omega_1 r_1)$

$= m\omega_1 r_1^2$

Final angular momentum $= m\omega_2 r_2^2$

Hence

$$m\omega_2 r_2^2 = m\omega_1 r_1^2$$

or

$$\omega_2 = \omega_1 \left(\frac{r_1}{r_2}\right)^2$$

$$= 2 \times \left(\frac{50}{10}\right)^2$$

Final angular velocity ω_2 = 50 rad/s

Assume $\frac{d^2r}{dt^2} = 0$

Inward radial acceleration of m $= r\omega^2$

Newtons second law for radial direction: $T = mr\omega^2$

Hence at ω_2; $T = 0.1 \times 0.1 \times 50^2$ N = 25 N

5-42

A projectile P weighing 20 lb is fired from a cannon at an angle of 37.5° from the horizontal and a muzzle velocity v_i = 1,295 ft/s. Calculate the angular momentum of the projectile when it reaches the maximum height about point O.

****

Solution

$$(v_x)_o = v_i \cos 37.5^\circ = (1295 \text{ ft/s})(0.793) = 1027.392 \text{ ft/s}$$

$$(v_y)_o = v_i \sin 37.5^\circ = (1295 \text{ ft/s})(0.608) = 788.346 \text{ ft/s}$$

The maximum height reached by the projectile is given by $+\uparrow \; (v_y)_p^2 = (v_y)_o^2 + 2a_c(s_y - s_o)$

OR, $0 = (788.346 \text{ ft/s})^2 + 2(-32.2 \text{ ft/s}^2)(y - 0)$

OR, $y = \dfrac{(788.346 \text{ ft/s})^2}{(64.4 \text{ ft/s}^2)} = 9650.456 \text{ ft}$

Now, $(v_x)_p = (v_x)_o = 1027.392 \text{ ft/s}$

Hence, $H_o = r(mv)$

$$= (9650.456 \text{ ft})\left(\frac{20 \text{ lb}}{32.2 \text{ ft/s}^2}\right)(1027.392 \text{ ft/s})$$

$$= 6.158 \times 10^6 \text{ lb.ft.s} \quad \text{Ans.}$$

WORK AND KINETIC ENERGY

5-43

The 20 pound weight is released from rest 0.5 feet above a linear spring. The spring constant is 50 lb/ft. Determine the maximum velocity reached by the weight during its descent.

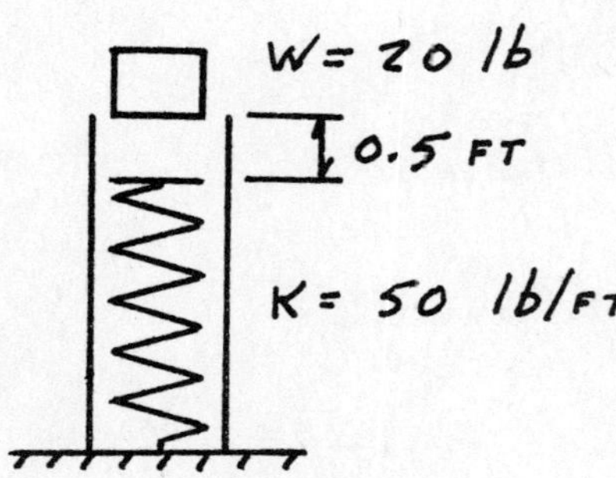

****

Note: The weight will strike the spring, and continue to accelerate (and thus its velocity will continue to increase) until the force in the spring equals the weight. At that time the instantaneous acceleration of the weight will be zero, and the velocity will be a maximum.

AT ②, $F_S = KX$

$20 = 50X$

$X = 0.4$ FT (THE DEFORMATION IN THE SPRING)

WORK OF WEIGHT $= Wh = 20(0.5+0.4) = 18$ FT·lb

WORK OF SPRING $= -\frac{1}{2}Kx^2 = -\frac{1}{2}(50)0.4 = -4$ FT·lb

KINETIC ENERGY OF WEIGHT $T_② = \frac{1}{2}mV_{MAX}^2 = \frac{1}{2}\left(\frac{20}{32.2}\right)V_{MAX}^2$

" " " " $T_① = 0$

USING PRINCIPLE OF WORK-ENERGY:

$T_① + U_{①\to②} = T_②$ $\qquad 0 + (18-4) = \frac{1}{2}\left(\frac{20}{32.2}\right)V_{MAX}^2$

$V_{MAX} = 6.71$ FT/S

5-44

In a piston cylinder assembly, the piston initially lies at 2 ft from the end of the cylinder and the pressure inside the cylinder is atmospheric. It is pushed to the left side at constant speed, to a position of 0.2 ft. During this motion of the piston, the temperature in the cylinder is kept constant by cooling the cylinder. Assuming that the gas inside the cylinder is ideal (for which pV=mRT holds, where R is a constant and m is the mass of the gas), determine the required work for this compression. The piston area is 1 ft^2 and the atmospheric pressure is 14.7 psia.

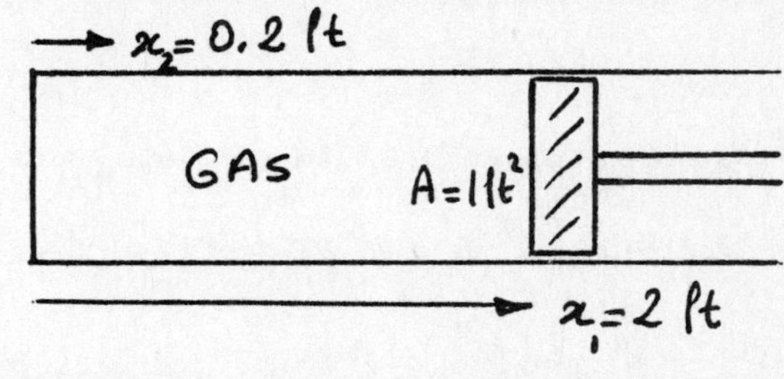

****

If $PV = mRT$ holds, for constant temperature we can write

$PV = P_1V_1 = P_2V_2$ etc. where 1,2 are the states of the gas.

$\therefore \quad P = P_1V_1/V = P_1.(x_1.A)/(x.A) = P_a.x_1/x \,, \quad x_1 = 2\text{ ft}.$

pressure of the gas at a position x.

The force acting on the piston at a position x :

P, P_a, F

$$F = p.A - P_a.A = (p - P_a).A$$ (piston is moving with constant speed, $\therefore$ $a = 0$)

$$F = \left(\frac{P_a.x_1}{x} - P_a\right).A = P_a\left(\frac{x_1}{x} - 1\right).A$$

$\therefore$, required work is

$$W_{12} = \int_2^{0.2} F.dx = \int_2^{0.2} P_a.A.\left(\frac{x_1}{x} - 1\right)dx = P_a.A \left| x_1 \ln x - x \right|_2^{0.2}$$

$$= (14.7 \times 144) \times (1) \times \left[(2 \ln 0.2 - 0.2) - (2 \ln 2 - 2)\right]$$

$$= -5938 \text{ lbf-ft}$$

(this work must be given to the piston).

5-45

A 50-lb block is acted upon by an 80 lb force as shown. If the block is initially at rest, determine its velocity after it has travelled a distance of 10 ft. Assume coefficient of kinetic friction between the block and the plane to be 0.2.

F = 80 lb
30°
V

****

Applying the work and kinetic energy principle,

Initial kinetic energy of block + work done by external force system on the block = Final kinetic energy of the block

$$0 + \int_0^{10} \left(80\cos 30^\circ - 0.2\left(50 + 80\sin 30^\circ\right)\right) dx = \frac{1}{2}\,\frac{50}{32.2}\,V^2$$

$$\frac{25}{32.2}V^2 = (80 \times 0.866 - 0.2 \times 90)(10)$$

$$V^2 = 660.486, \quad \text{or,} \quad V = 25.7 \text{ ft/sec}$$

5-46

A 30 lb block is released from rest on a smooth inclined plane as shown. If the spring constant k = 72 lb/ft, determine how far the spring would be compressed.

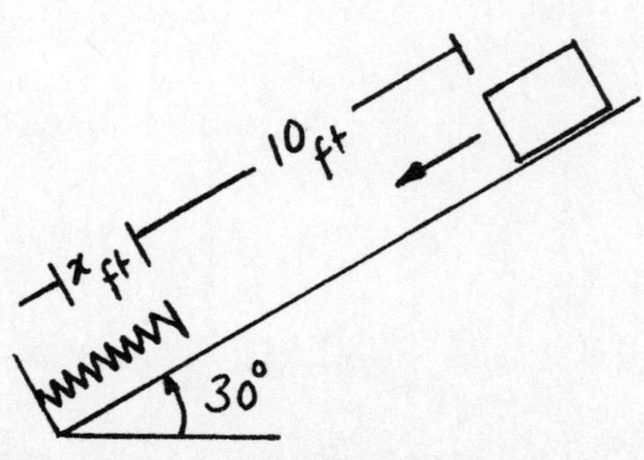

****

Initial kinetic energy of the particle + Work done by the external force system on the particle = Final Kinetic energy of the particle

$$0 + \left(\frac{1}{2}F\right)(10+x) - \frac{1}{2}kx^2 = 0$$

$$\frac{1}{2}(30)(10+x) - \frac{1}{2}(72)x^2 = 0$$

or, $2.4x^2 - x - 10 = 0$

solving, $x = 2.26$ ft

5-47

A 100 lb block is acted upon by a force F = 200 lb. Determine the final velocity of the block after it has travelled a distance of 10 ft up the inclined plane as shown. The block starts its motion from rest.

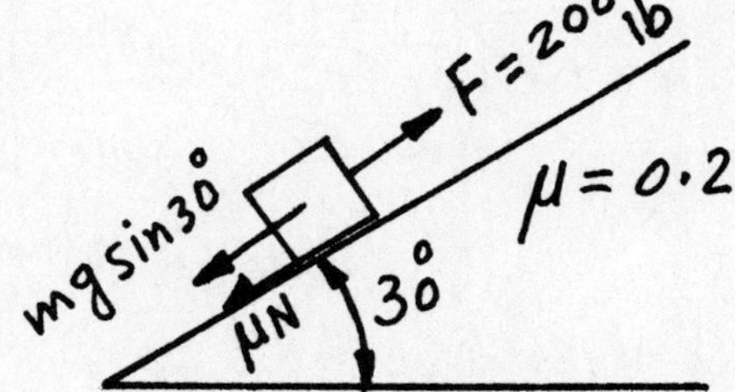

****

Consider the x-coordinate along the inclined plane

Initial kinetic energy of block + Total work done by external forces on the block = Final Kinetic energy of block

$$0 + \int_0^{10} (F - \mu N - mg\sin 30^\circ)\,dx = \frac{1}{2} m V^2$$

$$\Big[(200 - 0.2 \times 100 \times \cos 30^\circ - 100 \sin 30^\circ)x\Big]_0^{10} = \frac{1}{2}\frac{100}{32.2}V^2$$

Solving, one obtains $V = 29.2$ ft/sec

5-48

A particle of mass m is attached to a light inextensible string and swung about a point O on a frictionless table top, with angular velocity ω_o and radius r_o. If the string is slowly pulled down through a small hole O in the table, so that the new radius of rotation becomes r, determine the work done in pulling the string.

****

Since no friction, torque on m about the axis of rotation is zero. Hence the angular momentum about that axis is conserved;

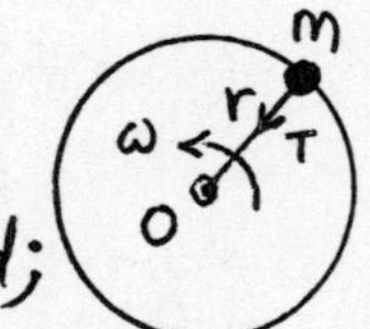

$$r\,(m r \omega) = r_o\,(m r_o \omega_o)$$

or,

$$\omega = \omega_o \frac{r_o^2}{r^2}$$

Neglecting $\frac{d^2r}{dt^2}$, the inward radial acceleration is $\omega^2 r$. From Newton's Second law, tension T in the string is given by

$$T = m\omega^2 r.$$

Substitute for ω;

$$T = m\left(\omega_o \frac{r_o^2}{r^2}\right)^2 r = \frac{m\omega_o^2 r_o^4}{r^3}$$

Work done in pulling the mass

$$W = \int_{r_o}^{r} T\,(-dr)$$

Note: r decreases in the positive direction of T; thus the negative sign in (−dr).

$$= -\int_{r_o}^{r} \frac{m\omega_o^2 r_o^4}{r^3}\,dr$$

$$= -\,m\omega_o^2 r_o^4 \left[-\frac{1}{2r^2}\right]_{r_o}^{r}$$

$$W = \frac{1}{2} m_o \omega_o^2 r_o^2 \left[\left(\frac{r_o}{r}\right)^2 - 1\right]$$

5-49

A car starts moving from rest at the point A. While it is passing through point B, its brakes are applied. Subsequently, the car slides along the track till it stops. The coefficient of friction is 0.25 . Calculate the velocity of the car at point B and the acceleration after the breaks are applied.

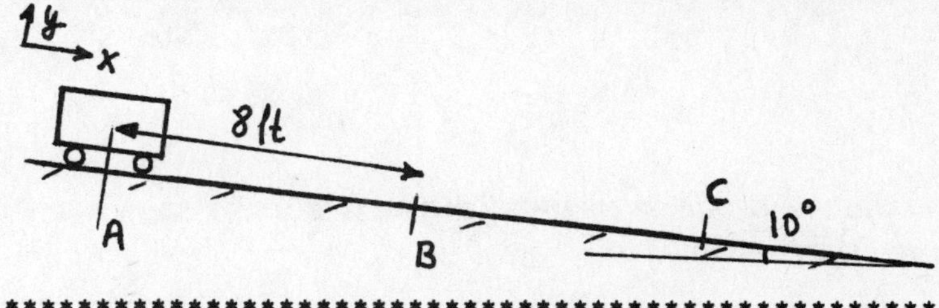

****

Energy method for points A and B :

$$T_A + U_{AB} = T_B \longrightarrow 0 + W.\,8.\,\sin 10^\circ = \frac{1}{2}.\frac{W}{g}.\,V_B^2$$

$$V_B = 9.46 \text{ ft/sec}$$

For B and C : $T_B + U_{BC} = T_C \longrightarrow \frac{1}{2} m V_B^2 + (W.\sin 10^\circ - \mu.N)\,\Delta x = 0$

$\Sigma F_y = 0$ gives $N = W\cos 10^\circ$, $\therefore\ \frac{1}{2} m V_B^2 + (W.\sin 10^\circ - W.\cos 10^\circ.\mu)\,\Delta x = 0$

$m = W/g$

10° W N

$$\frac{1}{2}.\frac{9.46^2}{32.2} + (\sin 10^\circ - \mu \cos 10^\circ).\Delta x = 0$$

$$\Delta x = 19.14 \text{ ft}$$

For the motion between B and C :

$$V_C^2 = V_B^2 + 2.a.\Delta x \longrightarrow 0 = 9.46^2 + 2.a.(19.14) \rightarrow a = -2.336 \text{ ft/sec}^2$$

uniformly decelerating motion.

CHECK : For the motion between B and C :

$$\Sigma F_x = m.a_x \longrightarrow W.\sin 10^\circ - \mu N = m.a_x$$

$$W.\sin 10^\circ - W.\cos 10^\circ.(0.25) = \frac{W}{g}.a_x$$

$$a_x = -2.336 \text{ ft/sec}^2$$

WORK AND KINETIC ENERGY: CONSERVATIVE FORCES

5-50

A particle is acted upon by a force which has the following expression:

$\underset{\sim}{F} = x(y^2+z^2)\underset{\sim}{i} + y(x^2+z^2)\underset{\sim}{j} + z(x^2+y^2)\underset{\sim}{k}$

Is this force conservative?

****

Solution :

A conservative force must satisfy the following relations :

$$\frac{\partial F_x}{\partial y} = \frac{\partial F_y}{\partial x} \quad ; \quad \frac{\partial F_y}{\partial z} = \frac{\partial F_z}{\partial y} \quad ; \quad \frac{\partial F_z}{\partial x} = \frac{\partial F_x}{\partial z}$$

In this problem,

$$F_x = x(y^2+z^2) \quad ; \quad F_y = y(x^2+z^2) \quad ; \quad F_z = z(x^2+y^2)$$

Taking the derivatives, we obtain :

$$2xy = 2xy$$

$$2yz = 2yz$$

$$2xz = 2xz$$

All relations are satisfied. Therefore this is a conservative force.

5-51

The 20 lb block has a spring attached to it (the spring constant is 4 lb/ft). The spring has an unstretched length of 4.0 ft. The block is pushed against the spring and released from rest at point A.

Determine the reaction of the block on the smooth curved surface when it gets to point A'.

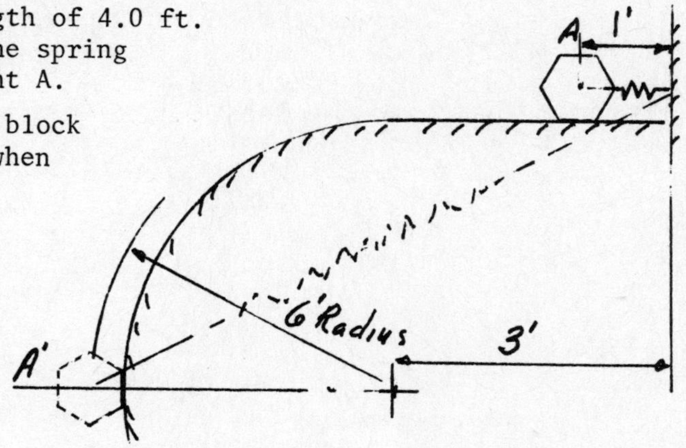

****

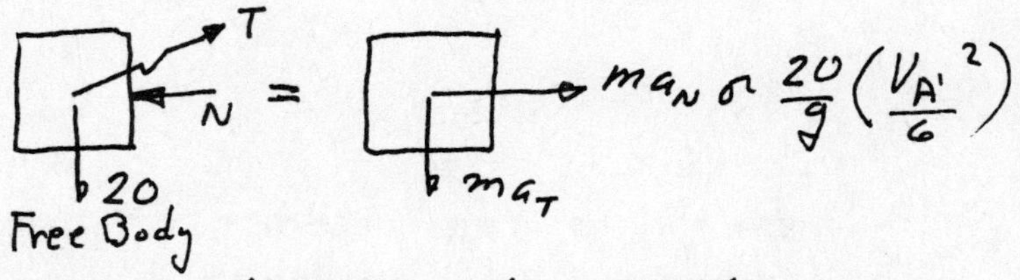

Use Work and Energy to get $V_{A'}$

draw a Force-distance diagram for the Spring

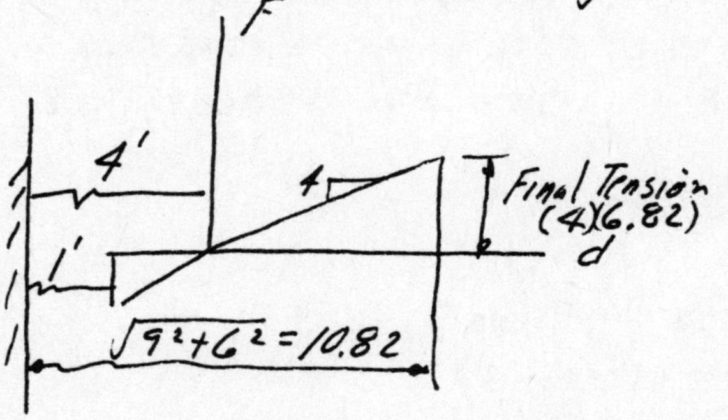

Area = Work of Spring

$= \frac{1}{2}4(3)^2 - \frac{1}{2}4(6.82)^2$

$= -74.93$

$U_{1-2} + T_1 = T_2$

$6(20) - 74.93 = \frac{1}{2}\frac{20}{g}V_{A'}^2$

$V_{A'}^2 = 145.1$

From Free Body

$\overset{+}{\leftarrow} \Sigma F_x$

$N - \frac{9}{10.82}(4)(6.82) = \frac{20}{g}\frac{145.1}{6}$

$N = 22.68 - 15.02$

$= 7.66$ lb

5-52

A particle of mass m is held against a spring of constant k. The spring is deflected a distance δ. The particle is resting at the top of a quarter-circular surface of radius R. The particle is released and slides along the frictionless surface. Derive a criterion involving m, k, δ, and R that will determine whether the particle leaves the surface before it reaches the bottom.

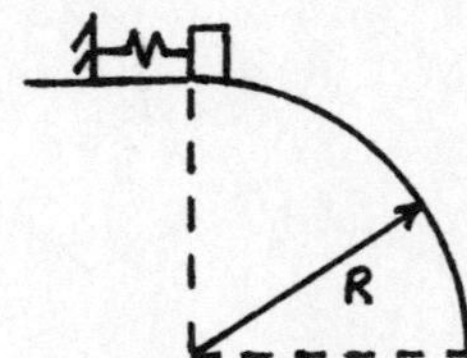

****

CONSIDER A FREE BODY DIAGRAM OF THE PARTICLE AFTER IT HAS MOVED THROUGH AN ANGLE θ

$$\Sigma F_n = m a_n$$

$$N - mg\cos\theta = \frac{-mv^2}{R}$$

WHERE V IS THE VELOCITY OF THE PARTICLE AT THAT INSTANT

THE VELOCITY CAN BE RELATED TO θ BY A CONSERVATION OF ENERGY ANALYSIS. LET THE POTENTIAL ENERGY DATUM BE AT THE BOTTOM OF THE SURFACE. ALL FORCES ARE CONSERVATIVE FORCES

$$\cancelto{0}{T_1} + V_1 = T_2 + V_2$$

$$\tfrac{1}{2}K\delta^2 + mgR = \tfrac{1}{2}mv^2 + mgR\cos\theta$$

$$mv^2 = 2\left[\tfrac{1}{2}K\delta^2 + mgR(1-\cos\theta)\right]$$

THUS

$$N - mg\cos\theta = -\tfrac{2}{R}\left[\tfrac{1}{2}K\delta^2 + mgR(1-\cos\theta)\right]$$

OR

$$N = mg(3\cos\theta - 2) - K\delta^2/R$$

THE PARTICLE WILL LEAVE THE SURFACE IF THE NORMAL FORCE IS ZERO. THUS WE SET

$$0 = mg(3\cos\theta - 2) - K\delta^2/R$$

$$\cos\theta = \frac{1}{3mg}\left(2mg + K\delta^2/R\right)$$

$$\theta = \cos^{-1}\left(2/3 + \frac{K\delta^2}{3mgR}\right)$$

A VALUE OF θ CAN BE FOUND ONLY IF

$$\frac{K\delta^2}{3mgR} < \frac{1}{3} \quad \text{OR} \quad \frac{K\delta^2}{mgR} < 1 \quad \text{OR} \quad K\delta^2 < mgR$$

5-53

A satellite of mass m is launched from earth with a velocity V_o making an angle of 30° with the horizontal. Find V_o so that the satellite is tangent at B to the orbit of radius 3R/2. (R: radius of earth).

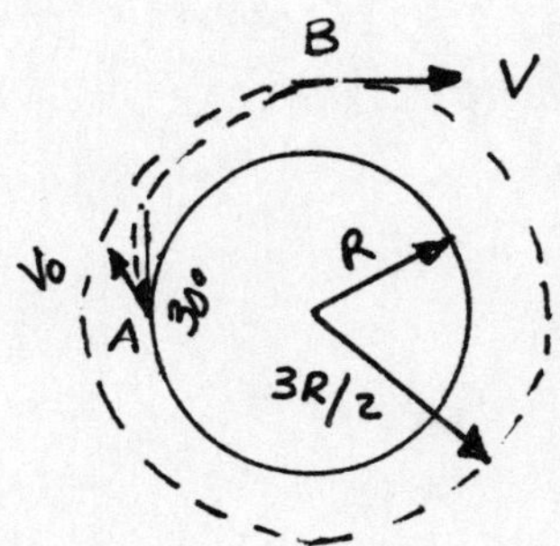

****

Conservation of energy:

$$T_A + V_A = T_B + V_B$$

$$\frac{1}{2} m v_o^2 - \frac{GmM}{R} = \frac{1}{2} m v^2 - \frac{GmM}{3R/2}$$

Conservation of angular momentum:

$$m v_o R \sin 60° = m v \frac{3R}{2} \quad \text{or} \quad v = \frac{1}{\sqrt{3}} v_o$$

$$GM = gR^2$$

Substituting, we obtain:

$$\frac{1}{2} m v_o^2 - \frac{gR^2 m}{R} = \frac{1}{2} m \left(\frac{1}{\sqrt{3}} v_o\right)^2 - \frac{2gR^2 m}{3R}$$

or $\quad V_o = \sqrt{gR}$.

5-54

A spring carries a body of 4 kg as shown in the figure. The spring has a constant of 200 N/m. Another body of 8 kg is hung above the 4 kg body at such a height that it just touches it. The wire carrying the second body is cut suddenly. Determine the maximum velocity of these two bodies during their motion and the maximum force acting on the bodies.

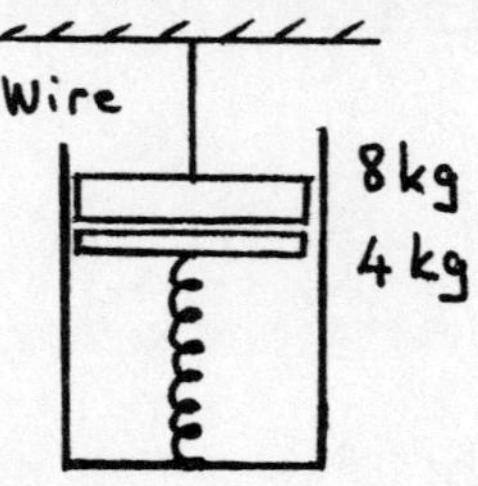

****

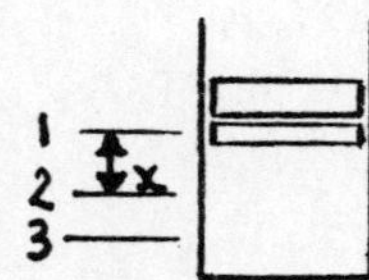

1 : initial position (system at rest)
2 : max. velocity position
3 : the system stops (at rest again).

Initial deflection of the spring : $\Delta x = F_0/k = 4 \times 9.8/200 = 0.197$ m

All forces are conservative and the principle of energy can be written as (for motion 1-2)

$$T_1 + V_1 = T_2 + V_2 \quad \text{or more specifically}$$

$$T_1 + V_{s_1} + V_{w_1} = T_2 + V_{s_2} + V_{w_2} \quad \text{if reference line is chosen at 2,}$$

$$0 + \frac{k}{2} \cdot \Delta x^2 + W \cdot x = \frac{1}{2} m V^2 + \frac{k}{2} \cdot (x + \Delta x)^2 + 0$$

$$W \cdot x = \frac{1}{2} m V^2 + \frac{1}{2} k x^2 + k \cdot x \cdot \Delta x \qquad m = \Sigma m_i = 12 \text{ kg}$$

$$12 \times 9.8\, x = \frac{12}{2} V^2 + \frac{200}{2} x^2 + 200\, x\, (0.196)$$

$$78.4\, x = 6 V^2 + 100\, x^2$$

Derivative w.r.t. x gives $78.4 = 12 V \frac{dV}{dx} + 200 x$ (*)

$V \frac{dV}{dx} = a$ and acceleration is zero when $V = V_{max}$ (at position 2)

$\therefore\ x_2 = 78.4/200 = 0.392$ m

∴ from equation (*) $78.4 \times 0.392 = 6v^2 + 100 \times 0.392^2 \rightarrow V_2 = 1.6$ m/sec

Maximum force occurs at position 3 (the system is at rest again)

$78.4\, x_3 = 6V_3^2 + 100.x^2$ with $V_3 = 0$ gives

$x_3 = 0.784$ m

∴ The maximum force $F = F_0 + x.k = 4 \times 9.8 + 0.784 \times 200 = 196$ N

5-55

A has a velocity of 15 ft/sec. to the right when it collides with B, which is at rest before impact. If the coefficient of restitution e is 0.6, determine the maximum compression of the spring.

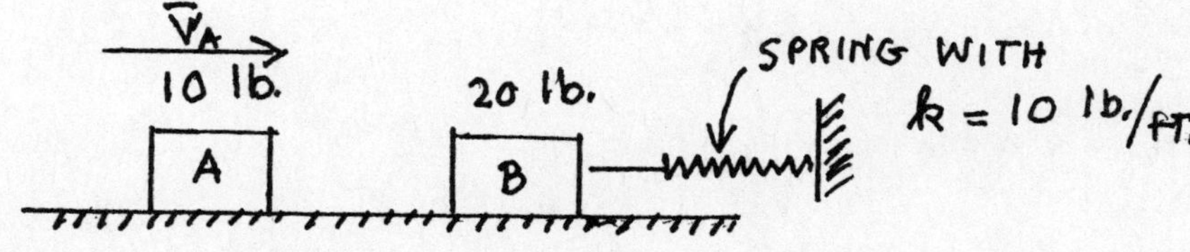

****

BY DIRECT IMPACT :

$$\frac{10}{g} \cdot 15 + \frac{20}{g} \cdot 0 = \frac{10}{g} \cdot V_1' + \frac{20}{g} V_2'$$

AND $.6 = \dfrac{V_2' - V_1'}{15}$. SOLVING, $V_2' = 8$ ft/sec.

BY CONSERVATION OF ENERGY $V_1 + T_1 = V_2 + T_2$ OR

$$\frac{1}{2} \cdot \frac{20}{g} \cdot 8^2 = \frac{1}{2} \cdot 10 \cdot x^2$$, (where x represents spring compression)

SOLVING, $x^2 = 3.975$ OR $x = 1.99$ ft.

5-56

A 10-lb load initially at rest at the position 1 starts sliding without friction on a vertical plane surface followed by a circular surface. Calculate the force exerted on the load at the positions 2 and 3.

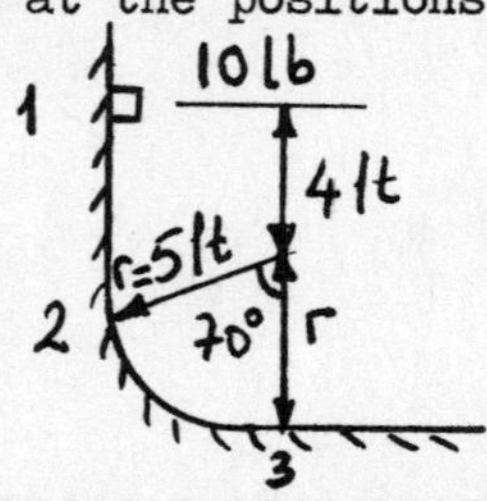

****

Energy method for points 1 and 2 :

$$T_1 + V_1 = T_2 + V_2 \longrightarrow W.\, h_{12} = \frac{1}{2}\,\frac{W}{g}.V_2^2$$

$$10\,(4 + 5.\cos 70) = \frac{1}{2}.\frac{10}{32.2}.V_2^2 \rightarrow V_2 = 19.18 \text{ ft/sec}$$

For point 2 : $\Sigma F_n = m.\dfrac{V_2^2}{r}$

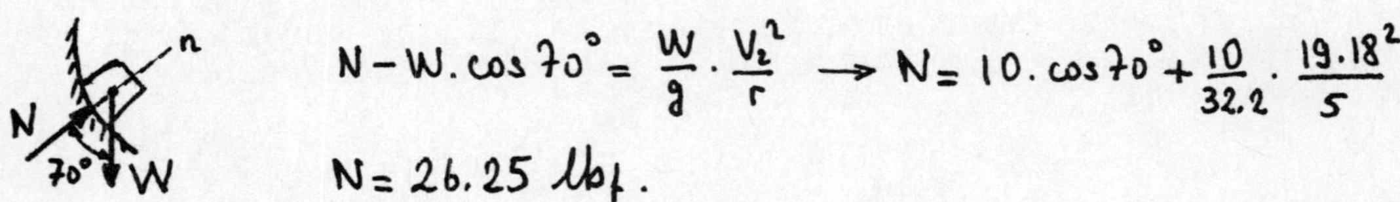

$$N - W.\cos 70° = \frac{W}{g}\cdot\frac{V_2^2}{r} \rightarrow N = 10.\cos 70° + \frac{10}{32.2}\cdot\frac{19.18^2}{5}$$

$N = 26.25$ lbf.

Energy method for points 1 and 3 :

$$T_1 + V_1 = T_3 + V_3 \longrightarrow W.h_{13} = \frac{1}{2}.\frac{W}{g}.V_3^2$$

$$10.(9) = \frac{1}{2}.\frac{10}{32.2}.V_3^2 \rightarrow V_3 = 24.08 \text{ ft/sec}$$

For point 1 : $\Sigma F_n = m.\dfrac{V_3^2}{r}$

W

N

$$N - W = m.\frac{V_3^2}{r} \longrightarrow N = 10 + \frac{10}{32.2}\cdot\frac{24.08^2}{5}$$

$N = 46$ lbf.

5-57

A pendulum consists of a bob weighing 5 pounds and a massless rod 48 inches long. The bob is raised to the position shown and released from rest. Find the tension in the rod when the rod is vertical.

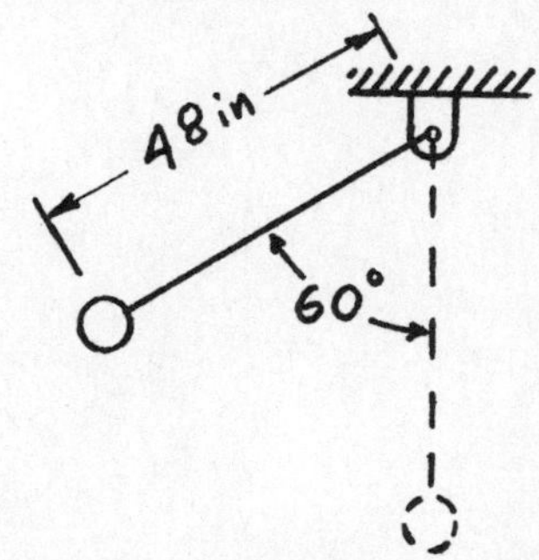

****

As bob swings, work = Δ kinetic energy

(4 ft, 60°, 2 ft, Δh, 4 ft)

$$\text{Work} = mg\Delta h$$

$$\Delta K.E. = \frac{1}{2}m(V_f^2 - V_i^2) = \frac{1}{2}mV_f^2$$

$$mg\Delta h = \frac{1}{2}mV_f^2 \quad \text{so} \quad V_f = \sqrt{2g\Delta h} = 11.3 \text{ fps}$$

At bottom of swing $\Sigma \bar{F} = (T - W)\bar{j} = m a_G \bar{j}$

(a_G, T, W)

Since bob is moving on a circular path,

$$a_G = \frac{V^2}{r} = \frac{(11.3)^2}{4} = 32$$

$$T = W + m a_G = 5 + \frac{5}{32}(32)$$

Answer $T = 10$ pounds

5-58

Suppose a force is given by $\vec{F} = i(y-x^2) + j(z-y^2) + k(x-z^2)$ and that its point of application moves from the origin to the point (1,1,1) (a) along the straight line OA or (b) along the curve:

$$c_2 = \begin{cases} x = t \\ y = t^2 \\ z = t^3 \\ 0 \le t \le 1 \end{cases}$$

Find the work done in these two cases.
(c) Is this force, $\vec{F}$, conservative or not?

****

(a) along the straight line OA; C_1: $x = y = z = t^{(*)}$ (using a parameter t) $0 \le t \le 1$

$W_{O\to A} = \int_{C_1} \vec{F}\cdot d\vec{r}$ $(*_1)$ $\vec{F}\cdot d\vec{r} = (y-x^2)dx + (z-y^2)dy + (x-z^2)dz$

where $d\vec{r} = dx\,i + dy\,j + dz\,k$

$= \int_0^1 3(t-t^2)dt$ $\overset{(*)}{=} [(t-t^2)+(t-t^2)+(t-t^2)]dt$

$\vec{F}\cdot d\vec{r} = 3(t-t^2)dt$

$= 3\left[\frac{t^2}{2} - \frac{t^3}{3}\right]_0^1 = 3\left[\frac{1}{2}-\frac{1}{3}\right] = 3\left[\frac{1}{6}\right]$

C_1: $W_{O\to A} = \frac{1}{2}$

(b) Along the path C_2:

$W_{O\to A} = \int_{C_2} \vec{F}\cdot d\vec{r}$ $\vec{F}\cdot d\vec{r} \overset{(*_1)}{=} (y-x^2)dx + (z-y^2)dy + (x-z^2)dz$

$\overset{(C_2)}{=} (t^2-t^2)dt + (t^3-t^4)2t\,dt + (t-t^6)3t^2dt$ (first term $= 0$)

$= \int_0^1 [2t^4 - 2t^5 + 3t^3 - 3t^8]dt$ $\vec{F}\cdot d\vec{r} = [2t^4 - 2t^5 + 3t^3 - 3t^8]dt$

$= \left.\frac{2t^5}{5} - \frac{2t^6}{6} + \frac{3t^4}{4} - \frac{3t^9}{9}\right|_0^1$

$= \frac{2}{5} - \frac{1}{3} + \frac{3}{4} - \frac{1}{3} = \frac{24-20+45-20}{60} = \frac{29}{60}$

If $\vec{F}$ was conservative, the work done would be the same over any path from O to A – hence this force is NOT conservative!

5-59

In a physics laboratory demonstration, a spherical ball whose mass is 2 kg is tied to a string so that the effective length between the center of sphere and the point of swing is 1.5 m. If the maximum swing angle ϕ from the vertical position is 72°, calculate the maximum speed of the sphere. Neglect any losses due to friction.

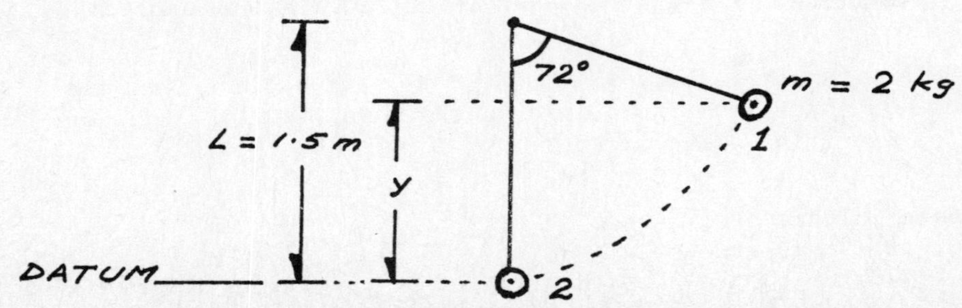

****

m

Total energy $E = T + V$

Hence, when T is maximum, V is minimum. Considering the point when the ball is vertical as datum, V_{min} would be obtained when $y = 0$

Also, $T_1 + V_1 = T_2 + V_2$

<u>At point 1</u>: $T_1 = \frac{1}{2} m v^2 = 0$, as $v_1 = 0$ (by inspection)

$$V_1 = m g y_1 = mg(L - L \cos 72°)$$
$$= (2\ kg)(9.81\ m/s^2)[(1.5\ m) - (1.5\ m)(0.309)]$$
$$= 20.345\ kg.\ m^2/s^2$$

<u>At point 2</u>: $T_2 = \frac{1}{2} m v^2 = \frac{1}{2}(2.5\ kg) v_2^2 = (1.25\ kg) v_2^2$

$$V_2 = m g y_2 = 0 \quad \text{as } y_2 = 0$$

Hence, $0 + 20.345\ kg.m^2/s^2 = (1.25\ kg)\, v_2^2 + 0$

OR, $v_2 = \sqrt{\dfrac{20.345\ kg.m^2/s^2}{1.25\ kg}}$

$$= 4.034\ m/s$$ <u>Ans.</u>

5-60

Rigid link AB has a mass of 15 kg and is connected at one end to a spring. The other end of the link is hinged. If the spring constant is 60 N/m, how much work is done in rotating the link by 90 degrees in the counter clockwise direction? The spring is initially stretched by 0.25 meters.

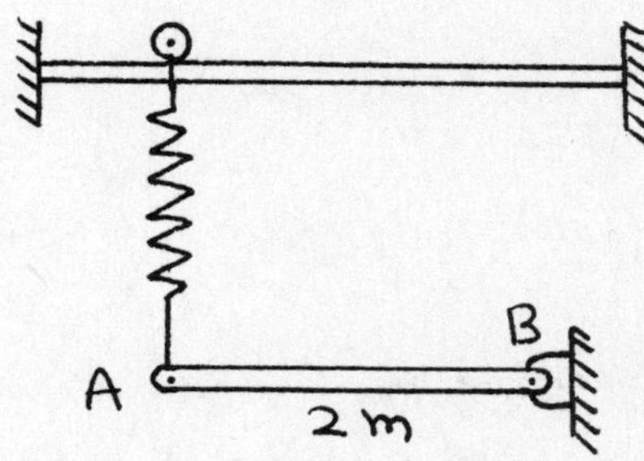

****

Work done in stretching the spring

$$W_1 = -\int_{0.25}^{2.25} KS\,dS = -\frac{1}{2}KS^2 \Big|_{0.25}^{2.25}$$

$$W_1 = -\frac{60}{2}\left[2.25^2 - 0.25^2\right] = -150\,N\text{-}m.$$

Work done in moving the center of mass of AB

$W_2 = mgh = 15(9.81)1 = 147.2$ N-m.

Total work done $= W_1 + W_2 = 147.2 - 150$

$= -2.8$ N-m.

or -2.8 Joule

WORK AND KINETIC ENERGY: TOTAL MECHANICAL ENERGY

5-61

A 40 lb block rests on a flat plate as shown. The spring has a stiffness of k = 100 lb/ft and is initially compressed 2 ft from position 2 to 1. If the block is released from rest at position 1, determine its velocity when it reaches position 2.

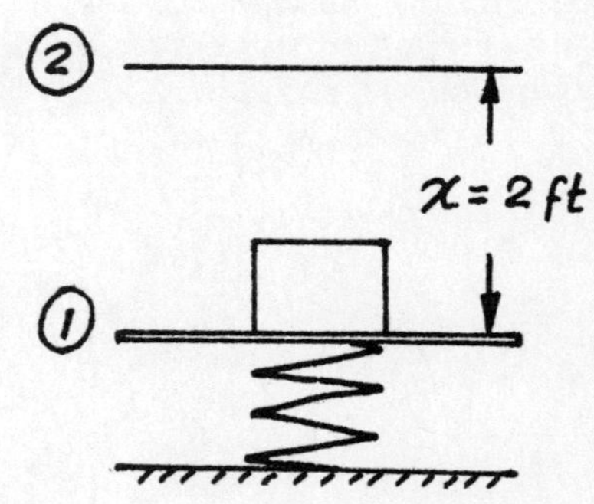

****

Total mechanical energy of system at position 1 = Total mechanical energy of system at position 2

$$\frac{1}{2}kx^2 = mgx + \frac{1}{2}mV^2$$

or, $$\frac{1}{2} \times 100 \times 4 = 40 \times 2 + \frac{1}{2}\frac{40}{32.2}V^2$$

Solving, we get $V = 13.9$ ft/sec

5-62

A roller coaster is released from rest and let go down an inclined plane leading into a "loop-the-loop" of radius r. Determine the minimum height h from which the roller coaster has to start in order to be able to make a complete revolution in the loop.

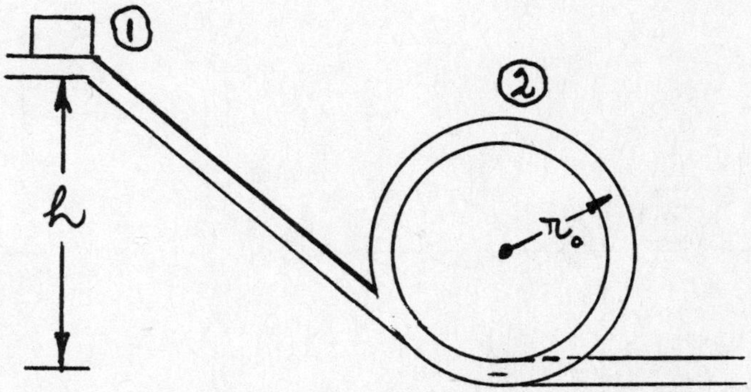

****

Conservation of energy between 1 and 2 gives

$$T_1 + V_1 = T_2 + V_2$$

or

$$mgh = \frac{mV_2^2}{2} + mg(2R) \quad \text{(datum at bottom)}$$

A free body diagram at 2 gives, or

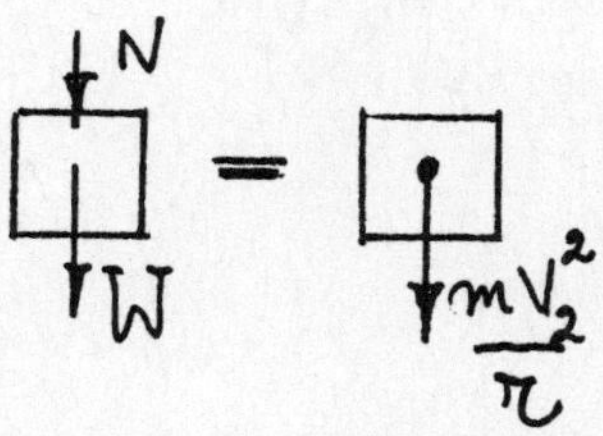

$$mg = \frac{mV_2^2}{r}$$

where the normal force N has been set equal to zero in order to determine the minimum h.

Eliminating V_2 between the two equations yields

$$h = 2.5r$$

5-63

Block B has a velocity V_o. If the coefficient of restitution between the two blocks is e, derive an expression for the height H that block A will attain. $m_A = 2\ m_B$.

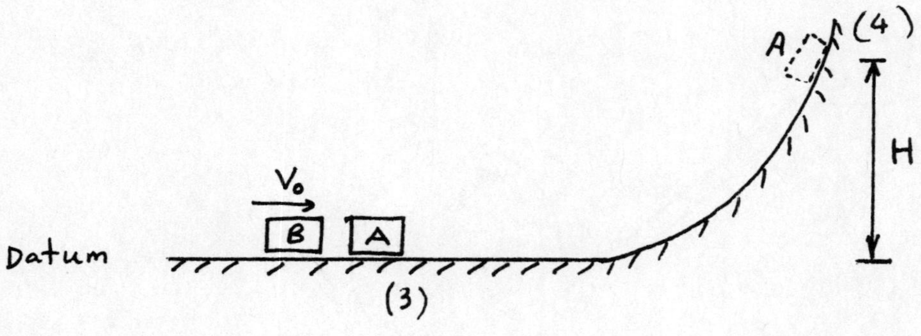

****

Before impact $v_A = 0$, $v_B = V_o$

After impact denote the velocities as v_A' and v_B'

Conservation of momentum :

$$m_A v_A + m_B v_B = m_A v_A' + m_B v_B'$$

$$2 m_B (0) + m_B V_o = 2 m_B v_A' + m_B v_B'$$

or $$2 v_A' + v_B' = V_o \qquad (1)$$

$$e(v_A - v_B) = v_B' - v_A'$$

or $$v_B' - v_A' = e(0 - V_o) \qquad (2)$$

Solving (1) and (2), we obtain :

$$v_A' = \frac{1}{3}(1+e) V_o$$

Conservation of energy for block A :

$$T_3 + V_3 = T_4 + V_4$$

$$\frac{1}{2} m_A v_A'^2 + 0 = 0 + m_A g H$$

$$H = \frac{1}{2g} v_A'^2 \quad \text{or} \quad H = \frac{1}{18g} (1+e)^2 V_0^2$$

5-64

A 500-kg block drops a distance of 5 m from level A and hits the unstressed spring at level B. The spring is compressed up to level C, a distance of 3 m as indicated. Determine the stiffness of the spring.

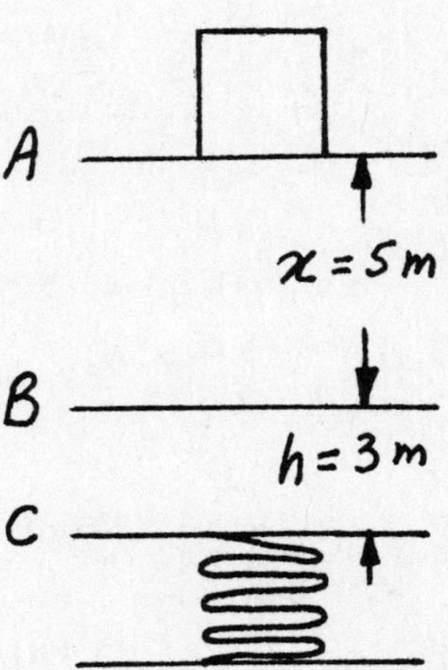

****

In this case, the block is acted upon by gravity which is a conservative force and the spring is acted upon by a force which is conservative as well. Therefore, potential energy released by block = potential energy absorbed by spring

or,
$$mg(x+h) = \frac{1}{2} k h^2$$
$$500 \times 9.81 (5+3) = \frac{1}{2} K (3)^2$$

Solving for, spring stiffness $= K = 8,720$ N/m

5-65

A train of joyride cars full of children in an amusement park is pulled by an engine along a straight, level track. It then begins to climb up a 5° slope. At a point B, 150 ft up the grade, the last car becomes uncoupled without the driver noticing, when its velocity is 20 m.p.h. If the car with its passengers weighs 1,700 lb and the track resistance is 2% of the car's weight, calculate (a) the total distance up the grade where the car stops at a point C, and (b) how far back on the level track the car will roll before it finally stops at a point D.

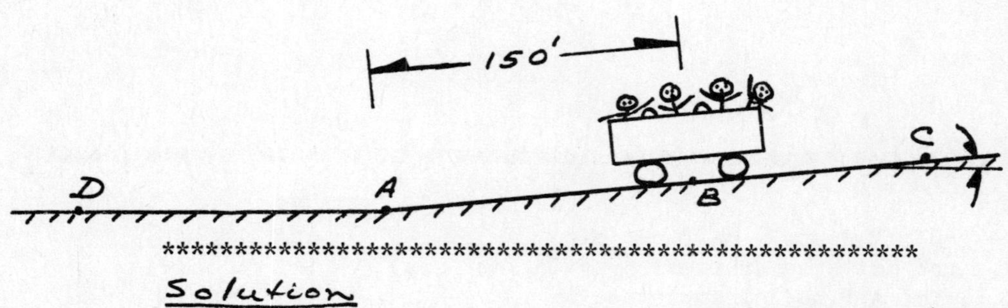

****

Solution

$$20 \text{ m.p.h.} = \frac{(20 \text{ m.p.h.})(5280 \text{ ft})}{(3600 \text{ s})} = 29.33 \text{ ft/s}$$

Let KE_1 = Initial K.E. when car becomes uncoupled

KE_2 = K.E. at point C = 0

KE_3 = K.E. at point D = 0

(a) $KE_1 = \frac{W v^2}{2g} = \frac{(1700 \text{ lb})(29.33 \text{ ft/s})^2}{2(32.2 \text{ ft/s}^2)} = 22708.434$ ft.lb

Positive work (PW) on car going up slope = 0

Negative work ($N_f W$) on car by friction

$= (F) W(BC) = (0.02)(1700 \text{ lb})(BC) = 34 (BC)$ ft.lb

Negative work ($N_g W$) on car by gravity

$= W(BC \sin 5°) = (1700 \text{ lb})(BC)(0.087) = 148.164 (BC)$ ft.lb

Now, $KE_1 + (PW) - (N_f W) - (N_g W) = KE_2$

OR, $[22708.434 + 0 - 34(BC) - 148.164(BC)]$ ft.lb $= 0$

OR, $BC = 124.659$ ft.

Hence, total distance travelled up the slope before car stops = (AB + BC)

$= (150 + 124.659)$ ft $= 274.659$ ft **Ans.**

(b) Positive work (U_p) by gravity

$= W(AC \sin 5°) = (1700 \text{ lb})(274.659 \text{ ft})(0.0871)$

$= 40694.786$ ft.lb

Negative work (U_n) by friction

$= FW(AC + CD) = (0.02)(1700 \text{ lb})(274.659 + AD)$ ft

$= [9338.406 + 34 (AD)]$ ft.lb

Again, $KE_2 + (U_p) - (U_n) = KE_3$

or, $\left[0 + 40694.786 - \{9338.406 + 34(AD)\}\right] ft.lb = 0$

or, $AD = 922.246$ ft. <u>Ans.</u>

5-66

The 2 kg. collar A was moving down the rod with a velocity of 4 meters/sec. when a force P was applied horizontally. If the coefficient of friction between the collar and rod is 0.25, determine the magnitude of the force P if the collar stopped after moving 1.5 meters more down the rod.

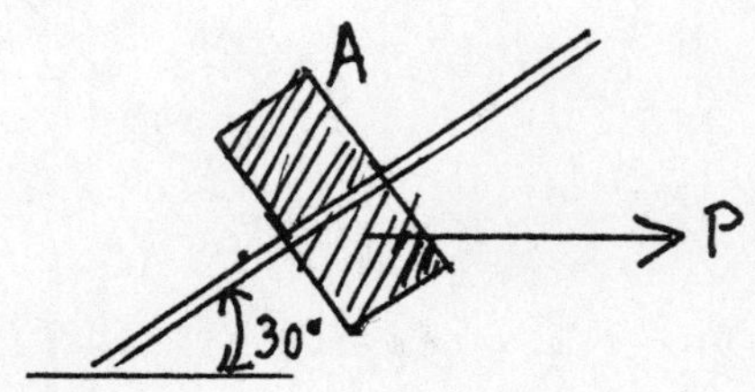

****

<u>FBD OF BLOCK OR COLLAR:</u>

$\Sigma F_y = 0$

$-2g\cos 30° - P\sin 30° + N = 0$

$\therefore N = 2g\cos 30° + P\sin 30°$

and $F' = \frac{1}{4}N = 4.248 + P/8$

BY WORK & ENERGY, $U = (2g\sin 30° - F' - P\cos 30°) \cdot \frac{3}{2}$

AND $U = \Delta T = -\frac{1}{2} \cdot 2 \cdot 4^2$

SUBSTITUTING FOR F' AND EQUATING: $P = 16.37$ N

5-67

A 20 lb. block slides on a rod as shown. Initially, at point 1, X_1= 0.5 ft. compressed, and X_2= 1.0 ft. stretched. Initial velocity =0, k=180 lb/ft. for both springs. The friction force between the block and the rod is constant at 10 lbs.
What is the velocity at point 2, after the block has slid 1.0 ft.?

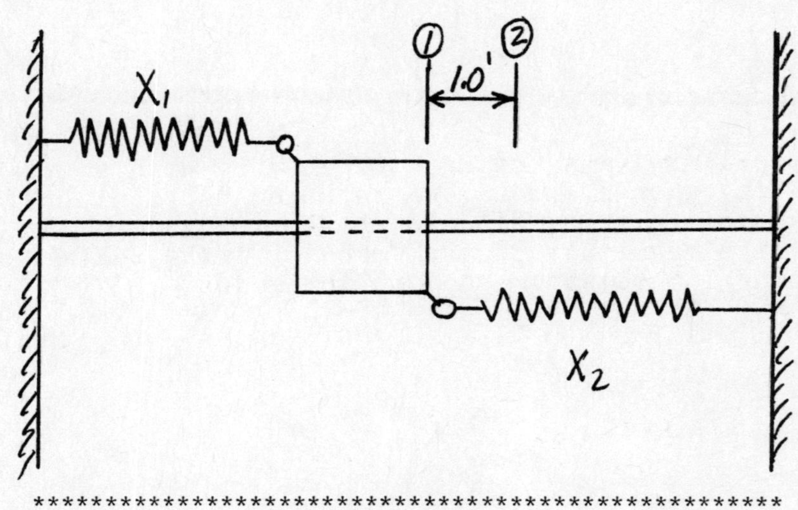

****

$$V_{1S} = \frac{1}{2}KX_1^2 + \frac{1}{2}KX_2^2 = \frac{1}{2}(180)(.5)^2 + \frac{1}{2}(180)(1.0)^2 = 112.5 \text{ ft-Lb.}$$

@ 2: $X_1 = 0.5'$, stretched; $X_2 = 0$

$$V_{2S} = \frac{1}{2}(180)(.5)^2 = 22.5 \text{ ft-Lb.}$$

$$W_f = -10(1) = -10 \text{ ft-Lb}$$

$$T_1 = \frac{1}{2}mV_1^2 = 0$$

$$T_2 = \frac{1}{2}mV_2^2 = \frac{1}{2}\left(\frac{20}{32.2}\right)(V^2)$$

$$T_1 + V_{1S} - W_f = T_2 + V_{2S}$$

$$0 + 112.5 - 10 = \frac{1}{2}\left(\frac{20}{32.2}\right)V^2 + 22.5$$

$$\boxed{V = 16.04 \text{ ft/SEC}} \quad \text{ANS.}$$

5-68

A toy truck travels around a 20-inch diameter vertical circular track as shown below. The mass center of the truck is 1.5 inches off the track. Determine the minimum velocity, V_o, which the truck must have at the bottom of the track in order to maintain contact with the track at all times. Assume frictionless contact.

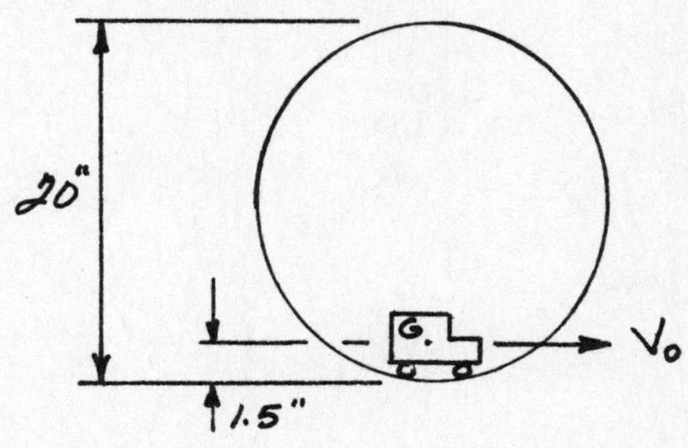

****

FBD of truck at top of track.

$$\oplus\downarrow \Sigma F_N = m a_N$$

$$W + N^{\,0} = m a_N$$

(NOTE: $N = 0$ since truck is barely making contact with track)

$$mg = m a_N$$

$$a_N = g = 32.2\ \frac{ft}{sec^2} = \frac{(V_{TOP})^2}{r}$$

$$V_{TOP} = \sqrt{\left(32.2\ \frac{ft}{sec^2}\right)\left(\frac{8.5}{12}\ ft\right)} = 4.78\ ft/sec$$

<u>CONSERVATION OF ENERGY</u>: (datum at bottom of track)

$$(T+V)_{BOTTOM} = (T+V)_{TOP}$$

$$\tfrac{1}{2} m V_0^2 + W \cdot 1.5\ IN = \tfrac{1}{2} m V_{TOP}^2 + W \cdot 18.5\ IN.$$

$$\tfrac{1}{2} \not{m} V_0^2 + \not{m}\left(32.2\ \frac{ft}{sec^2}\right)\left(\frac{1.5}{12}\ ft\right) = \tfrac{1}{2}\not{m}\left(4.78\ \frac{ft}{sec}\right)^2 + \not{m}\left(32.2\ \frac{ft}{sec^2}\right)\left(\frac{18.5}{12}\ ft\right)$$

$$\underline{\underline{V_0 = 10.68\ ft/sec}}$$

5-69

At position (1) the 3 lb. block is moving down the surface toward the spring with a velocity of 2 ft/sec. The spring, which has a stiffness of 10 lb/in, is initially unstretched. The coefficient of friction between the block and the surface is 0.3. Position (2) occurs when the spring has reached its maximum deflection. Position (3) occurs when the block reaches maximum height after being released by the spring.

a. What is the maximum deflection of the spring, Δ , after being hit by the block the first time ?

b. What is the maximum height, h, reached by the block after being released by the spring the first time ?

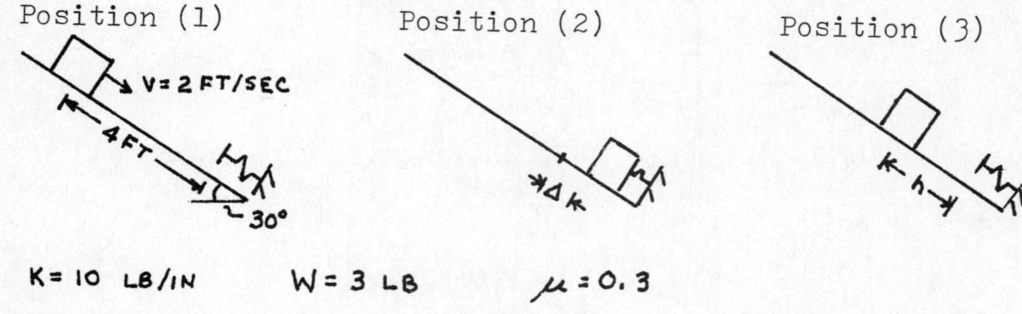

K= 10 LB/IN   W = 3 LB μ = 0.3

****

a. ENERGY IS NOT CONSERVED BETEEN POSITIONS (1) AND (2) DUE TO FRICTION. HOWEVER A WORK-ENERGY ANALYSIS CAN BE USED.

$$T_1 + U_{1\to 2} = T_2$$

T_1 = KINETIC ENERGY AT POSITION (1) = $\frac{1}{2} m v_1^2$

T_2 = KINETIC ENERGY AT POSITION (2) = $\frac{1}{2} m v_2^2 = 0$

(BLOCK HAS ZERO VELOCITY WHEN SPRING REACHES MAXIMUM DISPLACEMENT)

$U_{1\to 2}$ = WORK DONE BY ALL EXTERNAL FORCES ACTING ON THE BLOCK BETWEEN POSITIONS (1) AND (2)

W, N, F_s, $F = \mu N$

$$\Sigma F_n = 0$$

$$N = W \cos 30° = \frac{W\sqrt{3}}{2}$$

$$U_{1\to 2} = U_N + U_F + U_{F_s} + U_W$$

$U_N = 0$ (THE NORMAL FORCE IS ALWAYS

PERPENDICULAR TO THE PATH OF THE BLOCK)

$$U_F = -F(4\text{ FT} + \Delta) = -\mu \frac{W\sqrt{3}}{2}(4\text{ FT} + \Delta)$$

$$U_W = W(4\text{ FT} + \Delta)\sin 30^\circ = \frac{W}{2}(4\text{ FT} + \Delta)$$

$$U_S = \int_0^{\Delta} F_S\,dx = -\int_0^{\Delta} Kx\,dx = -\frac{1}{2}K\Delta^2$$

THUS

$$\frac{1}{2}mv_1^2 - \mu\frac{W\sqrt{3}}{2}(4\text{ FT} + \Delta) + \frac{W}{2}(4\text{ FT} + \Delta) - \frac{1}{2}K\Delta^2 = 0$$

SUBSTITUTING THE GIVEN VALUES

$$\frac{1}{2}\left(\frac{3}{32.2}\right)(2)^2 - (.3)\frac{(3\sqrt{3})}{2}(4+\Delta) + \frac{3}{2}(4+\Delta) - \frac{1}{2}(10)(12)\Delta^2 = 0$$

$$-120\Delta^2 + 0.721\Delta + 5.41 = 0$$

SOLVING FOR Δ

$$\Delta = \frac{-0.721 \pm \sqrt{(0.721)^2 + 4(120)(5.41)}}{-2(120)}$$

$$\Delta = 0.215\text{ FT},\ -0.209\text{ FT}$$

OBVIOUSLY Δ MUST BE >0 SO

$$\Delta = 0.215\text{ FT}$$

b. WORK-ENERGY CAN ALSO BE USED BETWEEN POSITIONS (2) AND (3)

$$T_2 + U_{2\to3} = T_3$$

$$T_2 = 0$$

$$T_3 = \frac{1}{2}mV_3^2 = 0$$

$$U_{2\to3} = U_N + U_F + U_{FS} + U_W$$

$$U_N = 0$$

$$U_F = -F(0.215 + h) = -0.3\frac{W\sqrt{3}}{2}(0.215 + h)$$

WHERE h IS MEASURED FROM THE UNSTRETCHED LENGTH OF THE SPRING

$$U_{FS} = \frac{1}{2}K\Delta^2 = \frac{1}{2}(120\text{ LB/FT})(0.215\text{ FT})^2$$

$$U_W = -W(0.215 + h)$$

THUS

$$-0.3\frac{3\sqrt{3}}{2}(0.215 + h) + \frac{1}{2}(120)(0.215)^2 - 3(0.215 + h) = 0$$

$$h = 0.548\text{ FT}$$

5-70

The bob of a pendulum starts in a vertical plane from the vertical position AB with an initial velocity V_o. The tension in the rigid rod becomes 7W when the angle is as shown. Find the initial velocity V_o if L = 6 ft. Use g = 32.2 ft/s^2.

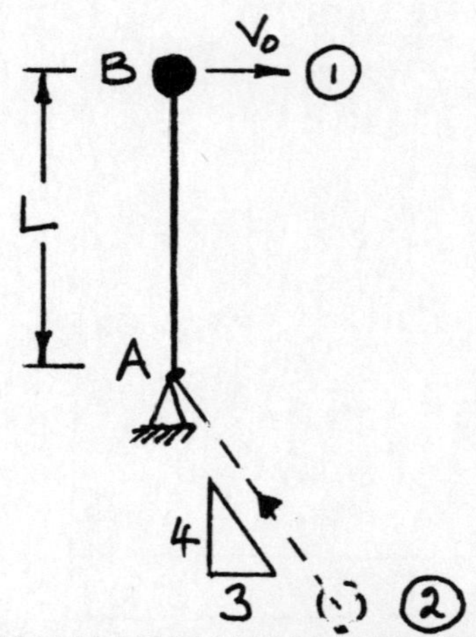

****

Newton's law of motion in the normal direction gives $7W - \frac{4}{5}W = \frac{mV_2^2}{L}$; $V_2^2 = \frac{31}{5}gL$

4, 3, 7W, W = mV_2^2/L, ma_t

for the freebody diagram at 2, shown.

Conservation of energy between 1 and 2 gives, taking V = 0 (datum) through A

$$\frac{1}{2}mV_o^2 + mgL = \frac{1}{2}mV_2^2 - \frac{4}{5}mgL$$

Elimination of V_2 between the two equations gives

$$V_o^2 = 2.6gL = (2.6)(32.2)(6)$$

$$\underline{\underline{V_o = 22.41 \text{ ft/s}}}$$

5-71

Using the Work-Energy Method, determine the distance, d, that the 13# block shown will move down the slope before coming to rest if the initial velocity is equal to 20 feet per second.

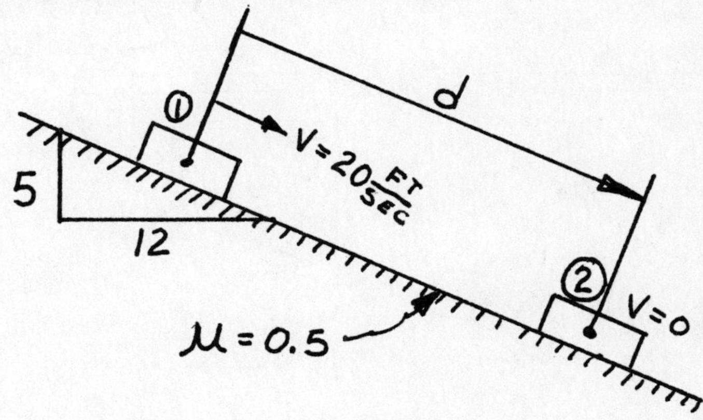

****

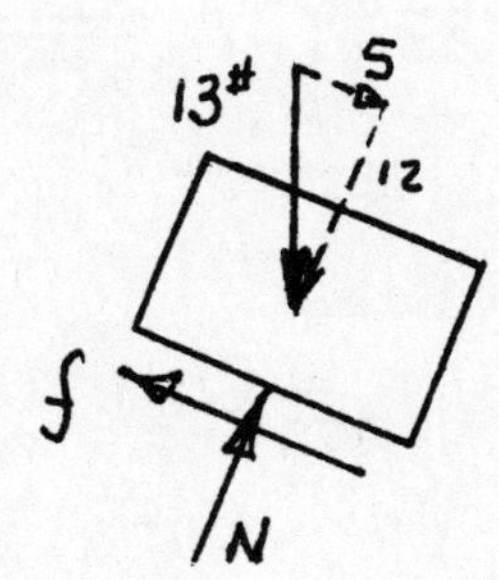

$$N = \frac{12}{13}(13^{\#}) = 12^{\#}$$

$$f = 0.5\,N = 0.5(12^{\#}) = 6^{\#}$$

$$(PE)_1 + (KE)_1 + GAINS = (PE)_2 + (KE)_2 + LOSSES$$

$$(13^{\#})(h_1) + \frac{1}{2}\left(\frac{13^{\#}}{32.2\,\frac{FT}{SEC^2}}\right)\left(20\,\frac{FT}{SEC}\right)^2 = 0 + 0 + (6^{\#})d$$

$$13^{\#}\left(\frac{5}{13}d\right) + 80.74 \text{ FT-lbs} = 6^{\#}d$$

$$80.74 \text{ FT-lbs} = 6^{\#}d - 5^{\#}d$$

$$d = \frac{80.74 \text{ FT-lbs}}{1^{\#}} = 80.74 \text{ FT}$$

5-72

A 10 lb. weight slides from rest on a frictionless rod from point P to point Q. The spring constant is k = 2 lb/ft and has an unstretched length of 1 foot. Find the speed of the weight at Q if (a) the path is a quarter circle, (b) the path is an ellipse.

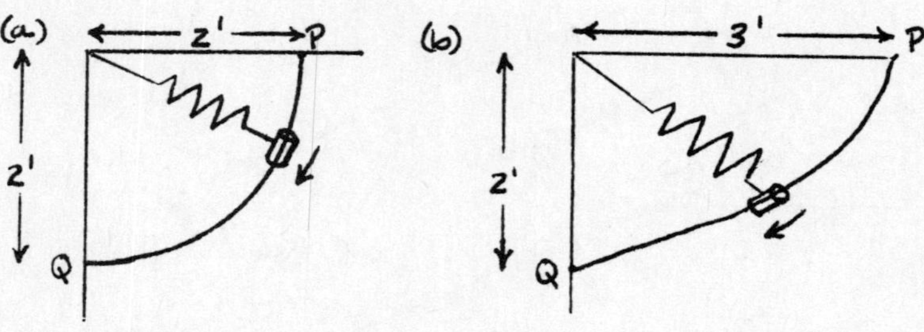

****

(a)

$$T_1 + V_1 = T_2 + V_2$$

$$0 + \begin{cases} Wh \\ \cancel{V_s} \end{cases} = \frac{1}{2}\left(\frac{W}{g}\right)v^2 + \begin{cases} 0 \\ \cancel{V_s} \end{cases}$$

$$\therefore \cancel{W}h = \frac{1}{2}\left(\frac{\cancel{W}}{g}\right)v^2$$

$$2hg = v^2$$

or $v = \sqrt{2hg} = \sqrt{2(2)32.2}$

$$\underline{v = 11.35 \text{ ft/sec}}$$

(b)

$$T_1 + V_1 = T_2 + V_2$$

$$0 + Wh + \frac{1}{2}Kx_P^2 = \frac{1}{2}\left(\frac{W}{g}\right)v^2 + \frac{1}{2}Kx_Q^2$$

(Recall: $V_{spring} = \frac{1}{2}Kx^2$)

but $x_P = 2$, $x_Q = 1$

$$10(2) + \frac{1}{2}(2)2^2 = \frac{1}{2}\left(\frac{10}{32.2}\right)v^2 + \frac{1}{2}(2)1^2$$

$$23 = 20 + 4 - 1 = \frac{5}{32.2}v^2 \quad \therefore v^2 = \frac{32.2 \cdot 23}{5}$$

$$\therefore v^2 = 148.12$$

$$\underline{\therefore v = 12.17 \text{ ft/sec}}$$

5-73

A spring loaded collar weighing 35 lb slides on a vertical frictionless shaft, as shown in the accompanying illustration. The height between the rest and final positions of the collar is 15 in, as is the distance between the shaft and the point of attachment of the spring's free end. The spring has an undeformed length of 7 in and a spring constant of 4.5 lb/in. Calculate the velocity of the collar at the final position on release from rest.

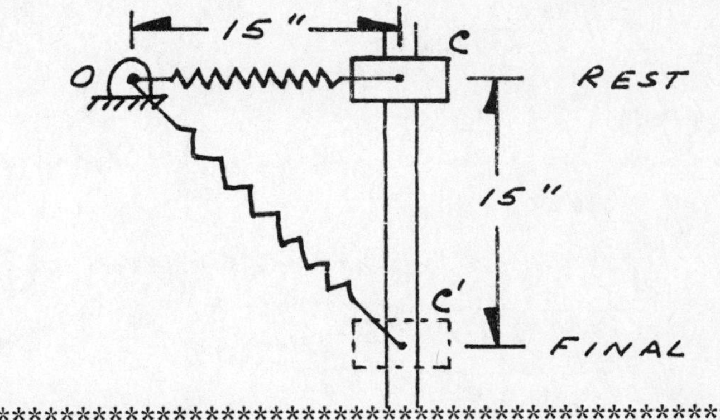

****

Solution

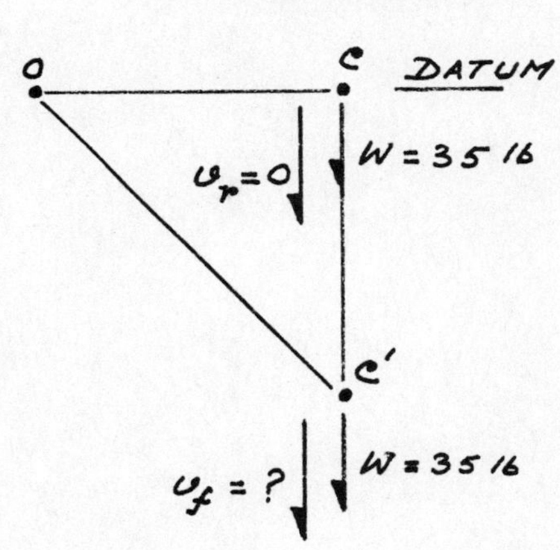

From geometry,

$$OC' = \sqrt{OC^2 + CC'^2}$$
$$= \sqrt{(15")^2 + (15")^2} = 21.213"$$

At rest position

(a) Potential energy:

Spring elongation

$$x_r = (15 - 7)" = 8"$$

and work done by spring

$$V_{s_r} = \frac{1}{2} k x_r^2$$
$$= \frac{1}{2}(4.5 \text{ lb/in})(8")^2$$
$$= 144 \text{ in. lb}$$

Taking datum at rest position, we get

$$V_r = V_{s_r} + V_g = (144 + 0) \text{ in. lb} = 144 \text{ in. lb}$$

(b) Kinetic energy:

$$T_r = 0 \quad \text{as} \quad v_r = 0.$$

At final position

(a) Potential energy:

Spring elongation $x_f = (21.213 - 7)" = 14.213"$

and work done by spring

$$V_{sf} = \frac{1}{2} k x_f^2$$
$$= \frac{1}{2}(4.5 \text{ lb/in})(14.213")^2 = 454.521 \text{ in.lb}$$
$$V_g = Wy = (35 \text{ lb})(-15") = -525 \text{ in.lb}$$

Thus, $V_f = V_{s_f} + V_g = (454.521 - 525)$ in.lb

$= -70.479$ in.lb

(b) Kinetic energy:

$$T_f = \frac{1}{2} m v_f^2$$

$$= \frac{1}{2}\left(\frac{35 \text{ lb}}{32.2 \text{ ft/s}^2}\right)(v_f \text{ ft/s})^2 = 0.543 v_f^2 \text{ ft.lb}$$

By applying the principle of conservation of energy at the rest and final positions, we get

$$T_r + V_r = T_f + V_f$$

OR, $0 + \left(\frac{144}{12}\right)$ ft.lb $= 0.543 v_f^2$ ft.lb $- \left(\frac{70.749}{12}\right)$ ft.lb

OR, $v_f = \sqrt{\dfrac{\left(\frac{144}{12} + \frac{70.749}{12}\right) \text{ft.lb}}{0.543 \text{ ft.lb}}} = \pm 5.737$ ft/s

Hence, $v_f = 5.737$ ft/s ↓ **Ans.**

6
KINETICS OF A SYSTEM OF PARTICLES

MOTION OF THE CENTER OF MASS

6-1

A 1 ton satellite is traveling at a velocity, relative to a particular coordinate system, of $\vec{V} = 1000i$(ft/sec). At t = 0, it is at the origin but breaks up into 3 parts A, B, and C of 400, 600, and 1000 lbs., respectively. When next observed 3 seconds later, the positions of A and B are $\vec{R}_A = 2000i + 1000j + 500k$ and $\vec{R}_B = -500i + 400j + 200k$. Where is C at t = 3?

****

We observe that there is no external force, hence the velocity of the mass center is $\vec{V} = 1000\,i$

at t = 3 sec: $\vec{R} = \vec{V}\cdot t = 3{,}000\,i$; then, since $m\vec{R} = \sum_{i=1}^{3} m_i \vec{R}_i$ $\quad (m = \sum_{i=1}^{3} m_i)$

$$m\vec{R} = m_A\vec{R}_A + m_B\vec{R}_B + m_c\vec{R}_c$$

$$\frac{2000}{g}[3{,}000\,i] = \frac{400}{g}[2000i + 1000j + 500k] + \frac{600}{g}[-500i + 400j + 200k] + \frac{1000}{g}\vec{R}_c$$

$$60{,}000\,i = 8{,}000\,i + 4{,}000\,j + 2{,}000\,k - 3{,}000\,i + 2400\,j + 1200\,k + 10\,\vec{R}_c$$

$$55{,}000\,i - 6{,}400\,j - 3{,}200\,k = 10\vec{R}_c$$

$$\therefore \vec{R}_c = 5{,}500\,i - 640\,j - 320\,k$$

6-2

Block A weighs 60 kilograms and block B weighs 20 kilograms. The pulley is weightless and frictionless. The system is released from rest. Find the acceleration of block A.

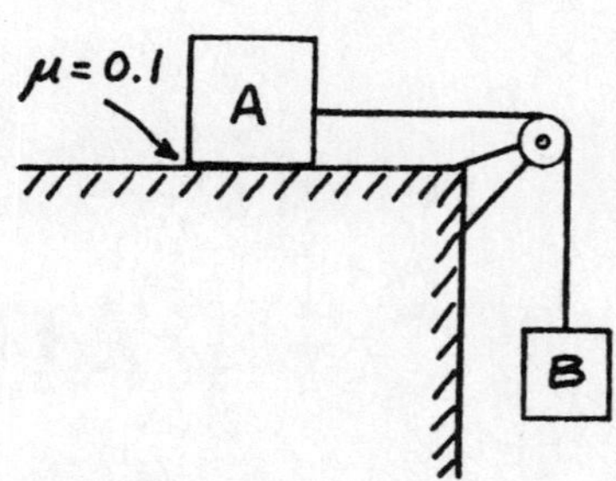

****

Block A

$$\Sigma \bar{F} = (T - f)\bar{i} + (R_N - W)\bar{j} = m\, a_{GA}\bar{i}$$

$W = 60\text{ kg}(9.8\text{ m/s}^2) = 588\text{ N}$, $R_N = W$, $m = 60$

slips so $f = \mu R_N = 0.1(588) = 58.8\text{ N}$

$$T - 58.8 = 60\, a_{GA} \qquad (\text{I.})$$

Block B

$$\Sigma \bar{F} = (T - W)\bar{j} = m\, a_{GB}(-\bar{j})$$

$m = 20$, $W = 20(9.8) = 196$

$$T - 196 = -20 a_{GB} \qquad (\text{II.})$$

Connecting string means $a_{GA} = a_{GB}$ (III.)

Solving (I.), (II.) and (III.) simultaneously

Answer

$$a = 1.72\text{ m/s}^2$$

6-3

A spring loaded device weighing 35.5 lb is ejected at a velocity of 250 ft/s in a horizontal direction, after which it splits up into two parts. Part 1 weighs 20 lb and travels at an angle of 37.5° to the left in a horizontal plane, while the other part travels to the right at an angle of 59.7° in a horizontal plane. Calculate the velocity of each part.

$v_1 = ?$ 1 20 lb 37.5°
$v_i = 250$ ft/s 35.5 lb x
59.7° 2 15.5 lb $v_2 = ?$

****

Solution

Linear momentum of the original device is conserved in the absence of an external force; hence

$$m v_i = m_1 v_1 + m_2 v_2$$

$$\overset{+}{\rightarrow} x : \left(\frac{35.5 \text{ lb}}{32.2 \text{ ft/s}^2}\right)(250 \text{ ft/s}) = \left(\frac{20 \text{ lb}}{32.2 \text{ ft/s}^2}\right) v_1 \cos 37.5° + \left(\frac{15.5 \text{ lb}}{32.2 \text{ ft/s}^2}\right) v_2 \cos 59.7°$$

or, $$275.621 = 0.492\, v_1 + 0.243\, v_2 \quad \ldots\ldots (1)$$

$$+\uparrow y : 0 = \left(\frac{20 \text{ lb}}{32.2 \text{ ft/s}^2}\right) v_1 \sin 37.5° - \left(\frac{15.5 \text{ lb}}{32.2 \text{ ft/s}^2}\right) v_2 \sin 59.7°$$

or, $$v_2 = \frac{0.378}{0.415} v_1 = 0.911\, v_1 \quad \ldots\ldots (2)$$

By substituting the value of v_2 in Eq.(1), we get

$$0.492\, v_1 + (0.243)(0.911\, v_1) = 275.621$$

$$\therefore \quad v_1 = \frac{275.621}{0.713} = 386.565 \text{ ft/s} \; \angle 37.5°$$

Ans.

and $$v_2 = (0.911)(386.565 \text{ ft/s}) = 351.774 \text{ ft/s} \; 59.7°$$

Ans.

IMPULSE AND MOMENTUM

6-4

A bird of 5 lb is hit from behind by a bullet of 0.1 lb while it is flying with a velocity of 50 ft/sec. Velocity of the bullet is 700 ft/sec and its direction forms an angle of 65° with the horizontal direction. After being hit, the bird falls to the ground in 2.3 seconds. Determine the distance from the point of hitting and the altitude the bird was flying at before it was hit.

5 lb

50 ft/sec

65°

700 ft/sec

****

Linear momentum is conserved during the impact:

$$m_b \vec{V}_b + m_B \vec{V}_B = (m_b + m_B)\vec{V}_F$$

$$0.1 \times 700 \times (\cos 65° \vec{i} + \sin 65° \vec{j}) + 5 \times 50 \vec{i} = (5 + 0.1)\vec{V}_F$$

$$54.82\vec{i} + 12.44\vec{j} = \vec{V}_F \quad \rightarrow \quad V_{Fx} = 54.82 \ , \quad V_{Fy} = 12.44 \text{ ft/sec}$$

For the x component of the motion:

$$x = V_{Fx} \cdot t + x_0 \quad \rightarrow \quad x - x_0 = \Delta x = V_{Fx} \cdot t = 54.82 \times 2.3 = 126.1 \text{ ft}$$

The bird falls to this distance after it is hit.

For the y component of the motion:

$$y = y_0 + V_{Fy} \cdot t - \frac{1}{2} g t^2 \longrightarrow \text{When it falls to the ground } t = 2.3 \text{ sec}, \ y = 0$$

$$0 = y_0 + 12.44 \times 2.3 - \frac{1}{2} \times 32.2 \times 2.3^2$$

$$y_0 = 56.56 \text{ ft}.$$

6-5

A 3 kg particle is moving with an initial velocity of 20 m/sec to the right when it explodes into two fragents. Immediately after the explosion, which takes place at the origin of the coordinate system shown the speed of the fragment A is observed to be 60 m/sec and the fragments are traveling in the direction shown.

a. What is the mass of fragment A ?
b. What is the velocity of fragment B ?
c. One second after the explosion is the mass center of the system located.

****

a) LINEAR MOMENTUM IS CONSERVED. THUS

$$m\vec{V}_0 = m_A\vec{V}_A + m_B\vec{V}_B$$

$$3\text{ Kg}(20\text{ m/sec})\vec{i} = m_A(60\text{ m/sec})\vec{j} + m_B(V_B\sqrt{2}/2\,\vec{i} - V_B\sqrt{2}/2)\vec{j}$$

THE CORRESPONDING SCALAR EQUATIONS ARE

$$60\text{ Kg m/sec} = m_B V_B \sqrt{2}/2 \qquad (1)$$

$$0 = m_A\, 60\text{ m/sec} - m_B V_B \sqrt{2}/2 \qquad (2)$$

THE MASS OF THE FRAGMENTS MUST SUM UP TO BE THE ORIGINAL MASS OF THE PARTICLE (MASS IS CONSERVED)

$$m_A + m_B = 3\text{ Kg} \qquad (3)$$

SUBSTITUTING (1) INTO (2) YIELDS

$$0 = 60\, m_A - 60 \longrightarrow m_A = 1\text{ Kg}$$

b) THEN FROM (3)

$$m_B = 2\text{ Kg}$$

FROM (1) $V_B = 60/\sqrt{2}$ m/SEC $= 42.4$ m/SEC

c) SINCE LINEAR MOMENTUM IS CONSERVED

$$\vec{V} = \text{CONSTANT} = 20\text{ m/SEC}\ \vec{i}$$

THUS AFTER 1 SECOND

$$\vec{r} = \vec{v}t = 20\text{ m}\ \vec{i}$$

6-6

Consider two bodies initially at rest as shown in the figure. Body 1 is on the slope and is released from rest at $\Theta = 60°$. Body 2 is hanged with a rope. After sliding without friction along the slope, body 1 hits the body 2. The impact is not exactly elastic and the coefficient of restitution is 0.8. Calculate the velocity of body 2 just after impact, the maximum tension in the rope and the height to which body 2 can rise.

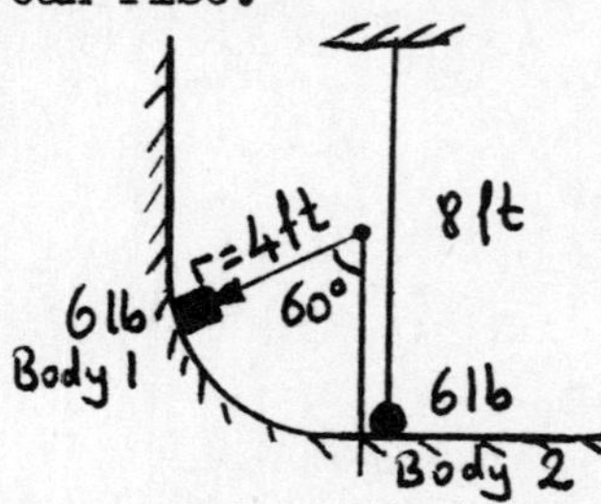

****

For the first part of this problem, we can use the energy method:

$$T_1 + V_1 = T_2 + V_2 \qquad \text{(all forces are conservative)}$$

$$0 + W(4 - 4\cos 60°) = \frac{1}{2}\frac{W}{g}V_1^2 \longrightarrow V_1 = 11.35 \text{ ft/sec}$$

velocity of body 1 when it falls to horiz. surface

For the second stage, the impact, we can use impulse and momentum method:

$$m_1 V_1 + m_2 V_2 = m_1 V_1' + m_2 V_2'$$

$$6 \times 11.35 + 0 = 6 \times V_1' + 6 \times V_2' \longrightarrow V_1' + V_2' = 11.35 \quad *$$

coefficient of restitution is given by $e = (V_2' - V_1')/(V_1 - V_2) = 0.8$

$$V_2' - V_1' = 0.8\,(11.35 - 0) \quad **$$

from * and **, $V_1' = 1.135$, $V_2' = 10.215$ ft/sec can be obtained.

For the third stage, when body 2 moves upward:

$$T_1 + V_1 = T_2 + V_2 \longrightarrow \frac{1}{2}\frac{W_2}{g}V_2'^2 + 0 = 0 + W_2 . h$$

$$\frac{1}{2} \cdot \frac{6}{32.2} \cdot (10.215)^2 = 6 . h \longrightarrow h = 1.62 \text{ ft.}$$

Maximum tension in the rope occurs at the minumum position of body 2:

Body 2: T (up), W_2 (down)

$$\Sigma F_n = m_2 \cdot \frac{V_2'^2}{\ell}$$

$$T - W_2 = \frac{W_2}{g} \cdot \frac{V_2'^2}{\ell}$$

$$T = 6 + \frac{6}{32.2} \cdot \frac{10.215^2}{8} = 8.43 \text{ lb}_f .$$

6-7

A bullet of mass 50 gm moving at velocity 50 m/s hits a particle of mass 0.5 kg resting on a frictionless surface. The particle splits into two parts A and B, and the bullet imbeds in B. The mass of fragment A is 250 gm. If the velocities of A and B make angles 60° and 30° respectively, to the direction of initial velocity of the bullet, determine the magnitudes of the velocities of the two fragments.

****

Conservation of momentum in the y direction;

$$0.25\, V_A \sin 60° - 0.3\, V_B \sin 30° = 0$$

Conservation of momentum in the x direction;

$$0.25\, V_A \cos 60° + 0.3\, V_B \cos 30° = 0.05 \times 50$$

simplify:

$$0.25\sqrt{3}\, V_A - 0.3\, V_B = 0$$

$$0.25\, V_A + 0.3\sqrt{3}\, V_B = 5.0$$

solve:

$$V_A = 5 \text{ m/s} \quad \angle 60°$$

$$V_B = 7.2 \text{ m/s} \quad \angle 30°$$

6-8

In a science fiction movie, a 1500 kg vehicle travelling in space at an initial velocity of 600 k m/s is obliged to simultaneously eject 2 objects in order to destroy 2 intruders approaching from 2 different directions, as shown in the accompanying illustration. If the final velocity of the parent vehicle after separation is given by $\mathbf{V}_f = (-0.762\mathbf{i} - 0.452\mathbf{j} + 0.694\mathbf{k})v_f$, calculate the velocities of the parent vehicle and the two objects after separation. Disregard the consequences of the two objects hitting the two intruders (!).

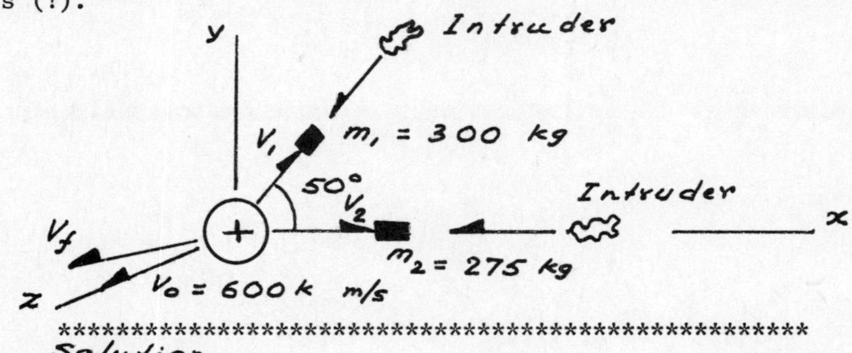

****

Solution

Let $m_t = m_p + m_1 + m_2 = 1500$ kg

OR, $m_p = (1500 - 300 - 275)$ kg $= 925$ kg

The total linear momentum of the system is conserved at the time of ejection (before the objects destroy the two intruders!), so that

$$L = m_t V_0 = m_p V_f + m_1 V_1 + m_2 V_2$$

OR,
$$(1500\text{ kg})(600\text{ m/s})\,k = (925\text{ kg})(-0.726i - 0.452j + 0.694k)v_f + (300\text{ kg})(\cos 50° i + \sin 50° j)v_1 + (275\text{ kg})\,i\,v_2$$

Separating the vector quantities in the x, y, and z directions, we get

$$x\ (i):\quad 0 = (925)(-0.762)v_f + (300)(\cos 50°)v_1 + (275)v_2$$
$$= -704.85\,v_f + 192.836\,v_1 + 275\,v_2 \quad \ldots\text{(1)}$$

$$y\ (j):\quad 0 = (925)(-0.452)v_f + (300)(\sin 50°)v_1$$
$$= -418.1\,v_f + 229.813\,v_1 \quad \ldots\text{(2)}$$

$$z\ (k):\quad (1500)(600) = (925)(0.694)v_f$$

OR,
$$v_f = \frac{(1500)(600)}{(925)(0.694)} = 1401.978 \text{ m/s}$$

By substitution in Eq. (2), we get

$$v_1 = \frac{(418.1)(1401.978)}{229.813} = 2550.626 \text{ m/s}$$

And by substitution in Eq. (1), we get

$$v_2 = \frac{(704.85)(1401.978) - (192.836)(2550.626)}{275} = 1804.842 \text{ m/s}$$

Hence, in vector form, we have

$$V_f = (-0.762\,i - 0.452\,j + 0.694\,k)\,1401.978 \text{ m/s}$$
$$= (-1068.307\,i - 633.694\,j + 972.972\,k) \text{ m/s} \quad \underline{\text{Ans.}}$$
$$V_1 = [(\cos 50°\,i) + (\sin 50°\,j)]\,2550.626 \text{ m/s}$$
$$= (1639.51\,i + 1953.892\,j) \text{ m/s} \quad \underline{\text{Ans.}}$$
$$V_2 = (1804.842\,i) \text{ m/s} \quad \underline{\text{Ans.}}$$

6-9

Consider 3 spheres each of mass m suspended in a line by strings of equal length. If the first sphere is released from some angle and hits the second sphere with velocity v_1, find the velocity, v_3^1 of the third sphere immediately after being struck by the second. Assume the coefficient of restitution is e. (Can you generalize to n balls?)

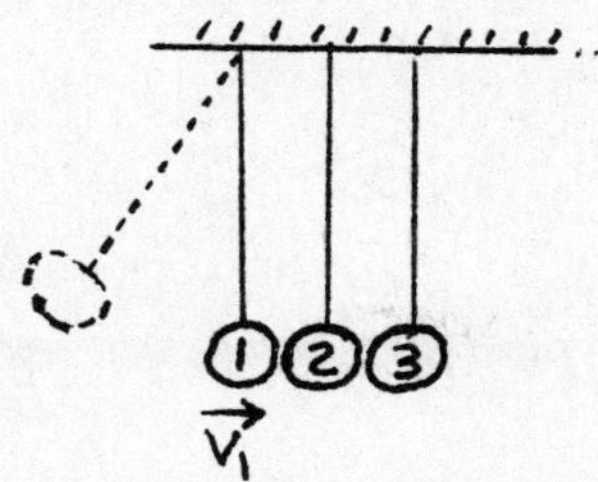

****

Consider the first two spheres:

① ② ⟶ x direction

V_1 (b/4 impact) $V_2 = 0$ (b/4 impact)

V_1' (velocity of ball 1 after impact) V_2' (Velocity of ball 2 after impact) etc...

Using conservation of momentum in the x-direction:

$$mV_1 + mV_2 = mV_1' + mV_2' \quad (V_2 = 0)$$

$$\therefore V_1 = V_1' + V_2' \Rightarrow V_1' = V_1 - V_2' \quad (*)$$

Also: (if e = coefficient of restitution)

$$V_2' - V_1' = e(V_1 - V_2) \quad (V_2 = 0)$$

$$\therefore V_2' = V_1' + eV_1 \overset{(*)}{=} V_1 - V_2' + eV_1$$

$$\Rightarrow 2V_2' = (1+e)V_1 \Rightarrow V_2' = \left(\frac{1+e}{2}\right)V_1 \quad (**)$$

Note: V_2' is also the velocity b/4 impact of ball 2 with ball 3; ∴ applying same logic to balls 2 & 3

hence, (**) ⇒ $V_3' = \left(\frac{1+e}{2}\right) V_2' = \left(\frac{1+e}{2}\right)\left(\frac{1+e}{2}\right) V_1$

or $V_3' = \left(\frac{1+e}{2}\right)^2 V_1$ — an obvious

induction argument now gives us a generalization:
– the velocity of the nth sphere, V_n', immediately after impact is $V_n' = \left(\frac{1+e}{2}\right)^{n-1} \cdot V_1$

Remark: If $e \cong 1$, the velocity of the 3rd sphere, after impact is the same as the initial velocity. Conservation of Energy says it will swing up approximately as high as the initial height of the 1st and this process will repeat in reverse. This is the basis for many commercial variations.

6-10

A 250-kg block with a velocity of 20 m/s is slid on to a stationary cart having a mass of 20 kg as shown. Determine the final velocity of the cart when the block has come to a complete halt on the cart. Assume the floor surface to be smooth.

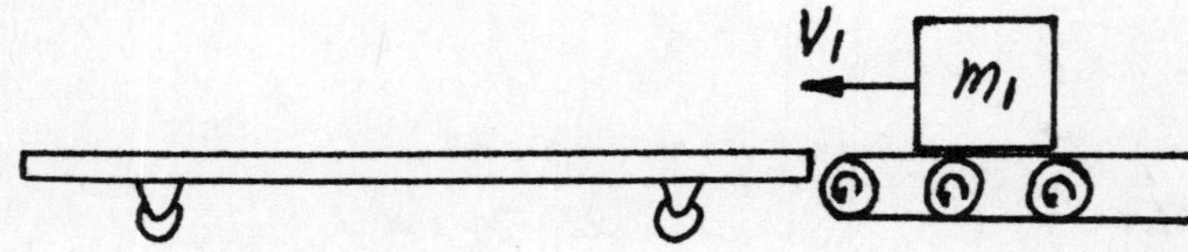

****

For the problem at hand one may write

Initial momentum of system + Impulse of external force system = Final momentum of system

$$m_1 V_1 + 0 = (m_1 + m_2) V$$

$$250 \times 20 = (250 + 20)\ V$$

or, $V = 250 \times 20 / 270 = 18.5$ m/s

6-11

A 30-ton railroad car with a velocity of 4 mph collides with a stationary 20-ton railroad car. The cars fail to couple and 20% of the energy is lost in the collision. Determine the velocities of each car after collision.

****

Let A = 30 Ton car and B = 20 Ton car.

Conservation of Linear Momentum: $\Sigma m v_1 = \Sigma m v_2$

$$m_A (v_A)_1 + m_B \cancelto{0}{(v_B)_1} = m_A (v_A)_2 + m_B (v_B)_2$$

$$\frac{30 \times 2000\text{ lb}}{32.2\text{ ft/sec}^2}(v_A)_1 = \frac{30 \times 2000\text{ lb}}{32.2\text{ ft/sec}^2}(v_A)_2 + \frac{20 \times 2000\text{ lb}}{32.2\ \frac{\text{ft}}{\text{sec}^2}}(v_B)_2$$

$$30\left(4\ \frac{\text{mi}}{\text{hr}} \times \frac{1\text{ hr}}{3600\text{ sec}} \times \frac{5280\text{ ft}}{\text{mi}}\right) = 30(v_A)_2 + 20(v_B)_2$$

$$17.61 = 3(v_A)_2 + 2(v_B)_2 \qquad \text{Eq. (1)}$$

Conservation of Energy: $.80\,T_1 = T_2$

$$.8\left(\frac{1}{2}\cdot\frac{30 \times 2000\text{ lb}}{32.2\text{ ft/sec}^2}\right)\left(5.87\ \frac{\text{ft}}{\text{sec}}\right)^2 = \frac{1}{2}\cdot\frac{30 \times 2000\text{ lb}}{32.2\ \frac{\text{ft}}{\text{sec}^2}}(v_A)_2^2 + \frac{1}{2}\cdot\frac{20 \times 2000\text{ lb}}{32.2\text{ ft/sec}^2}(v_B)_2^2$$

$$82.70 = 3(v_A)_2^2 + 2(v_B)_2^2 \qquad \text{Eq. (2)}$$

Substituting: $82.70 = 3\left(\frac{17.61 - 2v_{B2}}{3}\right)^2 + 2(v_B)_2^2$

$$10(v_B)_2^2 - 70.44(v_B)_2 + 62.0 = 0$$

$$(v_B)_2 = \frac{70.44 \pm \sqrt{(-70.44)^2 - 4(10)(62)}}{20} = 6.01 \text{ or } 1.03\text{ ft/sec}$$

if $(v_B)_2 = 1.03\ \frac{\text{ft}}{\text{sec}}$, then $(v_A)_2 = \frac{17.61 - 2(1.03)}{3}$

$(v_A)_2 = 5.18\text{ ft/sec}$ impossible!!!

$\therefore\ \underline{\underline{(v_B)_2 = 6.01\text{ ft/sec}}}$

$$(v_A)_2 = \frac{17.61 - 2(6.01)}{3} \qquad \underline{\underline{(v_A)_2 = 1.86\text{ ft/sec}}}$$

6-12

Two balls A and B of initial velocities V_o collide as shown in the figure. Find their velocities after impact. e=0.6, m_A = 3m, m_B = m.

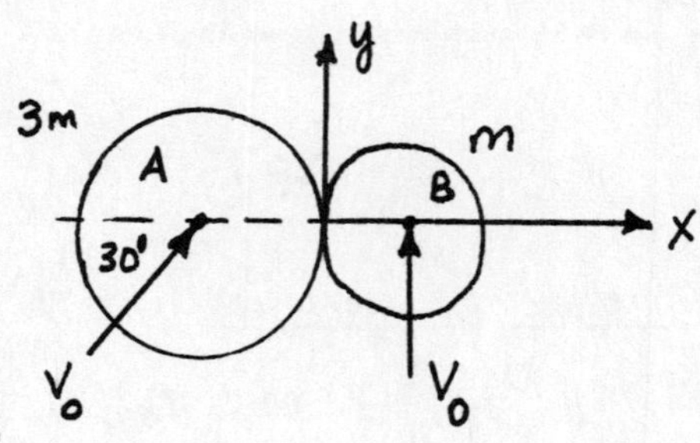

****

Assume $\underset{\sim}{v}_A'$ and $\underset{\sim}{v}_B'$ are the velocities after impact

Conservation of y-component of momentum for A:

$$m_A (v_A)_y = m_A (v_A')_y \;; \quad (v_A')_y = V_o \sin 30^\circ = \frac{V_o}{2}$$

Conservation of y-component of momentum for B:

$$m_B (v_B)_y = m_B (v_B')_y \;; \quad (v_B')_y = V_o$$

Conservation of x-component of momentum for the system:

$$m_A (v_A)_x + m_B (v_B)_x = m_A (v_A')_x + m_B (v_B')_x$$

$$3m\, V_o \cos 30^\circ + m\,(0) = 3m\,(v_A')_x + m\,(v_B')_x$$

or

$$3(v_A')_x + (v_B')_x = 3\,\frac{\sqrt{3}}{2}\,V_o \qquad (1)$$

The coefficient of restitution:

$$(v_B')_x - (v_A')_x = e\,[(v_A)_x - (v_B)_x]$$

or $(v_B')_x - (v_A')_x = 0.6\left[V_0 \frac{\sqrt{3}}{2} - 0\right]$ (2)

Solving (1) and (2) together, we obtain:

$(v_A')_x = 0.520\ V_0 \qquad (v_B')_x = 1.040\ V_0$

The velocities after impact:

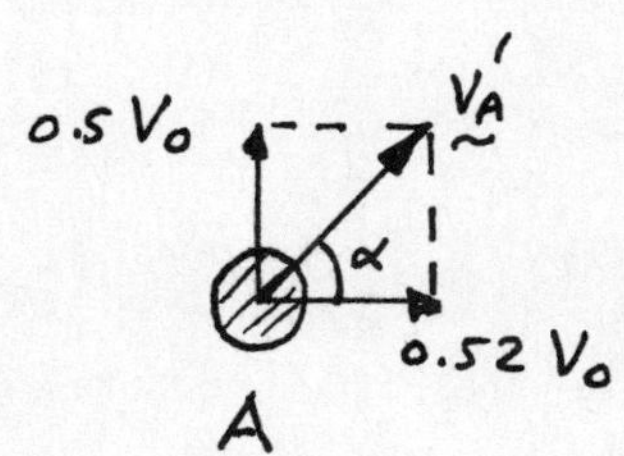

V_0, V_B', β, B, 1.040 V_0

$v_A' = \sqrt{(0.52\ V_0)^2 + (0.5\ V_0)^2}$

$= 0.721\ V_0$

$\tan\alpha = \dfrac{0.5\ V_0}{0.52\ V_0} = 0.962$

$\alpha = 43°\ 88$

$v_B' = \sqrt{V_0^2 + (1.04\ V_0)^2}$

$= 1.443\ V_0$

$\tan\beta = \dfrac{V_0}{1.04 V_0} = 0.962$

$\beta = 43°\ 88$

6-13

Air is received at a rate of 120 kg/sec by each side scoop of a jet plane. The engine consumes fuel at the rate of 3.5 kg/sec. The exhaust gases are discharged from the engine with a relative speed of 850 m/sec. Determine a) the thrust force developed by the engine, b) thrust horsepower of the engine.

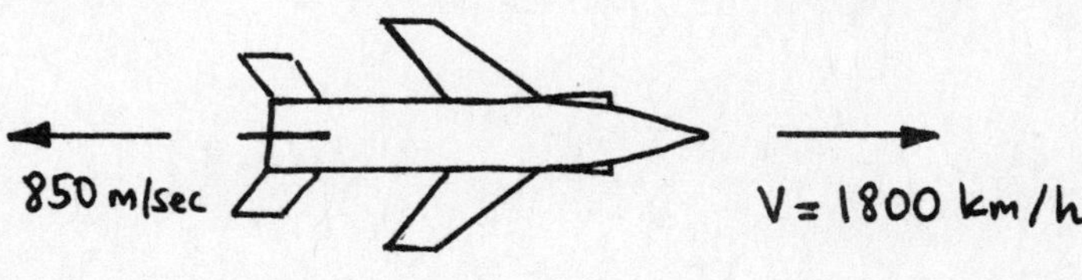

****

The method of impulse and momentum :

$$\Delta m_1 \cdot \vec{V_1} + \Sigma \vec{F} \cdot \Delta t = \Delta m_2 \cdot \vec{V_2} \qquad \text{or} \qquad \dot{m} = \Delta m / \Delta t \quad (\text{kg/sec})$$

$$\dot{m}_1 \vec{V_1} + T = \dot{m}_2 \vec{V_2} \quad , \qquad 1800 \text{ km/h} = 500 \text{ m/sec}$$

$$(2 \times 120) \times 500 + T = (2 \times 120 + 3.5) \times 850$$

thrust $T = 86975$ Newton

thrust horsepower $= T.V = 86975 \times 500 = 43.49 \times 10^6$ Watts

$= 43.49 \times 10^6 / 746 \quad = 58294$ horsepower

6-14

A 50 gram bullet is fired horizontally with a velocity of 300 m/sec. into a 4 kilogram block of wood which can move freely in the horizontal direction. Determine:

(a) the final velocity of the bullet.

(b) the ratio of the final kinetic energy of the block and bullet after impact to the initial kinetic energy of the bullet.

****

(a) BY DIRECT IMPACT $\quad (.05)\, 300 = 4.05\, V'$

$$V' = 3.70 \text{ m/sec}$$

(b) $(K.E)_1 = \frac{1}{2}(.05)(300)^2 = 2250$

$$(K.E)_2 = \frac{1}{2}(4.05)(3.70)^2 = 27.72$$

the ratio $\lambda = \dfrac{27.72}{2250} = 0.0123$

6-15

Ball A has a mass of 2 kg, B has a mass of 4 kg. The initial velocity of A is as shown, i.e., 2 m/sec to the right. Ball B is initially at rest. Assuming no friction and the coefficient of restitution e = 0.8, determine the final velocity and direction of balls A and B.

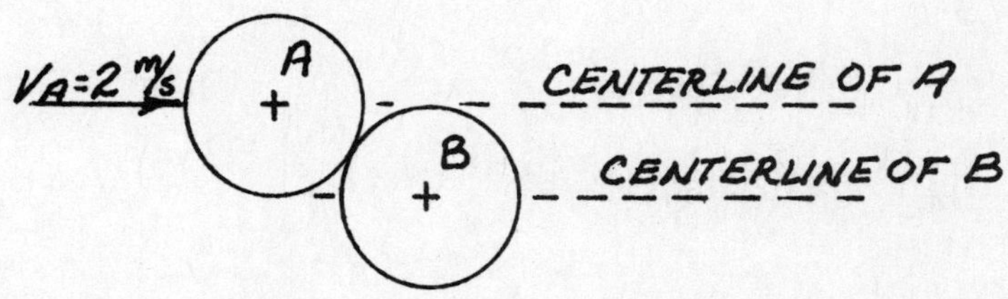

****

THE POINT OF IMPACT IS ON A LINE BETWEEN THE TWO CENTERS.

$\sin\theta = \frac{r}{2r} = \frac{1}{2}$, $\theta = 30°$

ALONG LINE OF IMPACT $2\cos 30°(2) + 0 = 2V'_{A_x} + 4V'_{B_x}$

$V'_{B_x} - V'_{A_x} = .8(2\cos 30° - 0)$, $V'_{A_x} = V'_{B_x} - 1.6\cos 30°$

$2\cos 30° = V'_{B_x} - 1.6\cos 30° + 2V'_{B_x}$

$3.6\cos 30° = 3V'_{B_x}$ $V'_{B_x} = 1.039$ m/s ↘ 30°

$V'_{A_x} = 1.039 - 1.6\cos 30° = -.346$ m/s

$V_{Ay} = V'_{Ay} = 2\sin 30° = 1$ m/s

$V_{By} = V'_{By} = 0$

V'_A = (.346, 1) = $\underline{\underline{1.058 \text{ m/s}}}$ ∠ 79.1°

$V'_B = \underline{\underline{1.039 \text{ m/s}}}$ ↘ 30°

6-16

The 20 lb block B is moving on a smooth surface with a velocity of 24 ft/sec to the left, when it is struck by a 5 lb smooth ball as shown. The coefficient of restitution is 0.5.

Determine
(a) The velocity of the ball after impact.
(b) The acceleration of block B when it passes point B'.

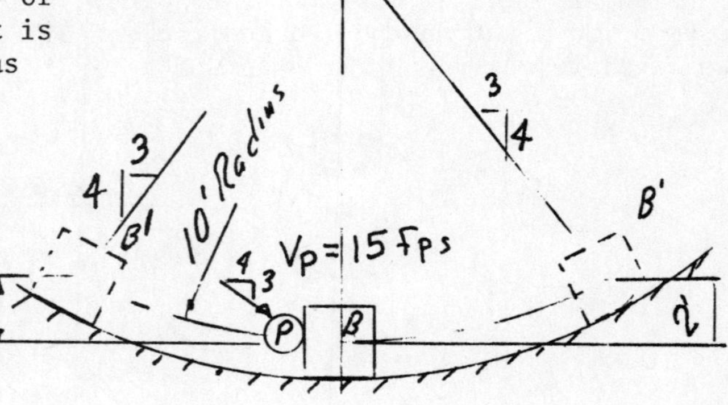

****

time (5, 20, N) + $\frac{5}{g}15$, $\frac{20}{g}(24)$ = $\frac{5}{g}V_{Px}$, $\frac{20}{g}V_B$, $\frac{5}{g}V_{Py}$

Since ball is smooth V_{Py} before Impact $= V_{Py}$ after impact

or $V_{Py} = \frac{3}{5}15 = 9\downarrow$

Linear Impulse $\xrightarrow{+}$

$$0 + \frac{4}{5}15\frac{5}{g} - \frac{20}{g}24 = \frac{5}{g}V_{Px} + \frac{20}{g}V_B$$

$$-84 = V_{Px} + 4V_B \quad \text{--- (1)}$$

$$e = -\frac{V_{B/P_x}\text{ (After Impact)}}{V_{B/P_x}\text{ Before Impact}} \qquad \xrightarrow{+} 0.5 = -\frac{(V_B - V_{Px})}{-24-12}$$

$$V_{Px} = V_B - 18 \quad \text{--- (2)}$$

Solve equations 1 & 2

$V_B - 18 = -4V_B - 84$ or $V_B = \overleftarrow{-13.2}$ and $V_{Px} = -31.2$

for part (a) $V_P = 9\downarrow \rightarrow 31.2$ or 32.5 ft/sec (slope 9 : 31.2)

Use Work and Energy to get velocity at point B'

$$U_{1-2} + T_1 = T_2$$

$$-2(20) + \frac{1}{2}\frac{20}{g}(13.2)^2 = \frac{1}{2}\frac{20}{g}V_{B'}^2 \quad \text{or } V_{B'}^2 = 45.4$$

W , N = $\frac{W}{g}a_N$, $\frac{W}{g}a_t$

$$a_N = \frac{V_{B'}^2}{R} \text{ or } \frac{45.4}{10} = 4.54 \text{ (slope 4 : 3)}$$

(b) $a_{B'} = 4.54$ (slope 4 : 3) $\rightarrow 19.32$ (slope 4 : 3)

$$= 19.85 \text{ ft/sec}^2$$

$\searrow + \; \Sigma F_T$

$$\frac{3}{5}W = \frac{W}{g}a_t$$

$$a_t = .6g$$

6-17

A ball of weight W_1 moving to the right with velocity V_1 strikes another ball of weight W_2 moving to the left with velocity V_2 as shown. If the coefficient of restitution between the two bodies is e, find a) the velocities of the two bodies after impact, and b) the magnitude of W_1 so that the weight W_1 comes to a complete stop immediately after impact.

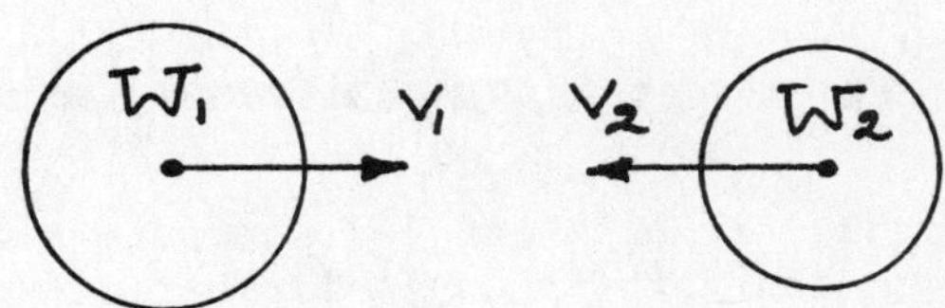

****

a) Since there is no external impulse the momentum of the entire system is conserved. Hence,

$$\frac{W_1}{g} V_1 + \frac{W_2}{g}(-V_2) = \frac{W_1}{g} V_1' + \frac{W_2}{g} V_2' \qquad (1)$$

where the primed velocities after impact are assumed positive to the right.

The restitution equation gives

$$e V_1 + V_1' = e(-V_2) + V_2' \qquad (2)$$

Solving (1) and (2) simultaneously gives

$$V_1' = \frac{W_1 V_1 - W_2\left[eV_1 + (1+e)V_2\right]}{W_1 + W_2}$$

and

$$V_2' = \frac{W_1\left[eV_2 + (1+e)V_1\right] - W_2 V_2}{W_1 + W_2}$$

b) For W_1 to come to a stop $V_1' = 0$. This gives

$$W_1 = W_2\left[e + (1+e)(V_2/V_1)\right]$$

Note that (b) has been readily calculated because V_1' and V_2' were obtained as general expressions, and not numerical.

MOMENTS OF FORCE AND MOMENTUM

6-18

A block weighing 56 lb is given an initial momentum such that when its center of gravity is 8 ft from the vertical pole its horizontal velocity is 9 ft/s, as shown in the accompanying illustration. Calculate (a) the velocity of the block at point B when its center of gravity is 6 ft from the pole and Δz = 2 ft, and (b) the vertical component of the velocity at B which is causing the block to fall downwards.

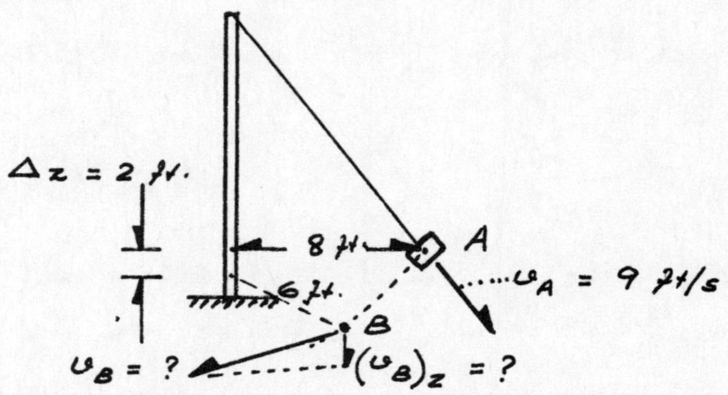

****

Solution

(a) $$T_A + V_A = T_B + V_B$$

By substitution, we get

$$\left[\frac{1}{2}\left(\frac{56\text{ lb}}{32.2\text{ ft/s}^2}\right)(9\text{ ft/s})^2\right] + 0 = \left[\frac{1}{2}\left(\frac{56\text{ lb}}{32.2\text{ ft/s}^2}\right)(v_B)^2\right] - \left[(56\text{ lb})(2\text{ ft})\right]$$

OR, $$v_B = 14.489\text{ ft/s}$$ Ans.

(b) $$(H_z)_A = (H_z)_B$$

By substitution, we get

$$(8\text{ ft})\left(\frac{56\text{ lb}}{32.2\text{ ft/s}^2}\right)(9\text{ ft/s}) = (6\text{ ft})\left(\frac{56\text{ lb}}{32.2\text{ ft/s}^2}\right)(v_B)_{hori.}$$

OR, $$(v_B)_{hori} = 12\text{ ft/s}$$

From geometry, we have

$$v_B^2 = (v_B)_h^2 + (v_B)_z^2$$

Hence $$(v_B)_z = \sqrt{v_B^2 - (v_B)_h^2}$$

$$= \sqrt{(14.489\text{ ft/s})^2 - (12\text{ ft/s})^2} = 8.119\text{ ft/s}$$ Ans.

6-19

The smooth homogeneous 100 lb bar is rotating in a vertical plane with an angular velocity of 2 rad/sec counterclockwise when it is struck by a 20 lb block which is moving to the left with a velocity of 10 ft/sec. The coefficient of restitution is 0.3. The block is on a smooth surface.

Determine the acceleration of the block when it reaches point B'.

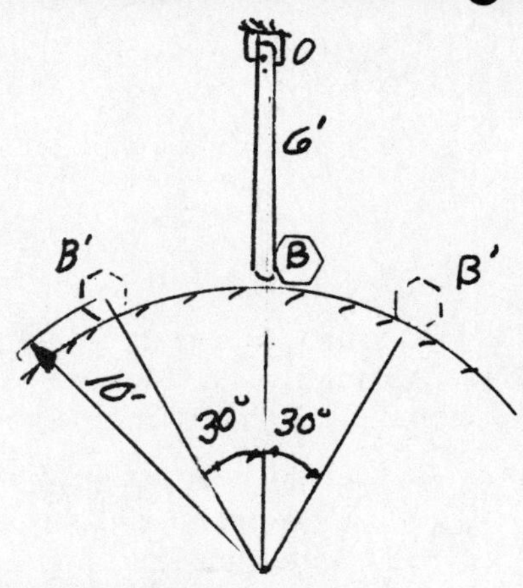

****

O_y, O_x, time, 100, 20, P, N

$\frac{100}{g}(3\omega)$, $\frac{1}{12}\frac{100}{g}6^2 2$, $\frac{20}{g}10$; $\frac{300}{g}\omega$, $\frac{300}{g}\omega$, $\frac{20}{g}V_{B_f}$

$\curvearrowleft \Sigma M_O(t)$

$$0+\frac{300}{g}(2)(3)+\frac{300}{g}(2)-\frac{20}{g}(10)(6) = \frac{300}{g}\omega(3)+\frac{300}{g}\omega+\frac{20}{g}V_{B_f}(6)$$

$$10 = 10\omega + V_{B_f} \quad (1)$$

Using coefficient of Restitution

$$e = -\frac{(V_{B/P})_f}{V_{B/P_i}} \text{ or } \xrightarrow{+} .3 = -\frac{(V_B-6\omega)}{-10-6(2)} \text{ or } -6.6 = 6\omega - V_B \quad (2)$$

Solve equations 1 and 2 $\quad -10\omega+10 = 6\omega+6.6$

$\omega = .213$ ↺ and $V_{B_f} = 7.875 \rightarrow$

Use Work and Energy to determine $V_{B'}$ $\quad U_{1-2}+T_1=T_2$

$$(10-\cos 30(10))(20)+\frac{1}{2}\frac{20}{g}(7.875)^2 = \frac{1}{2}\frac{20}{g}V_{B'}^2$$

$$V_{B'} = 12.178$$

$$\therefore a_{B_N} = \frac{(12.178)^2}{10} \text{ or } 14.83 \angle 30°$$

W, N, FBD, $\frac{W}{g}a_N$, $\frac{W}{g}a_t$

ΣF_T

$$\sin 30(W) = \frac{W}{g}a_t$$

$$a_t = \tfrac{1}{2}g$$

$$a_{B'} = 14.83 \angle 30° \rightarrow 16.1 \angle 30° \text{ or } \underline{\underline{21.89 \text{ ft/sec}^2}}$$

6-20

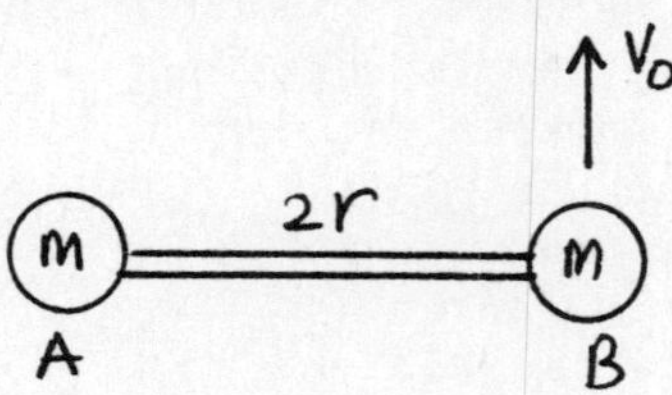

Two particles A and B of equal mass are connected by a rigid light rod of length 2r and placed on a frictionless horizontal surface. Mass B is suddenly imparted a horizontal velocity v_o perpendicular to AB.

(a) show that the centroid of the two-particle system will move in a straight line with speed $v_o/2$.

(b) Determine the angular velocity of the system soon after the motion begins.

****

Since there are no external forces in the horizontal direction (frictionless), both linear momentum along the plane and angular momentum about an axis perpendicular to the plane would be conserved.

(a) Let $\underline{\bar{V}}$ = velocity of centroid G

By conservation of linear momentum;

$$(m+m)\,\underline{\bar{V}} = m.\underline{0} + m v_o \underline{j}$$

{with the unit vectors $\underline{i}$ and $\underline{j}$ as shown

Hence

$$\underline{\bar{V}} = \tfrac{1}{2} v_o \underline{j}$$

This is rectilinear motion in the $\underline{j}$ direction (fixed), at constant velocity magnitude $\frac{1}{2}v_o$.

(b) Initial angular momentum about G

$$= 0.\underline{k} + r(m v_o)\,\underline{k}$$

Note: $\underline{i}, \underline{j}, \underline{k}$ is a right-handed orthogonal set of unit vectors.

If the angular velocity of the system as the motion begins is ω, velocity of A $= \frac{1}{2} V_0 \underline{j} - \omega r \underline{j}$

velocity of B $= \frac{1}{2} V_0 \underline{j} + \omega r \underline{j}$

Angular momentum about G $= -r.m(\frac{1}{2}V_0 - \omega r)\underline{k} +$

$r.m(\frac{1}{2}V_0 + \omega r)\underline{k} \quad = 2mr^2\omega \underline{k}$

Check: moment of inertia about G $\quad I_G = mr^2 + mr^2 = 2mr^2$

angular momentum $= I_G \omega \underline{k} = 2mr^2\omega\underline{k}$.

By conservation of angular momentum,

$$2mr^2\omega\underline{k} = mV_0 r\underline{k}$$

Hence

$$\omega = \frac{V_0}{2r}$$

6-21

A system consisting of three particles A,B and C is shown in the figure below. Calculate the angular momentum of this system about its mass center G and the point O (origin).Show that

$$\vec{H}_O = \vec{r}_G \times m\,\vec{V}_G + \vec{H}_G \qquad (\times : \text{vector product})$$

where $\quad m = \sum m_i$

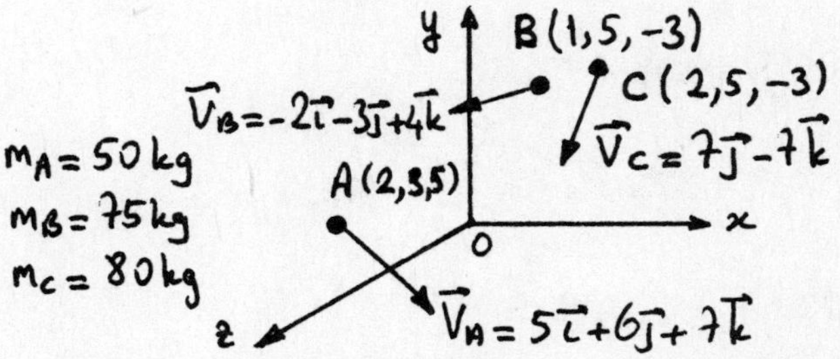

$m_A = 50$ kg
$m_B = 75$ kg
$m_C = 80$ kg

B(1,5,-3), $\vec{V}_B = -2\vec{i} - 3\vec{j} + 4\vec{k}$

C(2,5,-3), $\vec{V}_C = 7\vec{j} - 7\vec{k}$

A(2,3,5), $\vec{V}_A = 5\vec{i} + 6\vec{j} + 7\vec{k}$

****

$\vec{H}_O$: angular momentum about the origion O.
$\vec{H}_G$: angular momentum about the mass center G.
$\vec{r}_G$: position vector of G with respect to G.
$\vec{V}_G$: velocity of G.

$$\vec{H}_o = \sum \vec{r}_i \times m_i \vec{V}_i = 50 \begin{vmatrix} \vec{i} & \vec{j} & \vec{k} \\ 2 & 3 & 5 \\ 5 & 6 & 7 \end{vmatrix} + 75 \begin{vmatrix} \vec{i} & \vec{j} & \vec{k} \\ 1 & 5 & -3 \\ -2 & -3 & 4 \end{vmatrix} + 80 \begin{vmatrix} \vec{i} & \vec{j} & \vec{k} \\ 2 & 5 & -3 \\ 0 & 7 & -7 \end{vmatrix}$$

$$= -745\vec{i} + 1820\vec{j} + 1495\vec{k}$$

$$\vec{H}_G = \sum \vec{r}_i' \times m_i \vec{V}_i' = \sum \vec{r}_i' \times m_i \vec{V}_i$$

second choice is better, since velocities are already given in this problem; however, we still need $\vec{r}_i'$:

$$\vec{r}_i' = \vec{r}_i - \vec{r}_G$$

now, we need $\vec{r}_G$: $\quad m.\vec{r}_G = \sum m_i \vec{r}_i \qquad \sum m_i = m = 75+80+50 = 205$

$$\vec{r}_G = \frac{\sum m_i \vec{r}_i}{m} = \frac{50}{205}(2\vec{i}+3\vec{j}+5\vec{k}) + \frac{75}{205}(1\vec{i}+5\vec{j}-3\vec{k}) + \frac{80}{205}(2\vec{i}+5\vec{j}-3\vec{k})$$

$$\vec{r}_G = 1.634\vec{i} + 4.512\vec{j} - 1.049\vec{k}$$

$$\therefore \quad \vec{r}_A' = \vec{r}_A - \vec{r}_G = 0.366\vec{i} - 1.512\vec{j} + 6.049\vec{k}$$
$$\vec{r}_B' = \vec{r}_B - \vec{r}_G = -0.634\vec{i} + 0.488\vec{j} - 1.951\vec{k}$$
$$\vec{r}_C' = \vec{r}_C - \vec{r}_G = 0.366\vec{i} + 0.488\vec{j} - 1.951\vec{k}$$

$$\vec{H}_G = \vec{r}_A' \times m_A \vec{V}_A + \vec{r}_B' \times m_B \vec{V}_B + \vec{r}_C' \times m_C \vec{V}_C$$

$$= 50 \begin{vmatrix} \vec{i} & \vec{j} & \vec{k} \\ 0.366 & -1.512 & 6.049 \\ 5 & 6 & 7 \end{vmatrix} + 75 \begin{vmatrix} \vec{i} & \vec{j} & \vec{k} \\ -0.634 & 0.488 & -1.951 \\ -2 & -3 & -4 \end{vmatrix} + 80 \begin{vmatrix} \vec{i} & \vec{j} & \vec{k} \\ 0.366 & 0.488 & -1.951 \\ 0 & 7 & -7 \end{vmatrix}$$

$$= -1817.01\vec{i} + 2071.95\vec{j} + 908.57\vec{k}$$

$$\vec{V}_G = \frac{\sum m_i \vec{V}_i}{m} = \frac{50}{205}(5\vec{i}+6\vec{j}+7\vec{k}) + \frac{75}{205}(-2\vec{i}-3\vec{j}+4\vec{k}) + \frac{80}{205}(7\vec{j}-7\vec{k})$$

$$\vec{V}_G = 0.488\vec{i} + 3.097\vec{j} + 0.439\vec{k}$$

$$\vec{H}_o = m\vec{r}_G \times \vec{V}_G + \vec{H}_G = 205 \begin{vmatrix} \vec{i} & \vec{j} & \vec{k} \\ 1.634 & 4.512 & -1.049 \\ 0.488 & 3.097 & 0.439 \end{vmatrix} + \vec{H}_G$$

$$\vec{H}_o = -745\vec{i} + 1820\vec{j} + 1495\vec{k}$$

WORK AND KINETIC ENERGY

6-22

The 31 pound block is at rest on a frictionless surface when it is struck by a 1 pound bullet which is traveling at a speed of 320 feet per second. After impact, which is inelastic, the block slides up a ramp. The coefficient of friction between the block and the ramp is 0.25. How far along the ramp will it slide?

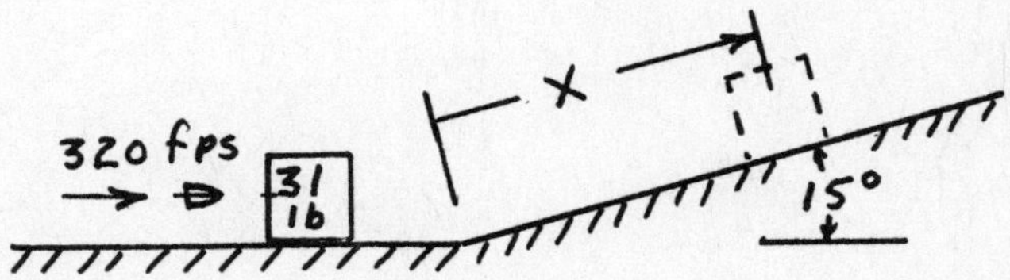

****

Impact → conservation of momentum

$$m_1 V_{1i} + m_2 V_{2i} = (m_1 + m_2) V_f$$

$$\left(\frac{1}{32}\right)(320) + \left(\frac{31}{32}\right)(0) = \left(\frac{1+31}{32}\right) V_f$$

$$V_f = 10$$

Slide → work = Δ kinetic energy

32, f, N

$$\Sigma F_N = 0 = N - 32 \cos 15°$$

So $N = 30.9$, for slide $f = \mu N = 7.73$

Work of weight $= -W\Delta h = -Wx \sin 15° = -8.28x$

Work of friction $= -fx = -7.73x$

Total work $= -16.0x$

$$\Delta K.E. = \frac{1}{2} m (V_f^2 - V_i^2) = \frac{1}{2}\left(\frac{32}{32}\right)(0^2 - [10]^2) = -50$$

$$-16x = -50$$

x = 3.13 feet

6-23

The 50 pound block shown below is released from rest with spring A compressed 6 inches. The coefficient of friction between the block and the inclined surface is 0.15. Determine the velocity of the block after it has compressed spring B 3 inches.

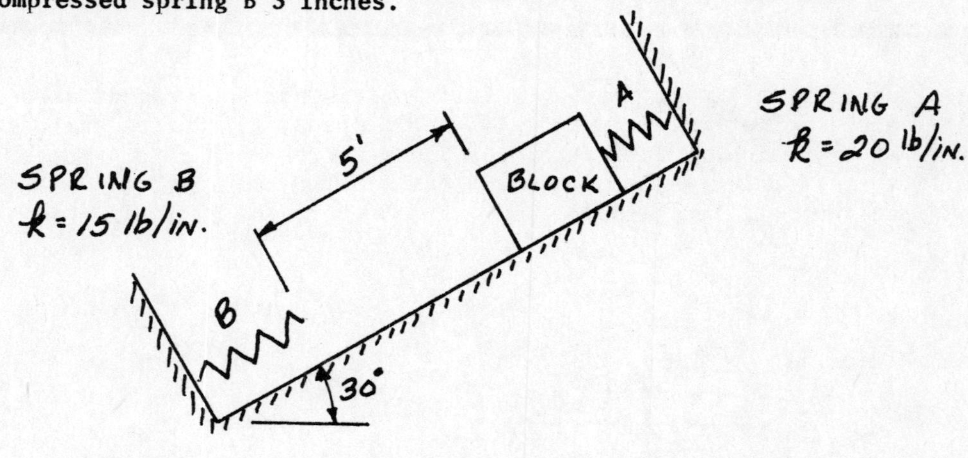

****

FREE BODY DIAGRAMS:

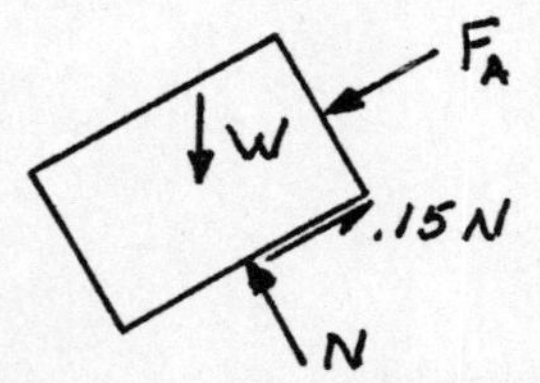

BLOCK IN CONTACT WITH SPRING A

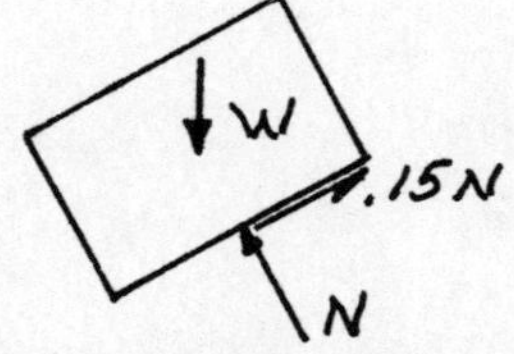

BLOCK BETWEEN SPRINGS A and B

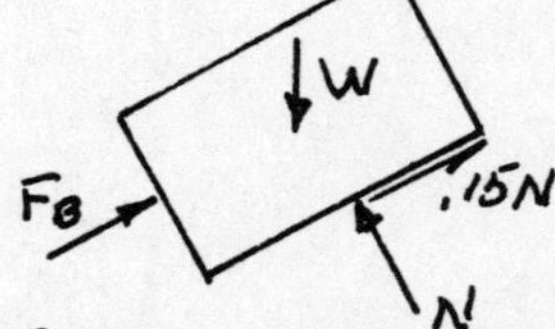

BLOCK IN CONTACT WITH SPRING B

PRINCIPLE OF WORK AND KINETIC ENERGY: $T_1 + U_{1-2} = T_2$

KINETIC ENERGY:

Initial Position: $V_1 = 0 \quad \therefore T_1 = 0$

Final Position: $T_2 = \frac{1}{2} m V_2^2 = \frac{1}{2} \frac{50}{32.2} V_2^2 = .776 V_2^2$

WORK:

FRICTION: $U_{1-2} = -.15N(60" + 3") = -9.45N$

$\oplus\!\uparrow \Sigma F_y = m a_y^{\nearrow 0} = 0 \qquad N - 50\cos 30 = 0$

$U_{1-2} = -9.45(50 \cos 30) = -409.2$ in.·lb

WEIGHT: $U_{1-2} = 50\,lb \sin 30\,(60'' + 3'')$
$= 1575\ in\cdot lb$

SPRING A: $U_{1-2} = \frac{1}{2} k_A (X_1^2 - X_2^2)$
$= \frac{1}{2}\left(\frac{20\,lb}{in}\right)\left[(6\,in)^2 - (0)^2\right] = 360\ in\cdot lb$

SPRING B: $U_{1-2} = \frac{1}{2} k_B (X_1^2 - X_2^2)$
$= \frac{1}{2}\left(\frac{15\,lb}{in}\right)\left[(0)^2 - (3\,in)^2\right] = -67.5\ in\cdot lb$

Substituting: $T_1 + U_{1-2} = T_2$

$$0 + (-409.2 + 1575 + 360 - 67.5)\ in\cdot lb = .776\ \frac{lb\cdot sec^2}{ft}\cdot V_2^2$$

$$(1458.3\ in\cdot lb)\left(\frac{1\,ft}{12\,in}\right) = .776\ \frac{lb\cdot sec^2}{ft}\cdot V_2^2$$

$$V_2^2 = 156.6\ \frac{ft^2}{sec^2} \qquad \underline{\underline{V_2 = 12.5\ ft/sec}}$$

6-24

A 10 kg block (A) rests on a smooth surface and the spring is compressed 0.1 meters. It is attached to an inextensible cord over frictionless, weightless pulleys to a 24 kg mass (B) as shown. If the system is released from rest and B travels 0.1 m, what is the velocity of A? The spring is attached to the objects at both ends and k = 490 N/m.

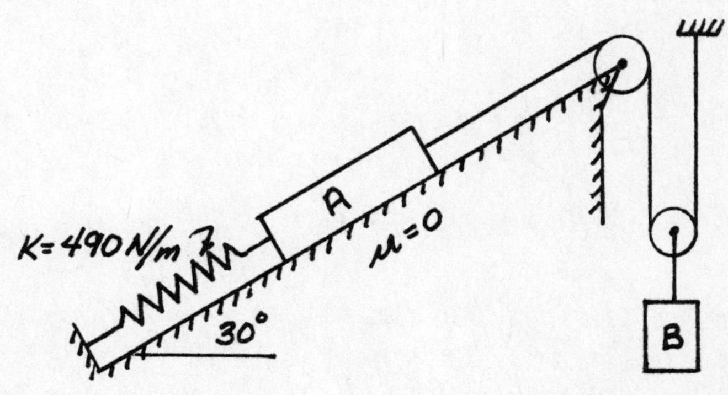

****

$$T_1 + V_1 = T_2 + V_2$$

$$0 + \frac{1}{2}(490)(.1)^2 = \frac{1}{2}(490)(.1)^2 + \frac{1}{2}(24)V_B^2 + \frac{1}{2}(10)V_A^2 + 10(9.81)\sin 30^\circ(.2) - 24(9.81)(.1)$$

$$0 = 12V_B^2 + 5V_A^2 - 13.73$$

BUT $V_B = \frac{V_A}{2}$, $0 = 12\left(\frac{V_A}{2}\right)^2 + 5V_A^2 - 13.73$

$$8V_A^2 = 13.73 \qquad V_A = 1.310 \text{ m/s}$$

6-25

The spring-mass system shown in the figure is released with the string taut, from the unstretched position of the spring (stiffness k). Neglecting friction, determine the velocity of mass B when it has moved through a distance of x. Assume $x < 4mg/k$.

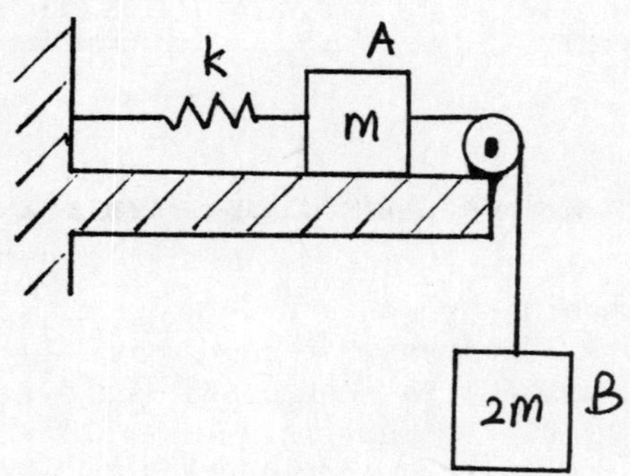

****

By compatibility (taut string),

velocity of A = Velocity of B = V

Initial kinetic energy = 0 (starting from rest)

Initial elastic potential energy = 0 (spring unstretched)

Initial gravitational potential energy = 0 (by properly chosing the datum).

Consider a displacement x of the two masses from the initial position, and corresponding velocity v.

Final kinetic energy $= \frac{1}{2}mv^2 + \frac{1}{2}2m\,v^2 = \frac{3}{2}mv^2$

Final elastic P.E. $= \frac{1}{2}kx^2$

Final gravitational P.E. $= -(2mg)x + 0 = -2mgx$

Since no work done due to friction, total energy is conserved;

$$\frac{3}{2}mv^2 + \frac{1}{2}kx^2 - 2mgx = 0 + 0 + 0$$

or, $$\frac{3}{2}mv^2 = 2mgx - \frac{1}{2}kx^2$$

Hence $$v = \sqrt{\frac{4mgx - kx^2}{3m}}$$ Note: $x < \frac{4mg}{k}$

6-26

The 10 kg cylinder shown is released from rest. Determine the velocity of G:

(a) after it has made 2 revolutions (by work and energy).
(b) after 2 seconds have elapsed (by impulse and momentum).

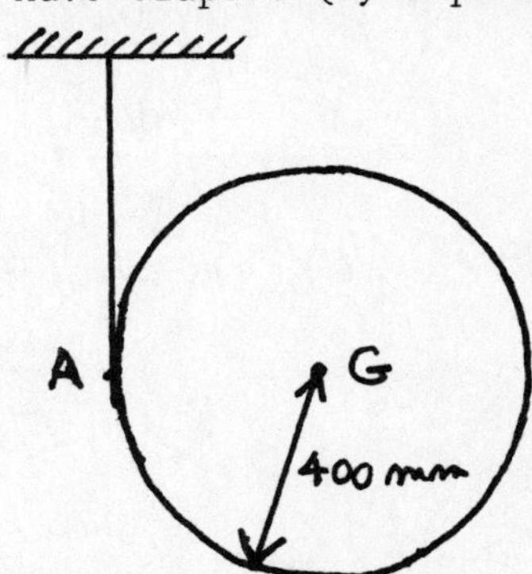

****

(a) distance G drops is $4\pi r = 1.6\pi$

$$\therefore\ 10g\,(1.6\pi) = \frac{1}{2}\cdot 10\cdot V_G^2 + \frac{1}{2}\left(\frac{1}{2}\cdot 10\cdot .4^2\right)\left(\frac{V_G}{.4}\right)^2$$

SOLVING $V_G^2 = 65.75$ OR $V_G = 8.11$ m/sec.

(b) $\sum M_A \cdot t = I_A \cdot \omega = I_A \cdot V_G/.4$

$$10g\,(.4)\,2 = \left[\frac{1}{2}\cdot 10\,(.4)^2 + 10\,(.4)^2\right]\cdot \frac{V_G}{.4}$$

SOLVING $V_G = \dfrac{78.48}{6} = 13.08$ m/sec.

6-27

The two blocks as shown are released from rest. Determine their velocity after they have travelled a distance of 10 ft. Assume the coefficient of friction between the blocks and the plane to be 0.2. Ignore the mass of the pulley.

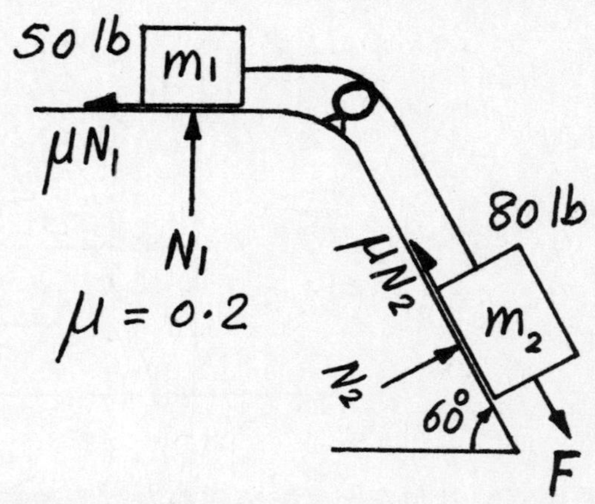

****

Initial kinetic energy of the system + Work done by the external forces on the system = Final kinetic energy of the system

$$0 + \int_0^{10} (F - \mu N_1 - \mu N_2)\,dx = \frac{1}{2}(m_1 + m_2)V^2$$

$$\left(80\cdot \sin 60^\circ - 0.2(50) - 0.2\,(80\cdot \cos 60^\circ))\,x\right]_0^{10} = \frac{1}{2}(m_1+m_2)V^2$$

or, $\frac{1}{2}(m_1 + m_2)V^2 = \frac{1}{2}\,\frac{130}{32.2}\,V^2 = 51.28(10)$

or, $V = 15.9$ ft/Sec

6-28

A 2 kg. block rests on a frictionless plane. It is attached, as indicated, to a 3 kg. weight and then to a fixed support. The cord is flexible and inextensible. The pulleys are frictionless and weightless. How far will the block on the plane travel before its speed is 4 m/sec?

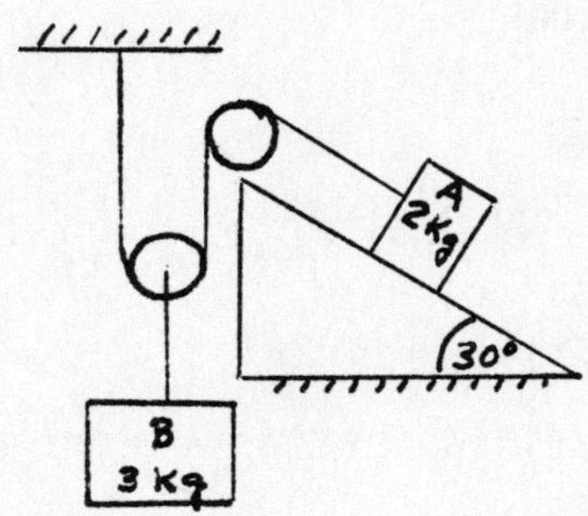

****

We note that the distance traveled by A and B satisfy $s_B = \frac{1}{2} s_A$ and the velocities satisfy $V_B = \frac{1}{2} V_A$

In this system $T_1 + U_{1\to 2}$ (work done going from state 1 to 2) $= T_2$ (*)

We observe: $T_1 = 0$, Final $T_2 = \frac{1}{2}(3)V_B^2 + \frac{1}{2}(2)V_A^2$

$$= \frac{1}{2}(3)2^2 + \frac{1}{2}(2)4^2$$

$$T_2 = 6 + 16 = 22 \text{ (N-m)}$$

Assuming motion down the plane:

[The work done is that due to the force of gravity on blocks A and B]

$$U_{1\to 2} = (2g)\cdot s \sin 30^\circ - 3g\left(\frac{s}{2}\right)$$

$$= 2(9.8)\cdot\frac{s}{2} - 3(9.8)\cdot\frac{s}{2}$$

$$= 9.8\left[1 - \tfrac{3}{2}\right]s$$

$$U_{1\to 2} = -4.9s$$

30°, s, h, 30°

$\Rightarrow h = \frac{s}{2}$

$\therefore$ (*) $\Rightarrow$ $T_1 + U_{1\to 2} = T_2$

$$0 - 4.9s = 22$$

$$s = -\frac{22}{4.9} \text{ m}$$

$s = -4.49$ m ($\therefore$ up the plane and our original assumption was incorrect.)

6-29

A ball of mass m is dropped on a plate of mass m from a height h. Assuming that the impact is perfectly plastic, determine the energy lost during impact, and the maximum displacement of the plate in terms of m, h, k and g.

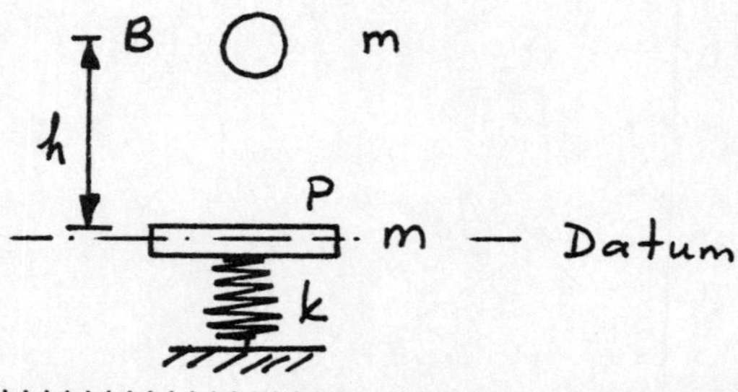

****

Solution :

Velocity of the ball before impact :

$$\frac{1}{2} m v_B^2 = mgh \quad ; \quad v_B = \sqrt{2gh}$$

After impact the plate and the ball will move with the same velocity v' (perfectly plastic impact).

Conservation of momentum :

$$m v_B + m(0) = mv' + mv' \quad \text{or} \quad v' = \frac{v_B}{2}$$

$$v' = \sqrt{gh/2}$$

Kinetic energy of the system before impact :

$$T_1 = \frac{1}{2} m v_B^2 + \frac{1}{2} m (0)^2 = mgh$$

Kinetic energy of the system after impact :

$$T_2 = \frac{1}{2}(2m)(v')^2 = \frac{1}{2} 2m \frac{gh}{2} = \frac{1}{2} mgh$$

Energy lost during impact :

$$\Delta T = T_1 - T_2 = mgh - \frac{1}{2} mgh = \frac{1}{2} mgh .$$

Maximum deformation:

At maximum deformation (δ) the system will come to rest. We will use the principle of conservation of energy.

$$T_A = \frac{1}{2} (2m) (v')^2 \qquad V_A = 0 + \frac{1}{2} k \delta_s^2$$

$$\delta_s = \frac{mg}{k} \quad \text{(static deformation)}$$

$$T_B = 0 , \qquad V_B = (V_g)_B + (V_e)_B = -2mg\delta + \frac{1}{2} k (\delta + \delta_s)^2$$

$$T_A + V_A = T_B + V_B$$

$$m \frac{gh}{2} + \frac{1}{2} \frac{m^2 g^2}{k} = -2mg\delta + \frac{1}{2} k \left(\delta + \frac{mg}{k}\right)^2$$

$$\frac{1}{2} k \delta^2 - mg\delta - mgh/2 = 0$$

$$\delta^2 - \frac{2mg}{k} \delta - \frac{mgh}{k} = 0$$

or $$\delta = \frac{mg}{k} + \sqrt{\frac{m^2 g^2}{k^2} + \frac{mgh}{k}}$$

6-30

A metal ball of weight W_1 is dropped from height h on a steel block of weight W_2 supported by a spring of stiffness k. The coefficient of restitution is e. a) Determine the height to which the ball rebounds, b) Set up the equation which gives the maximum deflection of the spring, and c) Find the mechanical energy lost on account of impact.

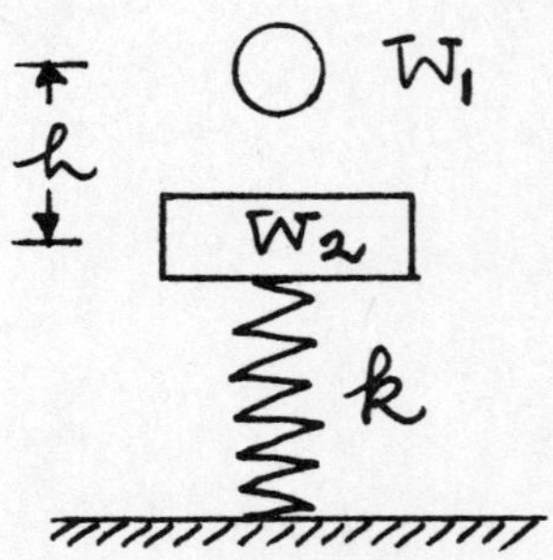

****

a) Conservation of energy: $mgh = \frac{1}{2} m v_1^2$

So that $v_1 = (2gh)^{1/2}$ just before impact.

Impact Phase:

Conservation of momentum for the entire system gives

$$\frac{W_1}{g}(2gh)^{1/2} = \frac{W_1}{g} v_1' + \frac{W_2}{g} v_2' \qquad (1)$$

The restitution equation gives

$$e(2gh)^{1/2} + v_1' = v_2' \qquad (2)$$

Solving (1) and (2) for v_1' and v_2' gives

$$v_1' = \frac{(W_1 - W_2 e)(2gh)^{1/2}}{W_1 + W_2} \; ; \; v_2' = \frac{W_1(1+e)(2gh)^{1/2}}{W_1 + W_2}$$

Conservation of energy after rebound gives $v_1' = (2gh_1)^{1/2}$, solving for h_1

$$h_1 = (W_1 - W_2 e)^2 h / (W_1 + W_2)^2$$

b) Taking the datum at the position of static equilibrium, conservation of energy gives, $T_1 = \frac{1}{2}\frac{W_2}{g}v_2'$ (kinetic energy after impact), $(V_1)_{spring} = \frac{1}{2}k\left(\frac{W_2}{k}\right)^2$ (corresponding to static deformation from unstretched length) $(V_1)_{gravity} = 0$ (datum).

At the position of max. deformation

$T_2 = 0$, $(V_2)_{spring} = \frac{1}{2}k\left(\frac{W_2}{k} + \delta_{max}\right)^2$

$(V_2)_{gravity} = -W_2\,\delta_{max}$. Substituting into $T_1 + V_1 = T_2 + V_2$ and solving for δ_{max} results in

$$\delta_{max}^2 = \frac{2W_1^2 W_2 (1+e)^2}{k(W_1+W_2)^2}\,h$$

c) The kinetic energy before impact is $T_0 = \frac{1}{2}\frac{W_1}{g}v_1^2 = W_1 h$. The total kinetic energy after impact is $T' = \frac{1}{2g}(W_1 v_1'^2 + W_2 v_2'^2)$ substituting for v_1' and v_2' from part (a) T' becomes

$$T' = \frac{(1+n e^2)T_0}{1+n} \quad \text{where } n = \frac{W_2}{W_1}$$

If $e = 1$ (perfectly elastic impact) $T' = T_0$, if $e = 0$ (perfectly plastic impact) $T' = (1+n)^{-1}T_0$ as expected in both, $e < 1$ leads to $T' < T_0$ as anticipated.

6-31

Figure 1 shows a system of particles at time t_1. Figure 2 shows the same particles at time t_2.

a. What is the linear impulse exerted on the system between t_1 and t_2 ?

b. What is the total angular impulse during the time interval about the origin ?

c. At time t what is the kinetic energy of the system relative to a centroidal frame ?

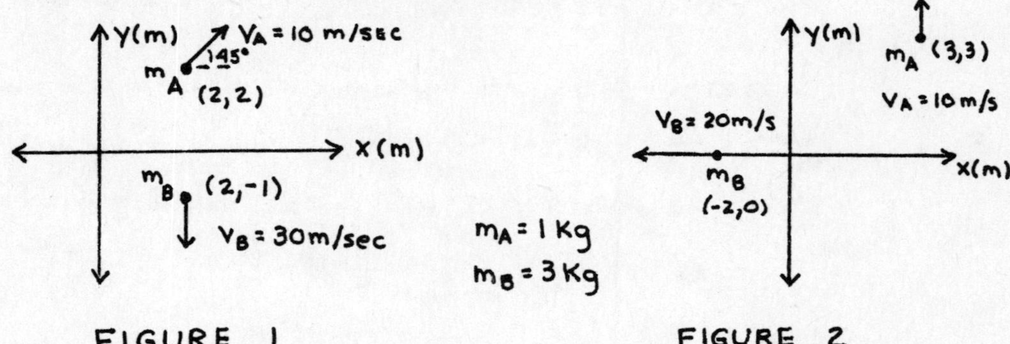

FIGURE 1 FIGURE 2

****

a) PRINCIPLE OF IMPULSE - MOMENTUM FOR A SYSTEM OF PARTICLES:

$$\left(\sum_{i=1}^{n} m_i \vec{v}_i\right)_{t=t_1} + \vec{I}mp_{1\to 2} = \left(\sum_{i=1}^{n} m_i \vec{v}_i\right)_{t=t_2}$$

FOR THIS PROBLEM

$$\vec{I}mp_{1\to 2} = m_A \vec{V}_{A_2} + m_B \vec{V}_{B_2} - m_A \vec{V}_{A_1} - m_B \vec{V}_{B_1}$$

$$= (1\text{ Kg})(10\text{ m/sec})\vec{j} + 3\text{ Kg}(-20\text{m/sec})\vec{i}$$
$$- (1\text{ Kg})(10\text{m/sec})(\sqrt{2}/2\,\vec{i} + \sqrt{2}/2\,\vec{j}) - 3\text{Kg}(30\text{ m/sec})(-\vec{j})$$

$$\vec{I}mp_{1\to 2} = 2.93\vec{i} + 22.9\vec{j}$$

b) PRINCIPLE OF ANGULAR IMPULSE - ANGULAR MOMENTUM ABOUT O

$$\vec{r}_{A_1} \times m_A \vec{V}_{A_1} + \vec{r}_{B_1} \times m_B \vec{V}_{B_1} + \overrightarrow{AngImp}_{1\to 2} = \vec{r}_{A_2} \times m_A \vec{V}_{A_2} + \vec{r}_{B_2} \times m_B \vec{V}_{B_2}$$

$$\overrightarrow{AngImp}_{1\to 2} = (3\vec{i} + 3\vec{j})\text{ m} \times (1\text{ Kg})(10\text{ m/sec}\,\vec{j})$$
$$+ (-2\vec{j}\text{ m}) \times (3\text{Kg})(20\text{ m/sec})(-\vec{i})$$
$$- (2\vec{i} + 2\vec{j})\text{m} \times (1\text{ Kg})(10\text{ m/sec})(\sqrt{2}/2\,\vec{i} + \sqrt{2}/2\,\vec{j})$$
$$- (2\vec{i} - \vec{j})\text{ m} \times (3\text{Kg})(30\text{ m/sec})(-\vec{j})$$
$$= 210\text{ Kg m}^2/\text{sec}\ \vec{K}$$

c) THE TOTAL KINETIC ENERGY OF A SYSTEM OF PARTICLES CAN BE WRITTEN AS

$$T = \tfrac{1}{2} m (\bar{\vec{v}} \cdot \bar{\vec{v}}) + \tfrac{1}{2} \sum_{i=1}^{n} m_i (\vec{v}_i{}' \cdot \vec{v}_i{}')$$

WHERE $m = \sum_{i=1}^{n} m_i$, $\bar{\vec{v}}$ IS THE VELOCITY OF THE MASS CENTER AND $\vec{v}_i{}' = \vec{v}_i - \bar{\vec{v}}$.

THE KINETIC ENERGY RELATIVE TO A CENTROIDAL FRAME IS SIMPLY

$$T' = \frac{1}{2} \sum_{i=1}^{n} m_i (\vec{v}_i{}' \cdot \vec{v}_i{}')$$

$$\bar{\vec{v}} = \frac{1}{m} \sum_{i=1}^{n} m_i \vec{v}_i$$

FOR THE PROBLEM AT HAND, AT $t = t_2$

$$\bar{\vec{v}} = \frac{1}{4\,\text{Kg}} \left[(1\,\text{Kg})(10\,\text{m/sec})\vec{j} + 3\,\text{Kg}\,(-20\,\text{m/sec})\vec{i} \right]$$

$$= 2.5\,\text{m/sec}\,\vec{j} - 15\,\text{m/sec}\,\vec{i}$$

$$\vec{v}_A{}' = \vec{v}_A - \bar{\vec{v}} = 10\,\text{m/sec}\,\vec{j} - 2.5\,\text{m/sec}\,\vec{j} + 15\,\text{m/sec}\,\vec{i}$$

$$= 15\,\text{m/sec}\,\vec{i} + 7.5\,\text{m/sec}\,\vec{j}$$

$$\vec{v}_B{}' = \vec{v}_B - \bar{\vec{v}} = -20\,\text{m/sec}\,\vec{i} + 2.5\,\text{m/sec}\,\vec{j} + 15\,\text{m/sec}\,\vec{i}$$

$$= -5\,\text{m/sec}\,\vec{i} + 2.5\,\text{m/sec}\,\vec{j}$$

$$T' = \tfrac{1}{2} \Big[(1\,\text{KG}) \left[(15)^2 + (7.5)^2 \right] \text{m}^2/\text{sec}^2$$

$$+ (3\,\text{Kg}) \left[(-5)^2 + (2.5)^2 \right] \text{m}^2/\text{sec}^2$$

$$= 187.5\ \text{Kg m}^2/\text{sec}^2$$

6-32

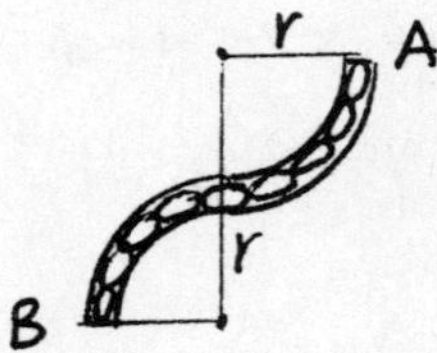

A smooth tube made of two 90° arcs of radius r as shown, is held vertically. A uniform chain of the same length is held inside the tube and released. Determine the velocity of the chain when its upper end ejects from the lower end of the tube.

****

Length of Chain = length of tube $= 2 \times \frac{\pi r}{2} = \pi r$

Initial Centroid of Chain = Centroid of tube $= G_1$

This falls to G_2 as the upper end of chain ejects from end B of the tube.

Vertical drop of the Centroid $= r + \frac{\pi r}{2}$

Initial kinetic energy of Chain $= 0$

Final K.E. of chain $= \frac{1}{2} m v^2$

Initial gravitational potential energy of Chain $= 0$ (G_1 = datum)

Final gravitational P.E. of Chain $= -mg\left(r + \frac{\pi r}{2}\right)$

No friction. By conservation of energy

$$\frac{1}{2} m v^2 - mg\left(r + \frac{\pi r}{2}\right) = 0$$

or,

$$v = \sqrt{gr(2+\pi)}$$

7
KINETICS OF RIGID BODIES

PLANE MOTION: MOMENT OF INERTIA

7-1

Determine the mass moment of inertia for the object shown with respect to the y-axis.

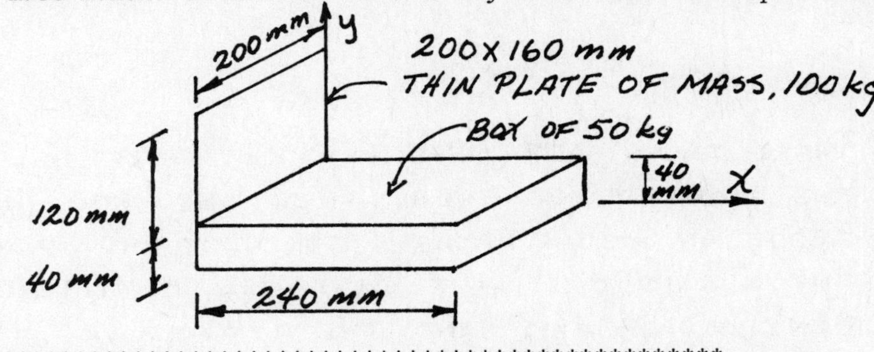

****

$$I_y = I_{y\,PLATE} + I_{y\,BOX}$$

$$= \frac{1}{12}(100)(.2)^2 + 100(.1)^2 + \frac{1}{12}(50)(.24^2 + .2^2)$$

$$+ 50\left[(.12^2 + .1^2)^{1/2}\right]^2$$

$$= .33 + 1.0 + .406 + 1.22 = \underline{\underline{2.96 \text{ kg m}^2}}$$

7-2

Four slender rods of mass m and length ℓ are welded together as shown below.

a. Calculate I_x.

b. Calculate I_z where the z axis is perpendicular to this page at C.

c. Calculate $I_{x'}$.

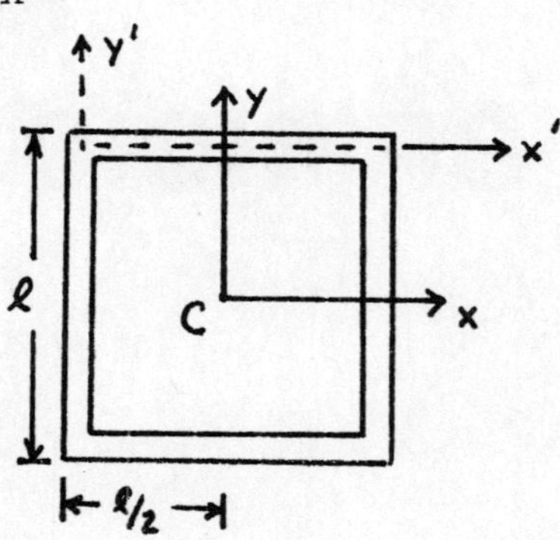

****

a) THE MOMENTS OF INERTIA OF A SLENDER ROD ABOUT ITS CENTROIDAL AXES ARE AS SHOWN BELOW

$$\bar{I}_x = \frac{1}{12} m L^2 = \bar{I}_z$$

$$\bar{I}_y = 0$$

THESE FACTS ARE USED IN THE TABLE BELOW. $\bar{I}_x$ IS THE MOMENT OF INERTIA OF THE ROD IN QUESTION ABOUT AN AXIS THROUGH ITS CENTROID PARALLEL TO THE x AXIS. d IS THE DISTANCE BETWEEN THIS CENTROIDAL AXIS AND THE x AXIS. $\bar{I}_x + md^2$ IS THE MOMENT OF INERTIA OF THE ROD, USING THE PARALLEL AXIS THEOREM, ABOUT THE x AXIS. THE TOTAL MOMENT OF INERTIA IS THEN JUST THE SUM OF THE 4 INDIVIDUAL MOMENTS OF INERTIA.

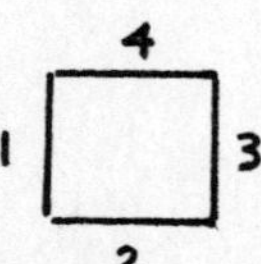

ROD	$\bar{I}_x$	d	$\bar{I}_x + md^2$
1	$\frac{1}{12}mL^2$	0	$\frac{1}{12}mL^2$
2	0	$L/2$	$mL^2/4$
3	$\frac{1}{12}mL^2$	0	$\frac{1}{12}mL^2$
4	0	$L/2$	$mL^2/4$

$$I_x = \frac{1}{12}mL^2 + m\frac{L^2}{4} + \frac{1}{12}mL^2 + mL^2/4 = \frac{2mL^2}{3}$$

b. BY INSPECTION ONE SEES THAT $I_y = I_x$. I_z, IS BY DEFINITION, THE POLAR MOMENT OF INERTIA OF THE STRUCTURE AT C. THUS

$$I_z = I_x + I_y = \frac{2mL^2}{3} + \frac{2mL^2}{3} = \frac{4mL^2}{3}$$

c. THE PARALLEL AXIS THEOREM MAY BE USED TO CALCULATE $I_{x'}$ AS THE x AXIS IS AN AXIS THROUGH THE CENTROID OF THE STRUCTURE. TO THIS END

$$I_{x'} = I_x + m(L/2)^2$$

$$= \frac{2mL^2}{3} + \frac{mL^2}{4} = \frac{11mL^2}{12}$$

7-3

Calculate the moment of inertia I_x of a right circular cone formed by revolving the shaded area shown in the accompanying illustration around the x axis. Consider the density ρ of the material to be constant and express the result in terms of the total mass m of the cone.

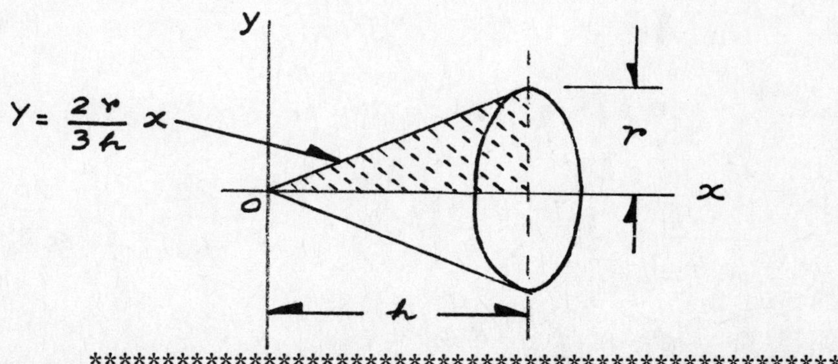

****

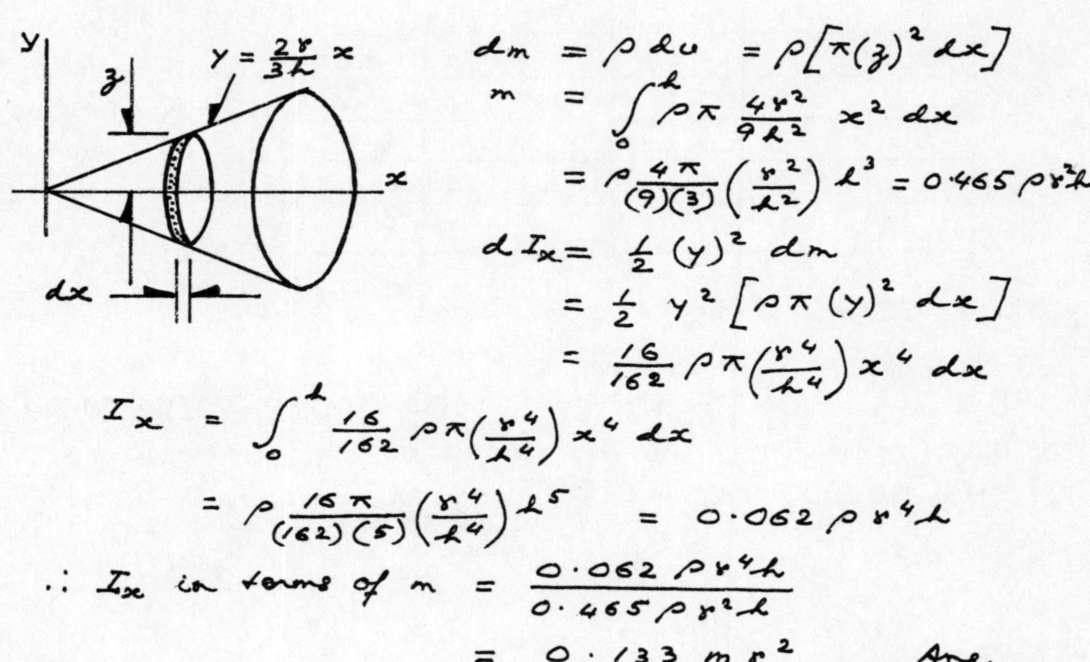

$$dm = \rho\, dv = \rho[\pi(y)^2\, dx]$$

$$m = \int_0^h \rho\pi \frac{4r^2}{9h^2} x^2\, dx = \rho \frac{4\pi}{(9)(3)}\left(\frac{r^2}{h^2}\right) h^3 = 0.465\,\rho r^2 h$$

$$dI_x = \frac{1}{2}(y)^2\, dm = \frac{1}{2} y^2 [\rho\pi(y)^2\, dx] = \frac{16}{162}\rho\pi\left(\frac{r^4}{h^4}\right) x^4\, dx$$

$$I_x = \int_0^h \frac{16}{162}\rho\pi\left(\frac{r^4}{h^4}\right) x^4\, dx = \rho\frac{16\pi}{(162)(5)}\left(\frac{r^4}{h^4}\right) h^5 = 0.062\,\rho r^4 h$$

$$\therefore I_x \text{ in terms of } m = \frac{0.062\,\rho r^4 h}{0.465\,\rho r^2 h} = 0.133\, m r^2$$ **Ans.**

7-4

A pendulum consists of a uniform 15-kg rod OA pinned at O and a 20-kg circular bob as shown. If the pendulum has a clockwise angular velocity of 10 rad/s at the instant shown, determine the moment of inertia and the angular momentum of the pendulum about the O-axis.

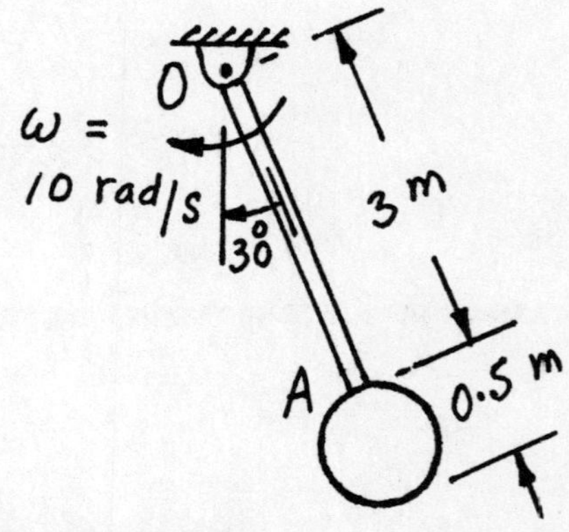

****

Moment of inertia of the system, I_O = moment of inertia of rod OA + moment of inertia of bob

$$\text{or, } I_O = \left(\frac{1}{12}\frac{15}{9.81}(3)^2 + \frac{15}{9.81}(1.5)^2\right) + \left(\frac{2}{5}\frac{20}{9.81}\left(\frac{1}{4}\right)^2 + \frac{20}{9.81}(3.25)^2\right)$$

$$\text{or } I_O = 26.172\ \text{kg}\cdot\text{m}^2$$

Angular momentum of pendulum, $H_o =$

$I_o \omega = 26.172 \times 10$

or, $H_o = 261.72$ N.m.s

7-5

Three homogeneous slender rods are riveted together in the shape indicated. If their mass is 0.2 slug/ ft. of length, find:
(a) the location of the center of gravity of mass, G
(b) mass moment of inertia about the y-axis
(c) mass moment of inertia about the x-axis
(d) mass moment of inertia about a vertical (y) axis through G.

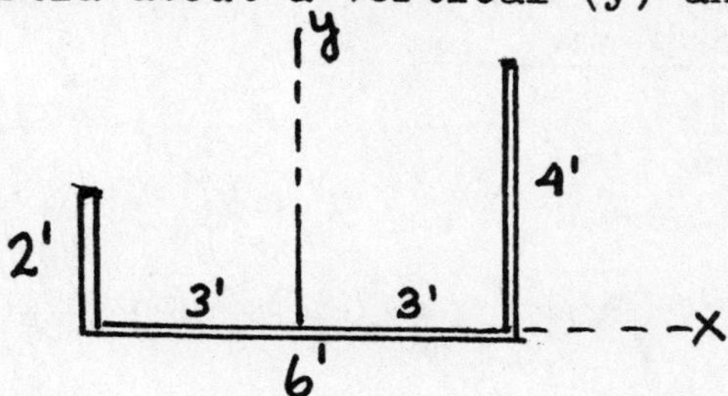

****

(a) MAKE THE FOLLOWING TABLE:

	L	$\bar{x}$	$\bar{y}$	M_y	M_x
	2'	−3	1	−6	2
	6'	0	0	0	0
	4'	3	2	12	8
TOTAL	12			6	10

then $\bar{x} = \frac{M_y}{\Sigma L} = \frac{1}{2}$ ft

$\bar{y} = \frac{M_x}{\Sigma L} = \frac{5}{6}$ ft

(b) $I_y = 6(.2)\,9 + \frac{1}{12}\cdot 6\cdot(.2)\cdot 6^2 = 14.4$ slug ft^2

(c) $I_x = \frac{1}{3}\cdot .2\,(.2)\,4 + \frac{1}{3}\cdot 4\,(.2)\cdot 4^2 = 4.8$ slug ft^2

(d) $(I_G)_y = I_y - md^2 = 14.4 - 12\,(.2)\left(\frac{1}{2}\right)^2 = 13.8$ slug ft^2

PLANE MOTION
FORCE AND MOMENT RELATIONSHIPS

7-6

The body is observed to rotate with an angular velocity of 5 rad/s and an angular acceleration of 3 rad/s^2, mass = 50 kg; find a) the pin reaction at A and, b) the mass moment of inertia about the mass center, G.

A COUPLE 50 N·m G 500 mm

****

R_y R_x 50 N·m 50 kg $\omega = 5$ rad/s $\alpha = 3$ rad/s^2

$$\vec{a}_G = -r\alpha\vec{j} - r\omega^2\vec{i} = -.5(3)\vec{j} - .5(5)^2\vec{i} = -1.5\vec{j} - 12.5\vec{i}$$

$$\Sigma F_x = ma_x \qquad R_x = 50(-12.5) = 625\text{ N} \leftarrow$$

$$\Sigma F_y = ma_y \qquad R_y - 50g = -50(1.5)$$

$$R_y = -50(1.5) + 50(9.81) = 415\text{ N} \uparrow$$

$$\Sigma M_o = I_o\alpha = \Sigma M_A = I_A\alpha$$

$$-50(9.81).5 + 50 = I_A(-3) \qquad I_A = 65.1\text{ kg m}^2$$

$$I_G = I_A - md^2 = 65.1 - 50(.5)^2 = \underline{\underline{52.6\text{ kg m}^2}}$$

OR $\Sigma M_G = I_G\alpha$

$$-415(.5) + 50 = I_G(-3) \qquad I_G = 52.6\text{ kg m}^2$$

7-7

Block A moves to the left with a velocity of 9 ft/sec and an acceleration of 4 ft/sec^2 in the position shown. The 100 lb disk rolls without slipping (with a clockwise angular velocity of 6 rad/sec) on the block. A cable wrapped around the hub (of negligible weight) of the disk and is attached to the 80 lb body B.

Determine:

(a) The acceleration of the body B.

(b) The vertical component of the reaction between the disk and the block.

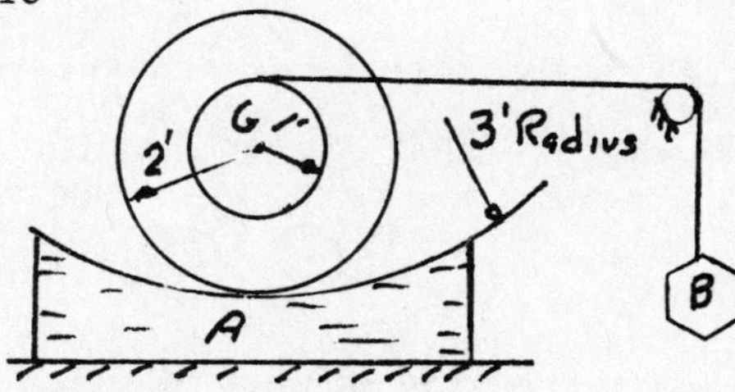

****

B' T 100 A' F N $\quad \frac{1}{2}\frac{100}{g}2^2\alpha \quad \frac{100}{g}a_{Gx} \quad \frac{100}{g}a_{Gy}$

Free Body

$a_G = a_{G/A'} \rightarrow a_{A'}$

$= 2\alpha,\ 2\omega^2 \rightarrow \overleftarrow{4} \rightarrow a_{A'y}$

$\vec{a}_{Gx} = 2\alpha - 4$

T 80 $\quad \frac{80}{g}a_B$

Free body $\quad$ Note: $a_B = a_{B'}$

$a_{B'} = a_{B'/G} \rightarrow a_G$

$= 1\alpha \rightarrow (2\alpha - 4) \rightarrow a_{Gy}$

$a_{B'} = \overrightarrow{3\alpha} - 4$

Using FBD of wt. B

$+\downarrow \Sigma F_y \quad 80 - T = \frac{80}{g}(3\alpha - 4)$

$T = 80 - \frac{80}{g}(3\alpha - 4)$

Using FBD of disk

$\curvearrowright + \Sigma M_{A'}$ $\quad$ m a d

$3T = \frac{1}{2}\frac{100}{g}2^2\alpha + \frac{100}{g}(2\alpha - 4)(2)$

$3\left(80 - \frac{240}{g}\alpha + \frac{320}{g}\right) = \frac{200\alpha}{g} + \frac{400}{g}\alpha - \frac{800}{g}$

$240g + 320 + 800 = (720 + 200 + 400)\alpha$

$\alpha = 6.7 \curvearrowright$

$a_B = 2(6.7) - 4$

$a_B = \underline{16.1\ \text{ft/sec}^2 \downarrow}$

For part b

Using FBD of disk

$\Sigma F_y \uparrow +$

$N - 100 = -\frac{100}{g}a_{Gy}$

$V_G = V_{G/A'} \rightarrow V_{A'}$

$= 2(6) \rightarrow \overleftarrow{9}$

$= 3 \quad a_{Gy} = \frac{3^2}{1} \uparrow$

$N = 100 - \frac{100}{g}(9)$

$\underline{N = 128\ \text{lb} \uparrow}$

7-8

A ladder used by a wall painter suddenly starts sliding from rest. The ladder weighs 55 lb. Find the reactions at the points A and B.

$\overline{AB} = 20$ ft.

50°

Kinematics: Choose coordinate axes as $y\uparrow \; \rightarrow x$.

$$\vec{a}_A = \vec{a}_B + \vec{a}_{A/B} \qquad \vec{a}_{A/B} = (\vec{a}_{A/B})_n + (\vec{a}_{A/B})_t$$

$$|(\vec{a}_{A/B})_n| = \overline{AB}.\omega_{AB} = 0 \quad \text{since } \omega_{AB} = 0$$

$$\vec{a}_{A/B} = (\vec{a}_{A/B})_t \qquad |(a_{A/B})_t| = \overline{AB}.\alpha_{AB}$$

$$\therefore -a_A\vec{j} = a_B\vec{i} + \overline{AB}.\alpha_{AB}.\cos 50^\circ(-\vec{j}) + \overline{AB}.\alpha_{AB}.\sin 50^\circ(-\vec{i})$$

$$\vec{i}: \quad 0 = a_B - \overline{AB}.\alpha_{AB}.\sin 50^\circ \quad \rightarrow a_B = 15.32\,\alpha_{AB}$$

$$\vec{j}: \quad -a_A = -\overline{AB}.\alpha_{AB}.\cos 50^\circ \quad \rightarrow a_A = 12.856\,\alpha_{AB}$$

For points G and A: $\vec{a}_G = \vec{a}_A + \vec{a}_{G/A}$ (see figure below)

$$|(\vec{a}_{G/A})_t| = \overline{AG}.\alpha_{AB} \qquad |(\vec{a}_{G/A})_n| = \overline{AG}.\omega_{AB} = 0 \qquad \overline{AG} = 10 \text{ ft}$$

$$\vec{a}_G = \vec{a}_A + \vec{a}_{G/A} = 12.856\,\alpha_{AB}(-\vec{j}) + \overline{AG}.\alpha_{AB}.\cos 40^\circ\,\vec{i} + \overline{AG}.\alpha_{AB}.\sin 40^\circ\,\vec{j}$$

$$\vec{a}_G = -6.428\,\alpha_{AB}\vec{j} + 7.66\,\alpha_{AB}\vec{i} \qquad a_{Gx} = 7.66\,\alpha_{AB},\; a_{Gy} = -6.428\,\alpha_{AB}$$

Kinetics: Newton's Second Law $\Sigma\vec{F} = m\vec{a}_G$

$$\vec{i}: (\overset{+}{\rightarrow}) \quad \Sigma F_x = m a_{Gx} \overset{(1)}{\rightarrow} N_A = \frac{W}{g} a_{Gx} = \frac{55}{32.2} \times 7.66\,\alpha_{AB} = 13.084\,\alpha_{AB}$$

$$\vec{j}: (+\uparrow) \quad \Sigma F_y = m a_{Gy} \overset{(2)}{\rightarrow} N_B - 55 = \frac{W}{g} a_{Gy} = \frac{55}{32.2} \times (-6.428)\,\alpha_{AB}$$

$$+\circlearrowleft \Sigma M_G = I_G.\alpha_{AB} \quad \rightarrow -N_A.\frac{\overline{AB}}{2}.\cos 40^\circ + N_B.\frac{\overline{AB}}{2}.\sin 40^\circ = \left(\frac{1}{12} m \overline{AB}^2\right).\alpha_{AB}$$

$$-7.66\,N_A + 6.428\,N_B = 56.936\,\alpha_{AB}$$

If $N_A = 13.084\,\alpha_{AB}$, $N_B = 55 - 10.98\,\alpha_{AB}$ are substituted from (1) and (2)

$$-13.084\,\alpha_{AB} \times 7.66 + 6.428\,(55 - 10.98\,\alpha_{AB}) = 56.936\,\alpha_{AB}$$

$$\alpha_{AB} = 1.5524 \text{ rad/sec}^2$$

$$\therefore \quad N_A = 13.084\,\alpha_{AB} = 20.31 \text{ lb}_f$$

$$N_B = 55 - 10.98\,\alpha_{AB} = 37.95 \text{ lb}_f$$

CHECK :

$$\Sigma M_K = m \cdot a_{Gx} \cdot \frac{\overline{AB}}{2} \cdot \cos 40^\circ + m \cdot a_{Gy} \cdot \frac{\overline{AB}}{2} \cdot \sin 40^\circ + I_G \cdot \alpha_{AB}$$

$$W \cdot \frac{\overline{AB}}{2} \cdot \sin 40^\circ = 55 \times \frac{20}{2} \times \sin 40^\circ$$

$$= \frac{55}{32.2} \times (7.66\,\alpha_{AB}) \times 10 \times \cos 40^\circ$$

$$+ \frac{55}{32.2} \times (6.428\,\alpha_{AB}) \times 10 \times \sin 40^\circ + 56.936\,\alpha_{AB}$$

$$\alpha_{AB} = 1.552 \text{ rad/sec}^2$$

7-9

Gear A weighs 48 lb and has a centroidal radius of gyration of 6 inches. Gear B weighs 16 lb and has a centroidal radius of gyration of 3 inches. Calculate the angular acceleration of gear B when a torque M of 60 inch-pounds is applied to the shaft of gear A.

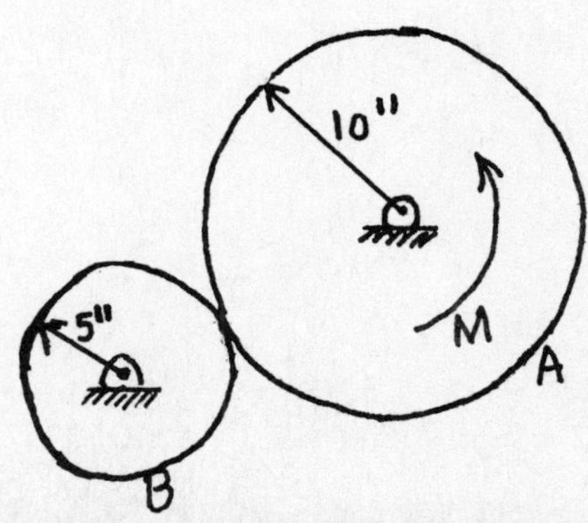

****

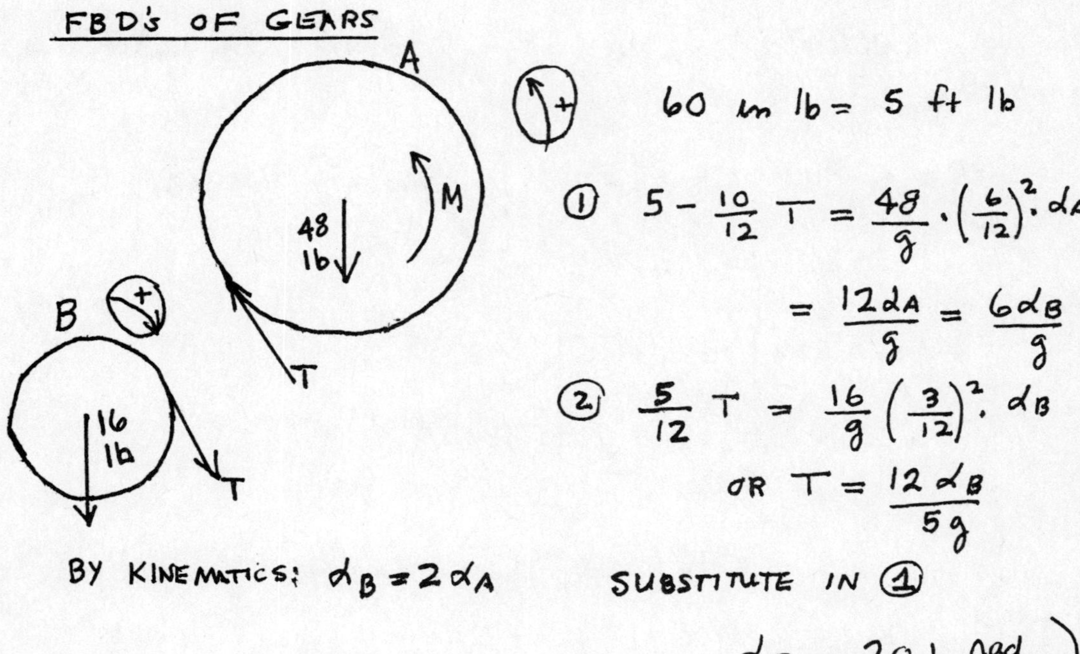

7-10

The 322 lb pulley has a radius of gyration of mass about a horizontal axis through its mass center of 1 foot. Determine the force P required to accelerate the mass center, G, upwards at 10 ft/sec^2. Assume both cords act vertically.

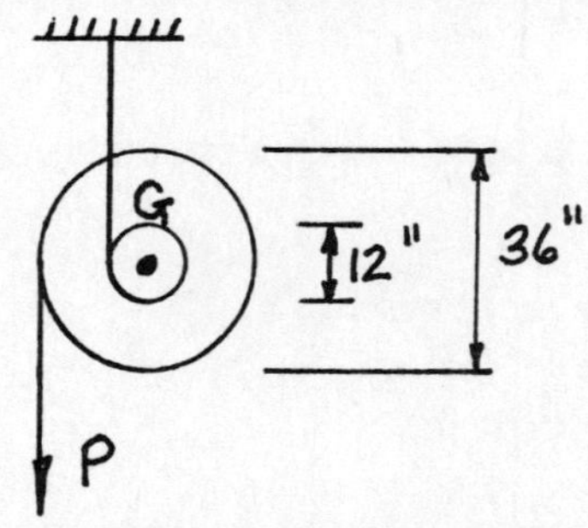

****

$$I = mk^2 = 10(1)^2 = 10 \text{ lb}\cdot\text{ft}\cdot\text{s}^2$$

$$a_G = r\alpha \quad , \quad \alpha = \frac{10}{1/2} = 20 \text{ rad/s}^2$$

$$\Sigma F_y = ma_y \quad , \quad T - P - 322 = \frac{322}{32.2}(10) = 100$$

$$\Sigma M_G = I_G\alpha \quad , \quad -\tfrac{1}{2}T + 1.5P = 10(\alpha) = 200$$

$P = T - 100 - 322 = T - 422$

$P = 3P - 400 - 422 \qquad P = \frac{822}{2} = \underline{\underline{411\ \text{lbs}}}$

7-11

The assembly shown accelerates to the right at 24 m/s^2. Determine the tension in the cable BC. The mass of AB is 20 kg and its mass center is at G. The dimensions are in meters.

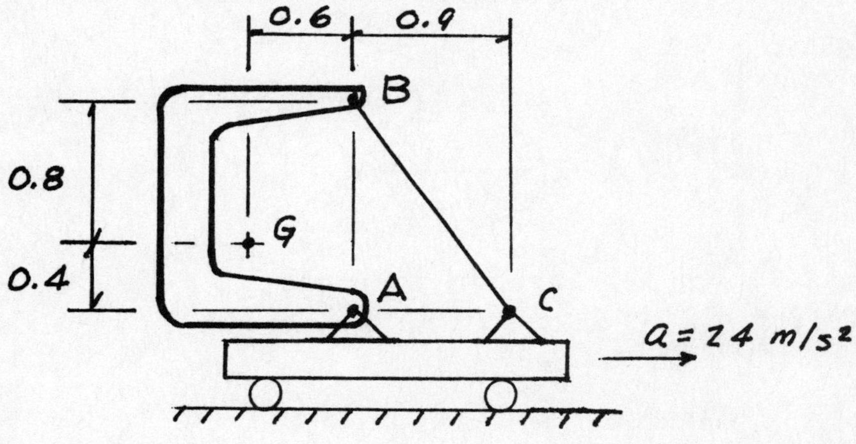

****

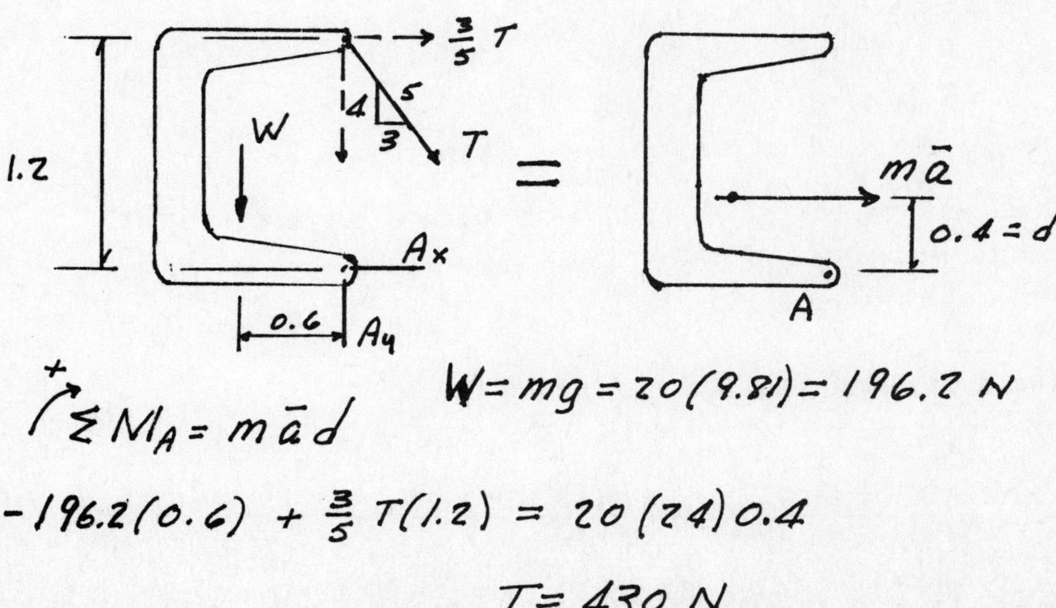

$$W = mg = 20(9.81) = 196.2\ \text{N}$$

$$\overset{+}{\curvearrowright} \sum M_A = m\bar{a}d$$

$$-196.2(0.6) + \frac{3}{5}T(1.2) = 20(24)0.4$$

$$T = 430\ \text{N}$$

7-12

A drum of radius 3 in. is attached to a disk of radius 6 in. The disk and drum have a total weight of 32.2 lbs and a centroidal radius of gyration of 1 ft. A cord is attached as shown and pulled with a force of magnitude 20 lb. The coefficient of friction between the disk and the floor is 0.4. Determine the angular acceleration of the disk.

$r_1 = 3\text{ in}$

$r_2 = 6\text{ in}$

$W = 32.2\text{ LB}$

$\bar{k} = 1\text{ FT}$

$\mu = 0.4$ G, r_1, r_2 $P = 20\text{ LB}$

****

FREE BODY DIAGRAM

EXTERNAL FORCES   EFFECTIVE FORCES

D'ALEMBERT'S PRINCIPLE

$$(\Sigma \overset{\curvearrowleft +}{M_c})_{ext} = (\Sigma \overset{\curvearrowleft +}{M_c})_{eff}$$

$$-Pr_1 = -\bar{I}\alpha - m\bar{a}\,r_2 \qquad (1)$$

$$(\Sigma F_x)_{ext} = (\Sigma F_x)_{eff}$$

$$P - F = m\bar{a} \qquad (2)$$

$$(\Sigma F_y)_{ext} = (\Sigma F_y)_{eff}$$

$$N - W = 0 \qquad (3)$$

ASSUME THAT SLIP DOES NOT OCCUR BETWEEN THE DISK AND THE FLOOR. THEN

$$\bar{a} = r_2\alpha \qquad (4)$$

SUBSTITUTING (4) INTO (1) YIELDS

$$Pr_1 = (\bar{I} + mr_2^2)\alpha \qquad \bar{I} = \bar{k}^2 m$$

$$\alpha = \frac{Pr_1}{m(\bar{k}^2 + r_2^2)} = \frac{(20\text{ LB})(\tfrac{1}{4}\text{ FT})}{\frac{32.2\text{ LB}}{32.2\text{ FT/SEC}^2}\left[(1\text{ FT})^2 + (\tfrac{1}{2}\text{ FT})^2\right]} = 4\ \frac{\text{RAD}}{\text{SEC}^2}$$

HOWEVER THE VALUE OF α OBTAINED IS DEPENDENT UPON THE ASSUMPTION OF NO SLIP. THAT ASSUMPTION MUST BE CHECKED. IF THE ASSUMPTION PROVES TO BE INVALID THE ANSWER IS INCORRECT AND THE PROBLEM MUST BE REWORKED ASSUMING SLIP OCCURS BETWEEN THE DISK AND THE SURFACE ($F = \mu N$ AND α AND $\bar{a}$ ARE INDEPENDENT). IF SLIP DOES NOT OCCUR THEN

$$F < \mu N$$

FROM (2)

$$F = P - m\bar{a} = 20\text{ LB} - \frac{32.2\text{ LB}}{32.2\text{ FT/SEC}^2}\left(\frac{1}{2}\text{ FT}\right)\left(4\text{ RAD/SEC}^2\right) = 18\text{ LB}$$

FROM (3)

$$N = W = 32.2\text{ LB}$$

IS $F < \mu N$?

$$18\text{ LB} < (.4)\,32.2\text{ LB} = 12.88 \qquad \underline{\text{No}}$$

THEN

$$F = \mu N = 12.88\text{ LB}$$

FROM (2)

$$\bar{a} = \frac{1}{m}(P - F) = 7.12\text{ FT/SEC}^2$$

FROM (1)

$$\alpha = \frac{1}{I}(Pr_1 + m\bar{a}r_2) = 6.85\text{ RAD/SEC}^2$$

7-13

A homogeneous block weighing 480 pounds is mounted on frictionless rollers at A and B. What is the maximum horizontal force, P, which can be applied at C without tipping the block, and what acceleration will result?

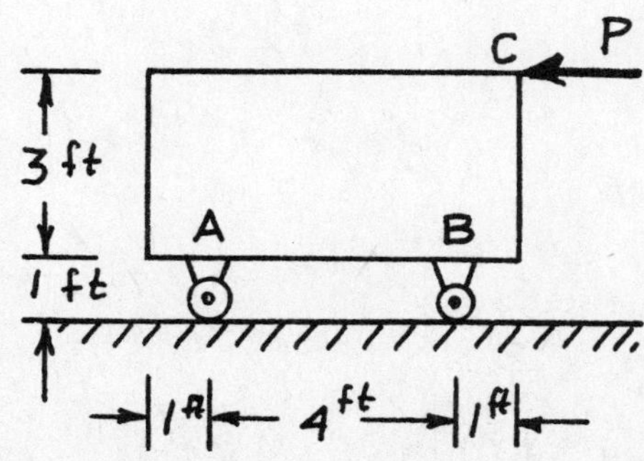

****

Just before it starts to tip, all weight is on front wheel.

$$\Sigma \bar{F} = (R_A - 480)\bar{j} - P\bar{i} = \frac{480}{32} a_G(-\bar{i})$$

"$\bar{j}$" $R_A = 480$, "$\bar{i}$" $P = 15 a_G$

Taking moments about mass center,

$$\Sigma T_G = -2R_A + \frac{3}{2}P = 0 \quad \text{(no rotation)}$$

$$P = \frac{4}{3} R_A = 640 \qquad a_G = \frac{P}{15} = 42.7$$

Answers

$P_{max} = 640$ lb $\qquad a_G = 42.7$ fps^2

7-14

A 1000-kg flywheel is acted upon by a couple on its inner hub as shown for a time interval of 10 seconds. If the flywheel is initially at rest, determine the linear velocity of its rim at the end of the time period. Assume the radius of gyration of the flywheel to be 0.8 m.

1 m; O; $\frac{1}{2}$ m; F = 500 N; F = 500 N

****

Initial angular momentum of flywheel + Angular impulse of external force system = Final angular momentum of flywheel

$$I_o \omega_1 + \int_0^{10} M_o \, dt = I_o \omega_2$$

$$0 + 500 \times \tfrac{1}{2} \times 10 = 1000 \times (0.8)^2 \times \omega_2$$

or, $\omega_2 = 3.91$ rad/sec

And rim velocity $= V = r\omega_2 = 1 \times 3.91 = 3.91$ m/s

7-15

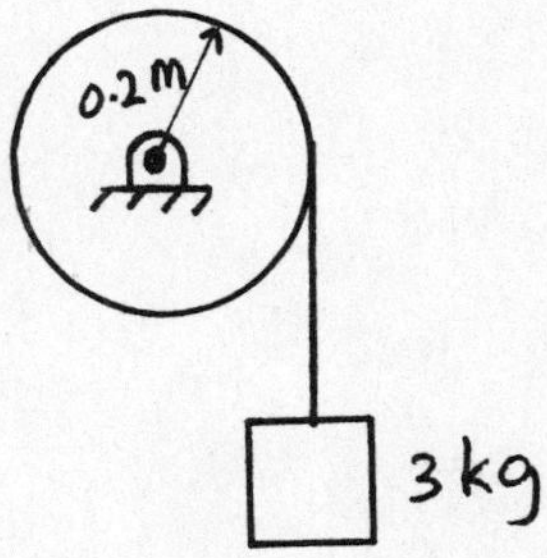

A cord wrapped around a wheel of radius 20 cm carries a 3 kg mass. The wheel is supported on frictionless bearings. When the system is released from rest, the mass moves 4 meters in 2 seconds. Determine the polar moment of inertia J of the wheel.

****

Let a = downward acceleration of the mass (m/s²)

Assuming the cord is kept taut,

angular acceleration of the wheel $= \dfrac{a}{\text{radius}} = \dfrac{a}{0.2}$ rad/s²

Using Newton's 2nd law

For Mass : Σ Force = mass x acceleration

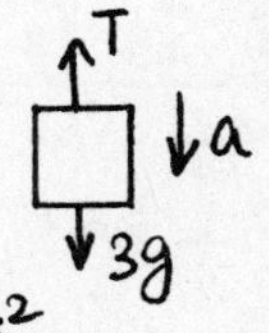

$$3g - T = 3a$$

$g = 9.81$ m/s²

For Wheel : Σ Torque = Moment of inertia x angular acceleration

$$0.2\,T = J \cdot \frac{a}{0.2}$$

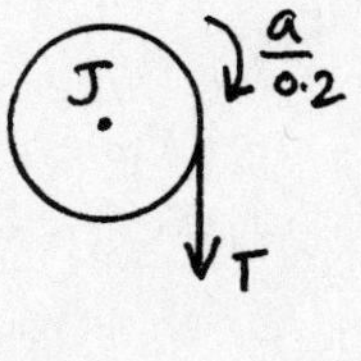

Eliminate tension T ;

$$3 \times 9.81 \times 0.2 = 3 \times 0.2a + J\frac{a}{0.2} \qquad \text{———(i)}$$

Note from (i) that acceleration a is a constant for this motion.

For rectilinear motion with constant acceleration a, starting from rest, distance moved is given by

$$x = \frac{1}{2} a t^2$$

With $x = 4\ \text{m}$ and $t = 2\ \text{s}$ we have

$$a = \frac{2 \times 4}{2^2} = 2\ \text{m/s}^2$$

Substitute in (i)

$$\underline{J = 0.47\ \text{kg m}^2}$$

7-16

A 15-kg bar is supported with two small pins which can slide freely in their tracks. A force of 800 N is applied to the pin A horizontally as shown in the figure. Calculate the forces acting on the pins A and B and the acceleration of the pin B. The length of the bar is 1.5 m.

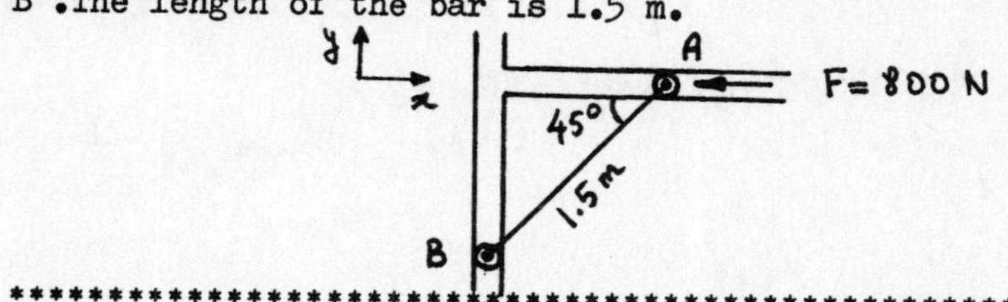

****

<u>Kinematics</u>: $\vec{a}_B = \vec{a}_A + \vec{a}_{B/A}$ (x, y chosen in the figure),

$$-a_B \vec{j} = -a_A \vec{i} + \overline{AB}.\alpha.\cos 45^\circ \vec{i} - \overline{AB}.\alpha.\sin 45^\circ \vec{j} \qquad \text{since } \omega = 0 .$$

$\vec{i}$: $0 = -a_A + \overline{AB}.\alpha.\cos 45^\circ \quad \rightarrow \quad a_A = \overline{AB}.\alpha/\sqrt{2}$

$\vec{j}$: $-a_B = -\overline{AB}.\alpha.\sin 45^\circ \quad \rightarrow \quad a_B = \overline{AB}.\alpha/\sqrt{2}$

A
G
$(\vec{a}_{G/A})_n = 0$
$(\vec{a}_{G/A})_t$
$(\vec{a}_{B/A})_n = 0$
B
45°
$(\vec{a}_{B/A})_t = \overline{AB}.\alpha$

For the mass center G and pin A : $\vec{a}_G = \vec{a}_A + \vec{a}_{G/A}$

$$\vec{a}_G = -\overline{AB}.\alpha/\sqrt{2}\ \vec{i} + (\overline{AB}.\alpha.\cos 45^\circ)/2\ \vec{i} - (\overline{AB}.\alpha.\sin 45^\circ)/2\ \vec{j}$$

$$\vec{a}_G = a_{Gx}\vec{i} + a_{Gy}\vec{j} = -0.53\alpha\vec{i} - 0.53\alpha\vec{j}, \quad |\vec{a}_G| = 0.7495\alpha$$

Kinetics : $\sum \vec{F} = m\vec{a}$: Newton's Second Law

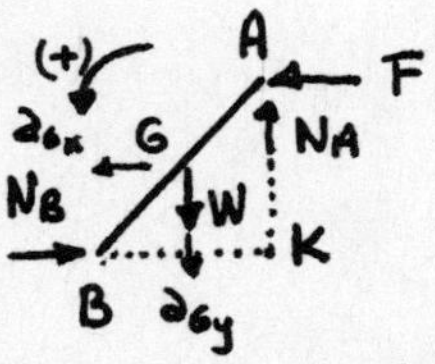

$$\sum F_x = m a_{Gx} \rightarrow N_B - F = m.a_{Gx} = 15(-0.53\alpha) \quad *$$

$$\sum F_y = m a_{Gy} \rightarrow -W + N_A = m\, a_{Gy} = 15(-0.53\alpha) \quad **$$

$$(+)\circlearrowleft \sum M_K = W.\frac{\overline{AB}}{2}.\cos 45° + F.\overline{AB}.\cos 45° = m.a_{Gx}.\frac{\overline{AB}.\cos 45°}{2} + m.a_{Gy}.\frac{\overline{AB}.\sin 45°}{2} + I_G.\alpha \quad , \quad I_G = \frac{1}{12} m l^2$$

$$(15\times 9.8)\times\frac{1.5}{2\sqrt{2}} + 800\times\frac{1.5}{\sqrt{2}} = 15\times(0.53\alpha)\times\frac{1.5}{2\sqrt{2}} + 15\times(0.53\alpha)\times\frac{1.5}{2\sqrt{2}} + \frac{15}{12}.(1.5)^2\alpha \text{ gives}$$

$\alpha = 82.39$ rad/sec^2

Therefore, from * and ** : $N_B = F - 15\times 0.53\alpha = 800 - 15\times 0.53\times 82.39 = 144.97$

$N_A = W - 15\times 0.53\alpha = 15\times 9.8 - 15\times 0.53\times 82.39 = -508.02$

$N_B \rightarrow$, $N_A \downarrow$.

$\therefore \quad a_B = \overline{AB}.\alpha.\sin 45° = 1.5\times 82.39\times\frac{1}{\sqrt{2}} = 87.39$ m/sec^2 $(\downarrow +)$

$|\vec{a}_A| = |\vec{a}_B| = 87.39$ m/sec^2 .

CHECK : Moment about the mass center

$$(+)\circlearrowleft \sum M_G = F.\frac{\overline{AB}}{2}.\sin 45° + N_A.\frac{\overline{AB}}{2}.\cos 45° + N_B.\frac{\overline{AB}}{2}.\cos 45° = I_G.\alpha$$

$$800\times\frac{1.5}{2}\times\frac{1}{\sqrt{2}} + N_A\times\frac{1.5}{2}\times\frac{1}{\sqrt{2}} + N_B\times\frac{1.5}{2}\times\frac{1}{\sqrt{2}} = \frac{15}{12}(1.5)^2\alpha$$

using the values obtained for N_A, N_B and α

$231.7 \simeq 231.7$.

7-17

End A of the 100 lb beam AB moves along the frictionless floor while end B is supported by a 4 ft. cable. Knowing that at the instant shown, end A is moving to the left with a constant velocity, determine the magnitude of the force P.

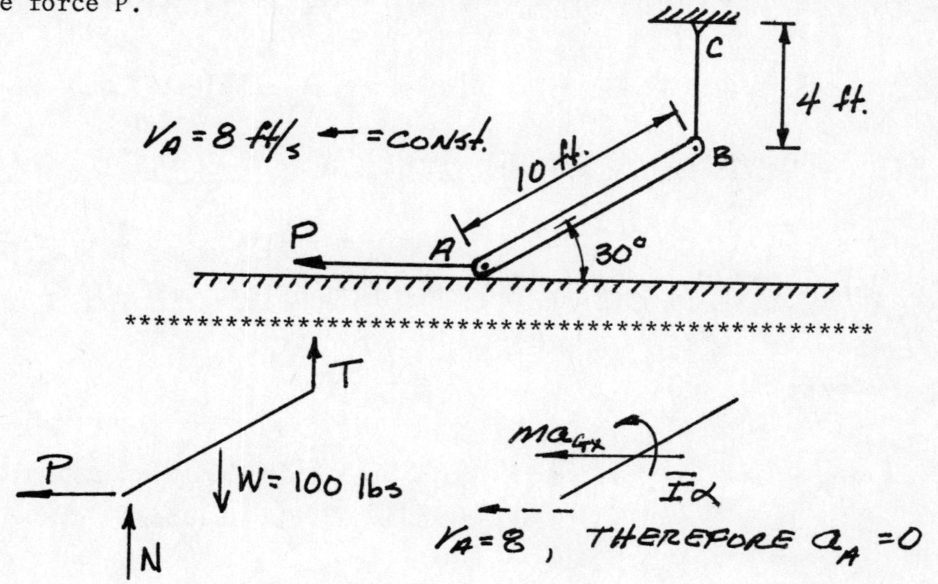

****

T
P
W = 100 lbs
N
ma_{Gx}
$\bar{I}\alpha$

$V_A = 8$, THEREFORE $a_A = 0$

SINCE V_A AND V_B ARE BOTH HORIZONTAL THE INSTANTANEOUS CENTER IS AT INFINITY, $\omega_{AB} = 0$, $V_B = 8$ ft/s

$$\omega_{CB} = \frac{8}{4} = 2 \text{ rad/s}$$

$$\Sigma F_x = ma \quad , \quad P = ma_{Gx}$$

FOR CB $\quad \vec{a}_B = r_{CB}\alpha_{CB}\vec{i} + r_{CB}\omega_{CB}^2\vec{j} = 4\alpha_{CB}\vec{i} + 4(2)^2\vec{j}$

$$\vec{a}_B = \vec{a}_A + \vec{a}_{B/A} = \alpha_{AB}\vec{k} \times \vec{r}_{B/A} - \omega_{AB}^2 \vec{r}_{B/A}$$

(with $\vec{a}_A = 0$ and $\omega_{AB} = 0$)

$$a_{Bx}\vec{i} + a_{By}\vec{j} = \alpha_{AB}\vec{k} \times (10\cos 30°\vec{i} + 10\sin 30°\vec{j}) - 0$$

$$\vec{a}_{By} = 16\vec{j} = 10(\cos 30°)\alpha_{AB}\vec{j} \qquad \alpha_{AB} = 1.848 \text{ rad/s}^2$$

$$\vec{a}_G = \vec{a}_A + \vec{a}_{G/A} = 0 + \alpha_{AB}\vec{k} \times \vec{r}_{G/A} - \omega_{AB}^2 \vec{r}_{G/A}$$

$$= \alpha_{AB}\vec{k} \times (5\cos 30°\vec{i} + 5\sin 30°\vec{j})$$

$$= 5\alpha_{AB}\cos 30°\vec{j} - 5\alpha_{AB}\sin 30°\vec{i}$$

$$a_{Gx}\vec{i} = -5\alpha_{AB}\sin 30^\circ \vec{i}$$

$$a_{Gx} = -5(1.848)\sin 30 = -4.62 \text{ ft/s}^2 = 4.62 \text{ ft/s}^2 \leftarrow$$

THEN $P = ma = \frac{100}{32.2}(4.62) = 14.35 \text{ lbs} \leftarrow$

7-18

The weight of the unbalanced wheel is 200 kilograms, and it has a radius of gyration about its mass center, G, of 0.30 meters. At the instant shown it is rotating at 6 radians per second in a clockwise direction. Find the forces acting at the pivot point C.

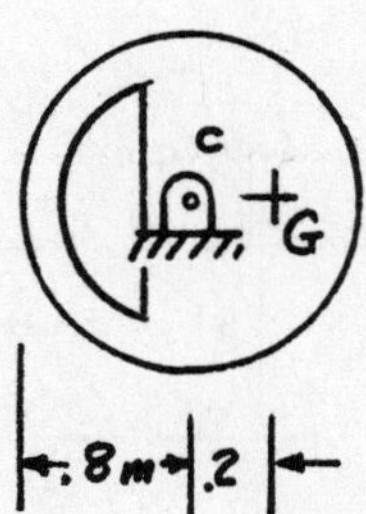

****

(Free-body sketch: α, C_x, C_Y, W)

$$W = 200\text{ kg}(9.8\text{ m/s}^2) = 1960 \text{ N}$$

$$I_c = I_G + md^2 = mk^2 + md^2 = 200([.3]^2 + [.2]^2) = 26$$

$$\Sigma M_c = .2(1960) = I_c\alpha = 26\alpha$$

$$\alpha = 15.08 \text{ rdps}^2$$

$$\bar{a}_G = -r\alpha\bar{j} - r\omega^2\bar{i} = -.2(15.08)\bar{j} - .2(6)^2\bar{i}$$

$$\Sigma\bar{F} = (C_x\bar{i} + C_y\bar{j}) - W\bar{j} = m\bar{a}_G$$

$$C_x = 200(-7.20), \quad C_y = 1960 - 200(3.02)$$

$$\boxed{C_x = -1440 \text{ N}, \quad C_y = 1356 \text{ N}}$$

7-19

A weightless cord is wrapped around a 5 ft diameter disk D and carries a block B weighing 75 lb at the free end, as shown in the accompanying illustration. The block causes the disk to rotate about its centroidal axis O, which is opposed by a constant bearing-friction moment of 10 lb.ft. If the moment of inertia I_o of the disk is 12 ft.lb.s^2, calculate (a) the angular acceleration α of disk D, (b) the linear acceleration a of block B, and (c) the tension T in the cord.

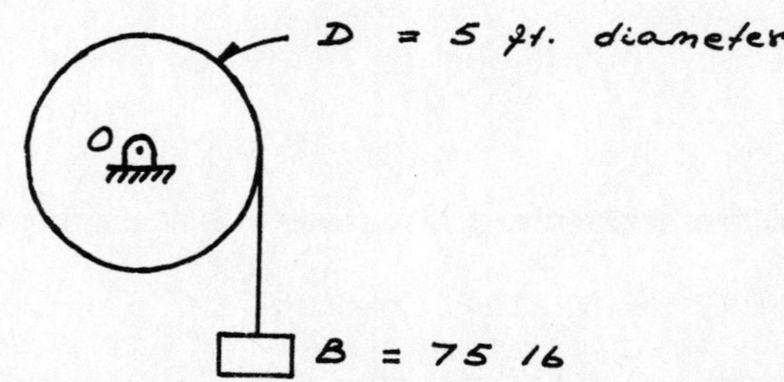

****

Solution

From the free-body diagram,

Unbalanced moment $= rT$ ↻

Inertia moment $= I\alpha$ ↺

Friction moment $= 10$ lb.ft. ↺

(a) For block B:

$$a = r\alpha\,;\quad F = ma = \frac{W}{g}a$$

$$\therefore\ F_i = \frac{W_B}{g}r\alpha$$

$$= \frac{(75\ \text{lb})(2.5\ \text{ft})}{(32.2\ \text{ft/s}^2)}\alpha = 5.822\alpha\ \text{lb.s}^2$$

$$\Sigma V = 0\,;\quad T + F_i = W_B$$

$$T = 75\ \text{lb} - 5.822\alpha\ \text{lb} \quad \ldots\ldots(1)$$

$$\Sigma M = 0\,;\quad rT = I\alpha + F_m$$

$$\text{or, } (2.5\ \text{ft})T = (12\ \text{ft.lb.s}^2)\alpha + 10\ \text{ft.lb.} \quad \ldots\ldots(2)$$

By solving simultaneous Eq. (1) & (2), we get

$$(2.5\ \text{ft})(75\ \text{lb} - 5.822\alpha\ \text{lb}) = (12\ \text{ft.lb.s}^2)\alpha + 10\ \text{ft.lb.}$$

$$\text{or, } \alpha = 7.437\ \text{rad/s}^2 \quad \underline{\text{Ans.}}$$

FREE-BODY DIAGRAM (labels: α, $I\alpha$, W, R, ω, Friction Moment, T, a, v, W_B, F_i)

(b) Linear acceleration: $a = r\alpha$

$$a = (2.5\ \text{ft})(7.437\ \text{rad/s}^2) = 18.592\ \text{ft/s}^2 \quad \underline{\text{Ans.}}$$

(c) Tension T:

Substituting the value of α in Eq. (1), we get

$$T = (75\ \text{lb}) - (5.822)(7.437\ \text{rad/s}^2) = 31.702\ \text{lb.} \quad \underline{\text{Ans.}}$$

Check :

By substituting the value of α in Eq.(2), we get

$$(2.5\ ft)\,T = (12\ ft.lb.s^2)(7.437\ rad/s^2) + 10\ ft.lb$$

$$\text{OR}\quad T = \frac{(12\ ft.lb.s^2)(7.437\ rad/s^2) + 10\ ft.lb}{(2.5\ ft)}$$

$$= 39.697\ lb. \qquad \underline{\text{Ans}}.$$

7-20

The yo-yo shown weighs 200 grams and has a radius of gyration of 20 cm about its mass center. The inner radius is 10 cm and the outer radius is 30 cm. If it is released from rest, how fast is it moving after 5 seconds?

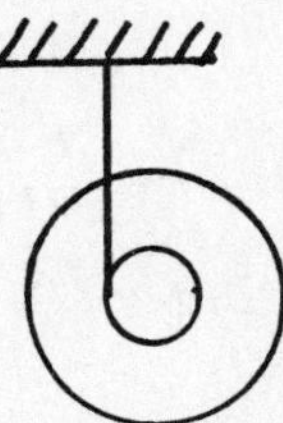

****

(T, ω, C, W)

$$W = \frac{200\,g}{1000\,g/kg}(9.8\,m/s^2) = 1.96\,N$$

$$I_c = m(k^2 + d^2) = 0.2([.2]^2 + [.1]^2) = 0.01\ kg \cdot m^2$$

Using impulse-momentum and instant center C,

$$(\Sigma \bar{T}_{IC})\Delta t = I_{IC}(\bar{\omega}_f - \bar{\omega}_i)$$

$$r_i W \Delta t = I_c(\omega_f - 0)$$

$$(.1)(1.96)(5) = 0.01\ \omega_f$$

$$\omega_f = 98\ rad/s$$

Using "rolling without slipping" $V_f = r\omega_f = (.1)\omega_f$

$$\boxed{V_f = 9.8\ m/s\ \text{downward}}$$

7-21

The wheel has an inner radius of 1 meter and an outer radius of 3 meters. It weighs 25 kilograms and has a radius of gyration about its mass center of 2 meters. The wheel rolls without slipping. Find the acceleration of block B which weighs 15 kilograms.

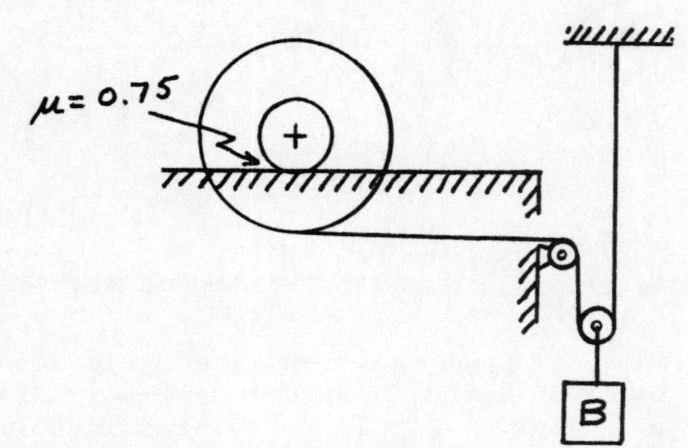

****

$$I_G = mk^2 = 25(2)^2,\; I_{IC} = I_G + md^2 = 25(2)^2 + 25(1)^2 = 125$$

$$W = mg = 25(9.8) = 245$$

$$\sum M_{IC} = 2T = I_{IC}\alpha \quad \text{or} \quad 2T = 125\alpha \quad \text{(I.)}$$

$$W = 15(9.8) = 147\ N$$

$$\sum \bar{F} = (2T - W)\bar{j} = m a_{GB}(-\bar{j})$$

$$2T - 147 = -15 a_{GB} \quad \text{(II.)}$$

Considering effect of pulley, $a_{GB} = \frac{1}{2} a_P$
Considering rolling without slipping, $a_P = 2\alpha$
so $a_{GB} = \alpha$ (III.)

Solve (I.), (II.), and (III.) by substitutions.

Answer
$a_{GB} = 1.05\ m/s^2$ (down)

7-22

A slender rod AB of length L and mass m is supported by a pin at point C. At this instant when the rod is in a vertical position, a force of magnitude P is applied to B as shown. Find:

(a) the angular acceleration for the rod
(b) The magnitude and direction of the reaction at C
(c) the magnitude and direction of the acceleration of G, the center of mass of the rod
(d) the magnitude and direction of the acceleration of A.

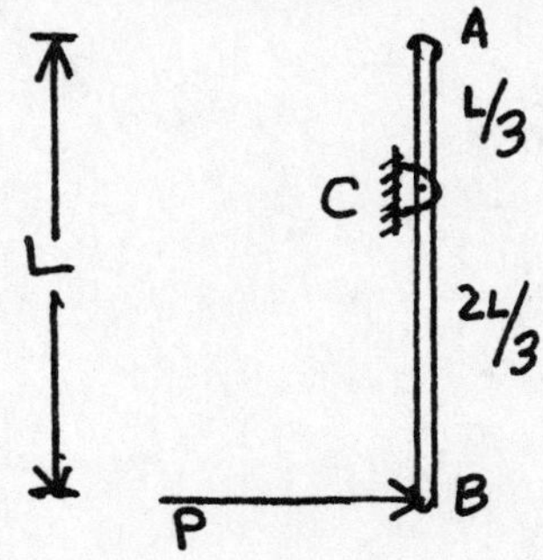

****

FBD OF ROD

C_y, C_x, mg, P

USE y, x, α

$$\Sigma F_x: \; P + C_x = m(a_G)_x$$

$$\Sigma F_y: \; C_y - mg = m(a_G)_y$$

$$\Sigma M_C: \; P \cdot \frac{2L}{3} = \left(\frac{1}{12} mL^2 + \frac{mL^2}{36}\right)\alpha$$

$$= \frac{1}{9} mL^2 \alpha$$

$$\text{then } \alpha = \frac{2PL}{3} \cdot \frac{9}{mL^2} = \frac{6P}{mL} \circlearrowleft$$

$$(a_G)_x = \frac{L}{6} \cdot \alpha = \frac{P}{m} \rightarrow$$

$$C_x = m \cdot \frac{P}{m} - P = 0$$

$$C_y = mg \uparrow$$

$$a_A = \frac{L}{3} \cdot \alpha = \frac{L}{3}\left(\frac{6P}{mL}\right) = \frac{2P}{m} \leftarrow$$

7-23

At the instant shown, block B is traveling on a smooth circular path with a velocity of 12 ft/sec to the right. Neglect the weight of block B. A 6 lb force is applied to the 20 lb non-homogeneous member AB. Member AB has a clockwise angular velocity of 2 rad/sec and a radius of gyration of 2 ft with respect to G.

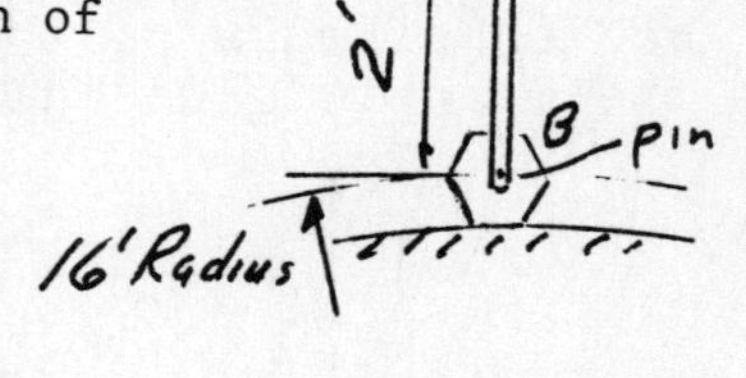

Determine:

(a) The acceleration of the block at the instant shown.

(b) The reaction on the block at the instant shown.

****

6 lb, 20 lb, N_B = $I_G\alpha$, $\frac{20}{g}a_G$, $m\bar{a} \approx 0$

Free Body

Since bar is in Plain Motion the ma vector has unknown (at G) direction. The ma vector for the block is zero since we neglect its weight.

$$a_G = a_{G/B} \rightarrow a_B$$

$$= \overset{2\alpha}{\overrightarrow{\quad}} \; 2(2)^2 \downarrow \rightarrow \frac{12^2}{16}\downarrow \rightarrow \overrightarrow{a_{B_T}}$$

$$a_G = 17\downarrow \rightarrow \overrightarrow{2\alpha} \rightarrow \overrightarrow{a_{B_T}}$$

substitute for the a_G

$$\curvearrowright \Sigma M_G \quad (1)6 = \frac{20}{g}(2)^2\alpha$$

$$\alpha = 2.415$$

$$\overrightarrow{\Sigma F_x}$$

$$6 = \frac{20}{g}(2\alpha + a_{B_T})$$

$$a_{B_T} = \frac{6g}{20} - 2(2.415)$$

$$a_{B_T} = \overrightarrow{4.83}$$

$$a_B = 9\downarrow \rightarrow \overrightarrow{4.83}$$

$$= \underline{\underline{10.2 \text{ ft/sec}^2}} \quad (4.83,\ 9)$$

For part b

$$\Sigma F_y \uparrow +$$

$$N_B - 20 = -\frac{20}{g}(17)$$

$$N_B = 20 - 10.56$$

$$N_B = \underline{\underline{9.44 \text{ lb}}}\uparrow$$

7-24

The homogeneous disk shown rolls without slipping with an angular velocity of 2 rad/sec and an angular acceleration of 3 rad/sec^2, both counterclockwise. The 10 ft homogeneous bar BC weighs 100 lb and has an angular velocity of 2 rad/sec clockwise in the position shown. The bar is connected to the disk at B by a smooth pin. Determine for this position the components of the pin reaction at B.

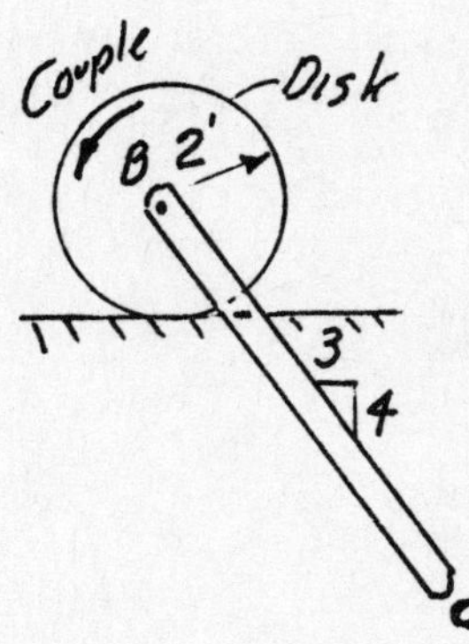

****

B, B_x, B_y, 100, C

Free Body

$= \; ma_G, \; I_G\alpha$

Since BC is in plain Motion the ma vector has an unknown slope. Must determine the a_G & α. Since the disk rolls without slipping $a_B = r\alpha$
$= 2(3)$ ←

$a_G = a_{G/B} \rightarrow a_B$ replace the acceleration by the three vectors

$= 5(2)^2 \; ↗ 5\alpha \rightarrow \overleftarrow{6}$

$\circlearrowleft + \Sigma M_B$ (m a d)

$$-\tfrac{3}{5}(5)100 = \tfrac{1}{12}\tfrac{100}{g}10^2\alpha + \tfrac{100}{g}5\alpha\,5 - \tfrac{100}{g}(6)4$$

$$3g = \tfrac{100}{12}\alpha + 25\alpha - 24$$

$$\alpha = -2.18 \quad \text{or} \quad \alpha = 2.18 \circlearrowright$$

$\overset{+}{\leftarrow} \Sigma F_x$

$$B_x = \tfrac{100}{g}6 - \tfrac{4}{5}\tfrac{100}{g}5(-2.18) + \tfrac{3}{5}\tfrac{100}{g}5(2)^2$$

$$= 18.63 + 27.08 + 37.26$$

$$= 83$$

$+\uparrow \Sigma F_y$

$$B_y - 100 = \tfrac{4}{5}\tfrac{100}{g}5(2)^2 + \tfrac{3}{5}\tfrac{100}{g}5(-2.18)$$

$$B_y = 100 + 49.7 - 20.3$$

$$= 129.4$$

$B = \overleftarrow{83} \rightarrow 129.4 \uparrow$ lb

7-25

A 100 kg cylindrical roller is initially at rest and is acted upon by a 300 Newton force as shown. Assuming the body rolls without slipping, determine:

(a) the velocity of G after 6 seconds.
(b) the friction force required to prevent slipping.

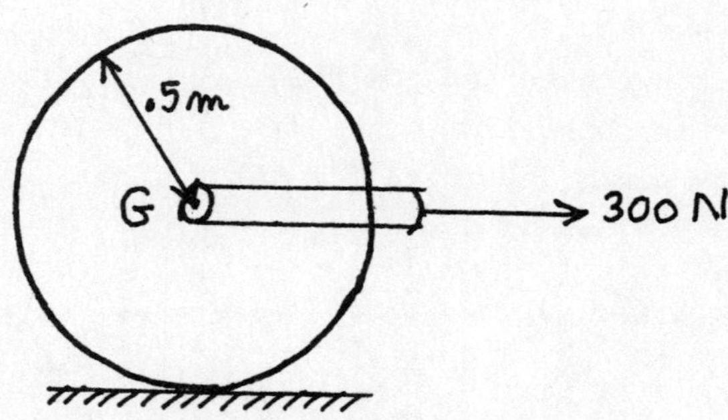

****

FBD OF CYLINDER

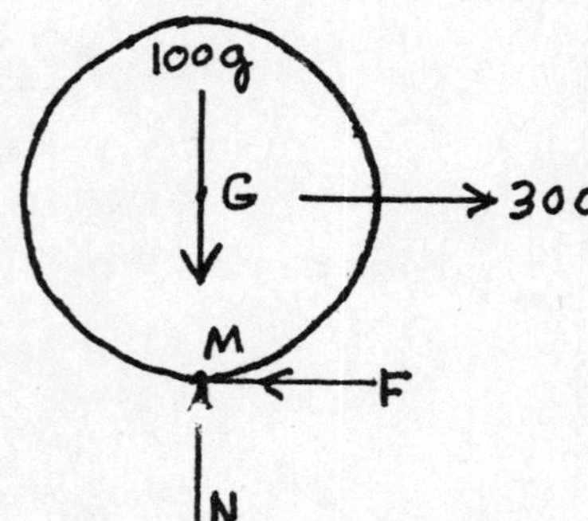

ROLLS WITHOUT SLIPPING: $\therefore V_G = \frac{1}{2}\omega$

$$\stackrel{+}{\circlearrowright} \Sigma M_M \cdot t = I_m \omega$$

$$300 \cdot \frac{1}{2} \cdot 6 = \left[\frac{1}{2} \cdot 100 \cdot \frac{1}{4} + 100 \cdot \frac{1}{4}\right] \omega$$

SOLVING: $\omega = 24$ rad/sec $\circlearrowright$

$$\Sigma F_x = 300 - F$$

$$\text{SO } (300 - F)\, 6 = 100 \cdot V_G = 100 \cdot \frac{24}{2}$$

$$F = 100 \text{ N}$$

$$\text{OR } \stackrel{+}{\circlearrowright} \Sigma M_G \cdot t = I_G \cdot \omega$$

$$6 \cdot \frac{1}{2} \cdot F = \frac{1}{2} \cdot 100 \cdot \frac{1}{4} \cdot \omega$$

$$F = \frac{12.5}{3}\,\omega = \frac{12.5}{3} \cdot 24 = 100 \text{ N}$$

7-26

The cylindrical body shown has a mass of 30 kg centered on its axis, a radius of gyration of 100 mm and a diameter of 500 mm. The coefficient of friction between the cylinder and ramp is 0.10.

(a) If released from rest, will the body slide, or roll without slipping on the surface? Verify your answer.

(b) Find the acceleration of G and the angular acceleration.

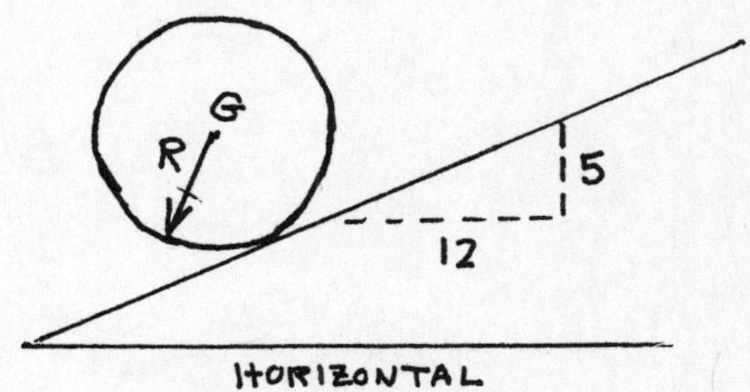

****

Let us assume no slipping.

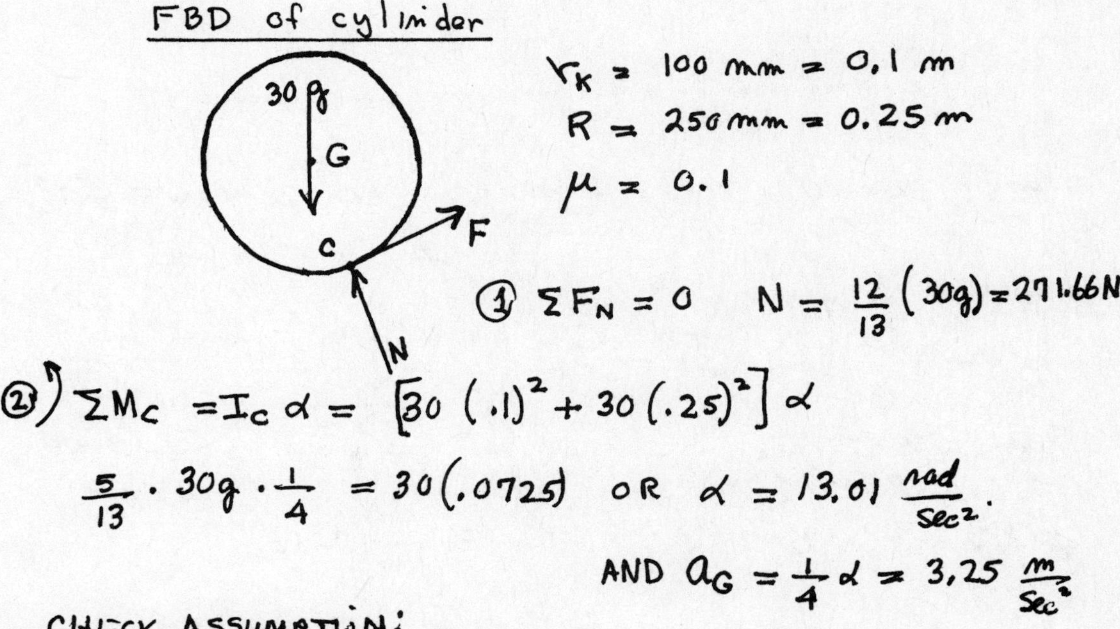

$r_k = 100\ mm = 0.1\ m$

$R = 250\ mm = 0.25\ m$

$\mu = 0.1$

① $\Sigma F_N = 0 \quad N = \frac{12}{13}(30g) = 271.66 N$

② $\circlearrowleft \Sigma M_C = I_C \alpha = [30(.1)^2 + 30(.25)^2]\alpha$

$\frac{5}{13} \cdot 30g \cdot \frac{1}{4} = 30(.0725)$ OR $\alpha = 13.01 \frac{rad}{sec^2}$.

AND $a_G = \frac{1}{4}\alpha = 3.25 \frac{m}{sec^2}$

CHECK ASSUMPTION:

$\Sigma F \parallel$ to plane: $30g\left(\frac{5}{13}\right) - F = 30 \cdot a_G$

SOLVING $F = 15.69 N$

BUT $F' = 0.1(N) = 27.17 N > 15.69 N$

$\therefore$ Assumption is verified.

7-27

Point D has a velocity of 4 ft/sec to the right when a force, P = 100 lb, is applied. The cord is wrapped around the 128.8 lb wheel, and the wheel rolls without slipping. The radius of gyration of mass of the wheel with respect to G is 2.0 ft.

Determine, for this position, the reaction exerted on the wheel by the curved plane.

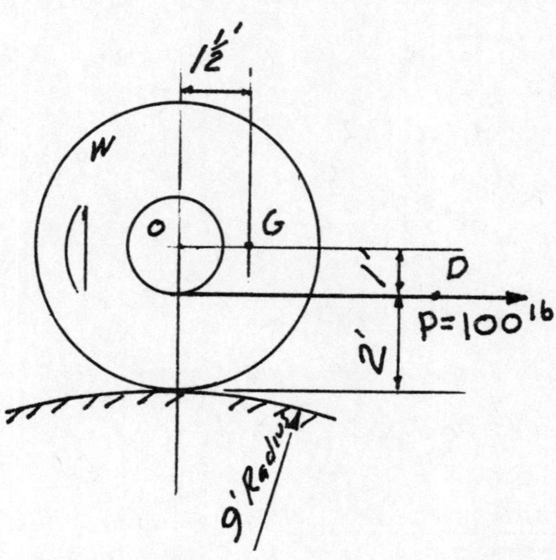

****

Instant center at point of contact ∴ $V_D = r_{ic}\omega$ $\omega = \frac{4}{2}$ or 2

$a_{o_x} = 3\alpha$ $a_{o_y} = \frac{V_o^2}{R}$ or $\frac{(3(2))^2}{12}$

$I_G\alpha = 4(2^2\alpha)$

$a_G = a_{G/O} \rightarrow a_o$

$= \frac{3}{2}(2)^2 \quad \frac{3}{2}\alpha \rightarrow 3\alpha \rightarrow 3$

128.8, 100, N, F — FBD; $4(6)$, $4(3)$, $4(3\alpha)$, $4(\frac{3}{2}\alpha)$ — EFD

From FBD $+\circlearrowleft \Sigma M_{ic}$

$100(2) + 128.8(\frac{3}{2}) =$

$4(2)^2\alpha + 4(3)\frac{3}{2} + 4(\frac{3}{2}\alpha)\frac{3}{2} + 12\alpha(3) - (24)(3)$

$\alpha = 7.33$

$+\uparrow \Sigma F_y$

$N - 128.8 = -4(\frac{3}{2}\alpha) - 4(3)$

$N = 128.8 - 6(7.33) - 12$

$N = 72.8$ lb $\uparrow$

$\xrightarrow{+} \Sigma F_x$

$F + 100 = 4(3)\alpha - 24$

$F = -100 + 12(7.33) - 24$

$F = -36$

Reaction = 72.8 lb $\uparrow$ + $\leftarrow$ 36 lb

7-28

A bar of uniform thickness is suspended with two wires as shown in the figure. The wire A is cut suddenly. For this moment, calculate the tension in the wire B, the angular acceleration of the bar and accelerations of the points G (mass center) and A.

$l = 10\,ft$, G, A, B, $W = 100\,lb$, x, y

****

Kinematics: $\vec{a}_G = \vec{a}_B + \vec{a}_{G/B} = 0 + \frac{l}{2}\alpha(-\vec{j}) \rightarrow \vec{a}_G = -\alpha\frac{l}{2}\vec{j}$

Kinetics:

$$\Sigma F_x = 0$$

$$(+)\uparrow \Sigma F_y = m.a_{G_y} \longrightarrow T - W = -m.a_{Gy} \quad (*)$$

$$\Sigma M_B = I_G.\alpha + m.a_{Gy}.\frac{l}{2} = I_B.\alpha$$

$$(m.g).\frac{l}{2} = \frac{1}{12}ml^2\alpha + m.(\alpha\frac{l}{2})\frac{l}{2} \longrightarrow \alpha = \frac{3g}{2l} = \frac{3\times32.2}{2\times10} = 4.83 \text{ rad/sec}^2$$

$$\therefore T = W - m.a_{Gy} = mg - m(\frac{\alpha l}{2}) = mg - \frac{3g}{2l}m\frac{l}{2} = \frac{mg}{4} = \frac{100}{4} = 25\ lbf$$

Acceleration of point A: $\vec{a}_A = \vec{a}_B + \vec{a}_{A/B}$

$$= 0 + l.\alpha.(-\vec{j}) = l.\frac{3g}{2l}.(-\vec{j}) = \frac{3g}{2}(-\vec{j}) = -48.3\vec{j} \text{ ft/sec}^2$$

$$\therefore \vec{a}_G = \frac{3g}{4}(-\vec{j}) = -24.15 \text{ ft/sec}^2.$$

CHECK: $\Sigma M_A = I_G.\alpha - m.a_{Gy}.\frac{l}{2}$

$$T.l - W\frac{l}{2} = \frac{1}{12}ml^2.\alpha + m.\frac{\alpha l}{2}.\frac{l}{2}$$

from(*), $T = W - m\,a_{Gy}$ is substituted here and after some arrangement

$$\alpha = \frac{3g}{2l}$$

7-29

The cylinder shown is homogeneous and has a mass of 60 kg. The coefficient of friction between the cylinder and the plane is 0.30. Find all unknown forces acting on the wheel immediately after it is released from rest and find the angular acceleration.

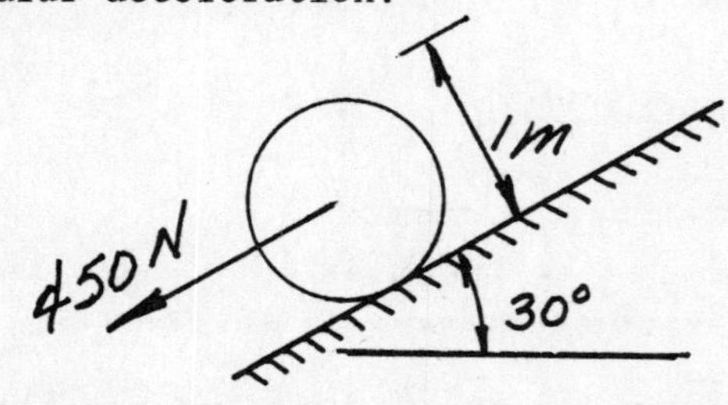

****

ASSUME NO SLIPPAGE

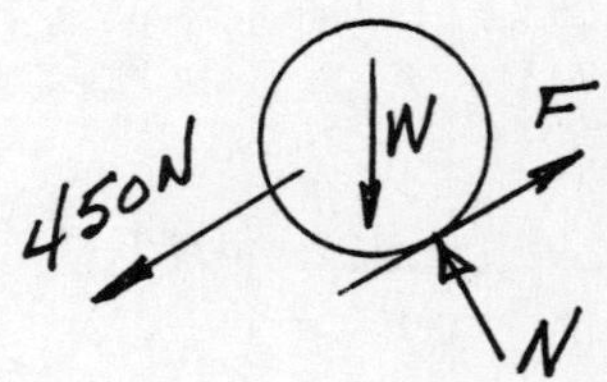

$\Sigma F_N = 0$

$60(9.81)\cos 30° = N = 510\,N$

MAX POSSIBLE FRICTION FORCE (ASSUME CONSTANT COEFFICIENT OF FRICTION)

$F_{MAX} = \mu N = .3(510) = 152.9\,N$

$\swarrow + \Sigma F_F = ma$, $a = .5\alpha$ FOR CENTER OF ROLLING SPHERE

$450 + 60(9.81)\sin 30 - F = 60(.5\alpha)$

$\Sigma M_G = \bar{I}\alpha$, $\bar{I} = \frac{1}{2}(60)(.5)^2 = 7.5\ kg \cdot m^2$

$F(.5) = 7.5\alpha$, $F = 15\alpha$

$744 - F = 30\alpha = 2F$ $\quad F = \frac{744}{3} = 248 > 152.9\,N$

NO GOOD

BODY IS SLIPPING, ASSUME $F = F_{MAX} = 152.9\,N$

$\alpha = \frac{F}{15} = \frac{152.9}{15} = \underline{\underline{10.2\ rad/s^2}}$

7-30

The 50 kg. uniform rod AB is released from rest in the position shown. Knowing that end B may slide free on the frictionless floor, determine (a) the angular acceleration of the rod (b) the tension in wire AC (c) the reaction at point B.

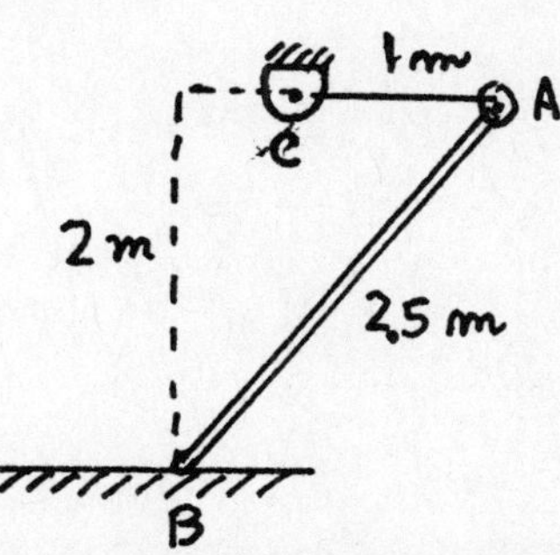

****

FBD OF ROD

T, A, G, 50 kg., B, N

USE: y, x, α

$$\Sigma F_x: \; -T = 50\,(a_G)_x$$

$$\Sigma F_y: \; N - 50g = 50\,(a_G)_y$$

$$\Sigma M_G: \; N \cdot \frac{3}{4} - T \cdot 1 = \frac{1}{12} \cdot 50 \cdot \left(\frac{5}{2}\right)^2 \cdot \alpha$$

4 unknowns, so we need to use relative acceleration:

$$\bar{a}_A = \bar{a}_B + \bar{a}_{B/A}$$

$$-a_A \bar{j} = -a_B \cdot \bar{i} + \frac{5}{2}\alpha\left(-\frac{4\bar{i}+3\bar{j}}{5}\right)$$

$$\therefore \; a_A = -3/2\,\alpha \; ; \; a_B = -2\alpha$$

then $\bar{a}_G = \bar{a}_A + \bar{a}_{G/A}$

$$(a_G)_x \bar{i} + (a_G)_y \bar{j} = -\frac{3}{2}\alpha \bar{j} + \frac{5}{4}\alpha\left(\frac{-4\bar{i}+3\bar{j}}{5}\right)$$

$$= -\alpha \bar{i} - 3/4\,\alpha \bar{j}$$

substituting back in the force equations:

$$T = 176.6 \text{ N} \; ; \; N = 358.1 \text{ N}$$

$$\alpha = 3.53 \text{ rad/s}^2$$

7-31

The uniform slender bar weighs 75 pounds and is six feet long. It is at rest on a smooth horizontal plane when a force P of 15 pounds is applied one foot from the left end. P is applied perpendicular to the bar and parallel to the plane. Determine the acceleration of end B of the bar immediately after P is applied.

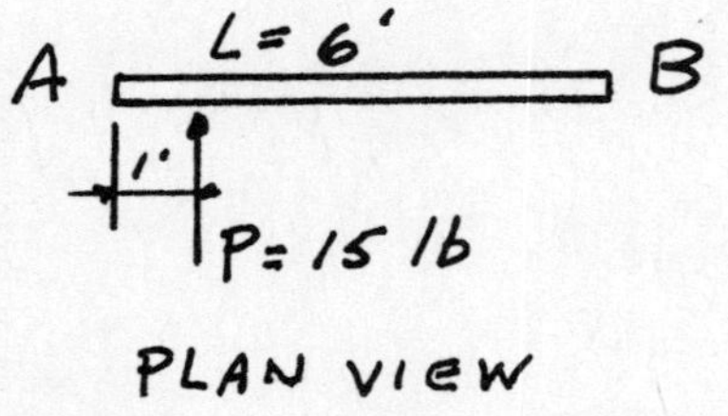

For a slender rod $\bar{I} = mL^2/12$

****

TO FIND a_G:

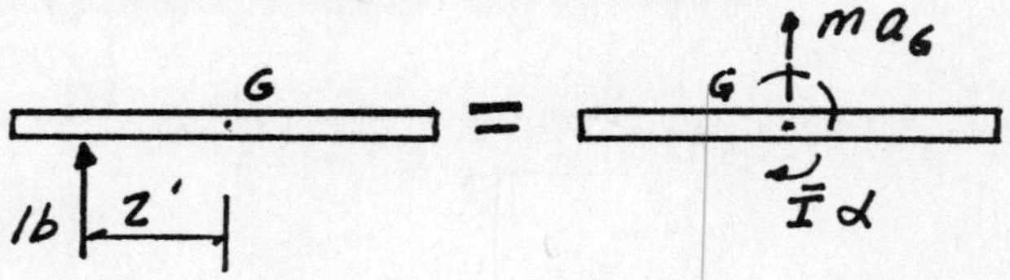

$$\uparrow \Sigma F = ma_G = \frac{W a_G}{g}$$

$$15 = \frac{75}{32.2} a_G$$

$$a_G = 6.44 \text{ FT/s}^2 \uparrow$$

TO FIND α:

$$+\curvearrowleft \Sigma M_G = \bar{I}\alpha = \frac{1}{12} mL^2\alpha = \frac{1}{12}\frac{W}{g}L^2\alpha$$

$$15(2) = \frac{1}{12}\left(\frac{75}{32.2}\right)6^2\alpha \Rightarrow \alpha = 4.293 \text{ rad/s}^2 \curvearrowright$$

$$\underline{a_B} = \underline{a_G} + \underline{a_{B/G}} = a_G\uparrow + (r\omega^2 \leftarrow + r\alpha\downarrow)$$

FOR $a_{B/G}$

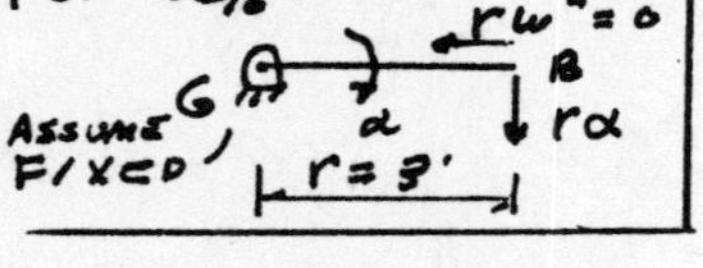

$$= 6.44\uparrow + 0 + 3(4.293)\downarrow$$

$$= +6.44 - 12.88 = 6.44 \text{ FT/s}^2 \downarrow$$

Note: Immediately after the force is applied the angular velocity ω of the bar will be zero. A graph of the accelerations for all points on the bar would be linear as shown here:

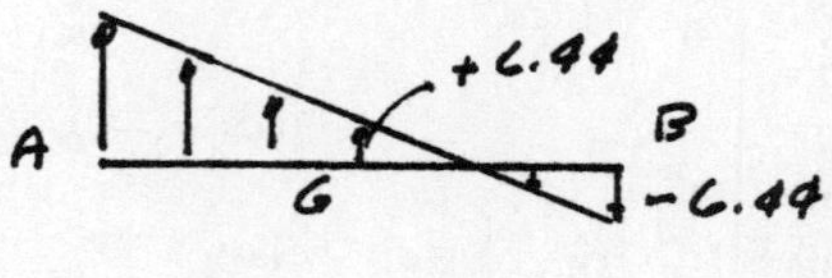

7-32

The 10 pound sphere rolls and slips on the horizontal plane. The coefficient of friction is 0.4. At the instant shown, the velocity of point A is 3 ft/sec to the right, and of point C is 1.2 ft/sec to the right. At this instant, find the angular acceleration of the sphere.

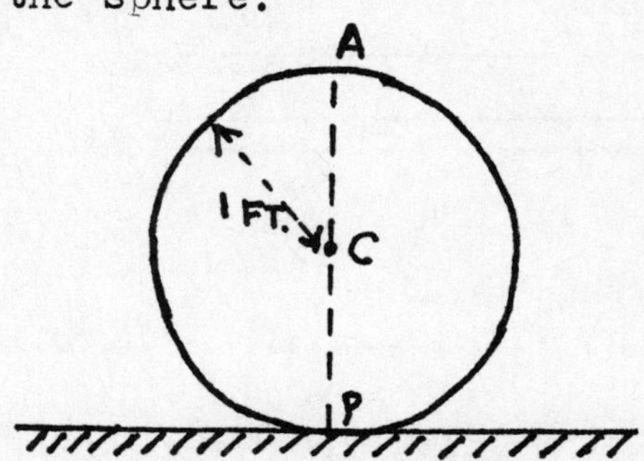

****

FBD OF SPHERE:

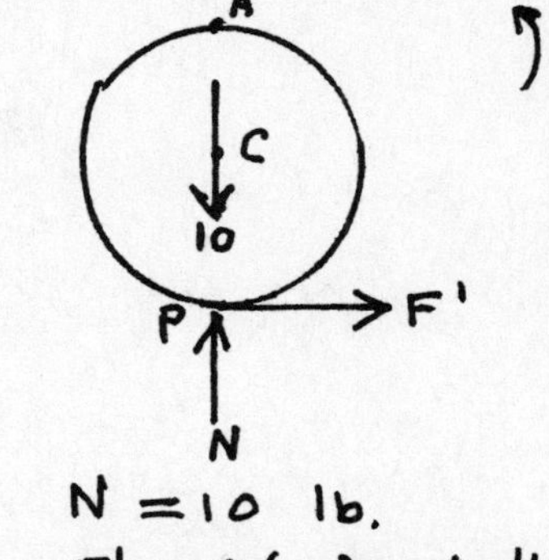

$N = 10$ lb.

$F' = .4(10) = 4$ lb.

Assume F' IN DIRECTION SHOWN

$$\curvearrowleft \Sigma M_C = I_C \cdot \alpha$$

$$4 \cdot 1 = \left(\frac{2}{5} \cdot \frac{10}{g} \cdot 1^2\right) \alpha$$

$$\alpha = g = 32.2 \text{ ft/s}^2$$

VELOCITY DIAGRAM

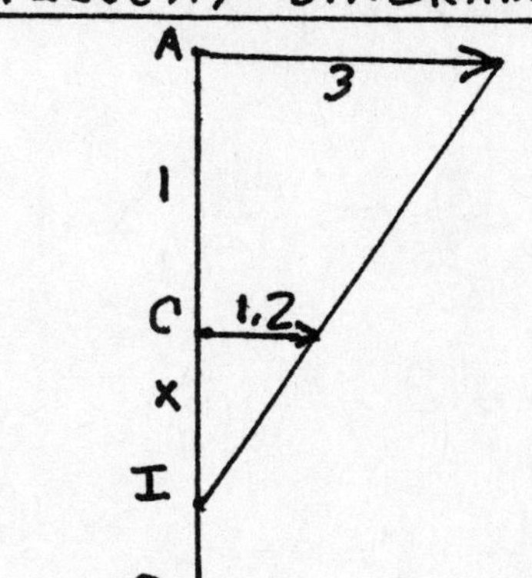

TO VERIFY DIRECTION OF F'

BY PROPORTIONS:

$$\frac{x}{1.2} = \frac{1+x}{3} \qquad x = \frac{2}{3} \text{ ft}$$

$$\text{then } \omega = \frac{V_C}{CI} = \frac{1.2}{2/3} = 1.8$$

$$\text{and } V_P = (IP)\omega = .6 \text{ ft/s} \leftarrow$$

$\therefore$ F' IS TO THE RIGHT

7-33

A uniform bar of weight W is simply supported at A and B. If the support at B is suddenly removed, determine the angular accelaration of the bar and the reaction at A.

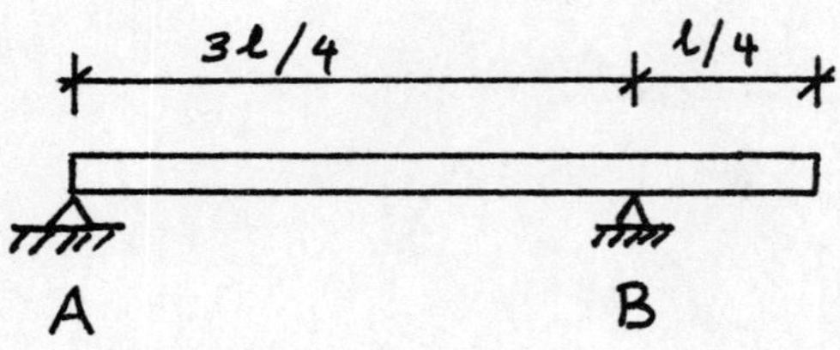

****

Solution :

A_x, A_y, W, $l/2$ $\equiv$ α, $\bar{a}$

$$+\circlearrowright \Sigma M_A = I_A \alpha \qquad W = mg$$

$$I_A = \bar{I} + m\left(\frac{l}{2}\right)^2 = \frac{1}{12} m l^2 + m \frac{l^2}{4} = \frac{1}{3} m l^2$$

$$mg\left(\frac{l}{2}\right) = \frac{1}{3} m l^2 \alpha \quad \text{or} \quad \alpha = \frac{3}{2}\frac{g}{l} \circlearrowright$$

$$\bar{a} = \alpha \frac{l}{2} = \frac{3}{2}\frac{g}{l} \cdot \frac{l}{2} = \frac{3}{4} g \downarrow$$

$$\overset{+}{\rightarrow} \Sigma F_x = m \bar{a}_x \qquad A_x = 0$$

$$+\downarrow \Sigma F_y = m \bar{a}_y \qquad W - A_y = m \bar{a}$$

$$mg - A_y = m \cdot \frac{3}{4} g \quad \text{or} \quad A_y = \frac{1}{4} mg = \frac{1}{4} W \uparrow$$

7-34

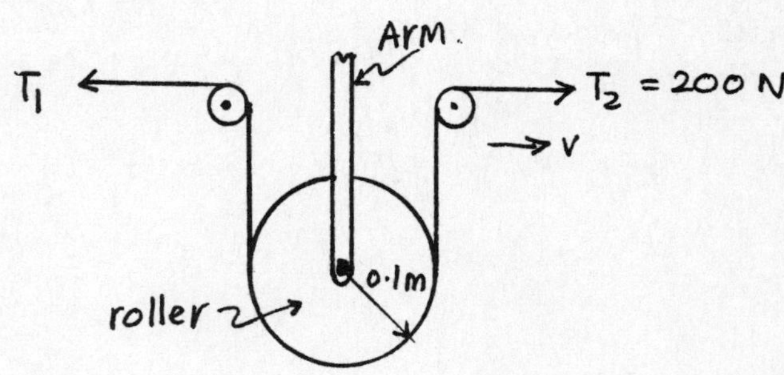

A belt-tensioning device is made of a tension roller of radius 0.1 m that is supported on frictionless bearings in the carriage arm. The drive side of the belt is maintained at tension T_2 = 200 N. If there is belt slippage when the roller is engaged and the arm held stationary, determine the initial acceleration of the roller.

Polar moment of inertia of the roller J = 0.1 kgm^2.

Coefficient of kinetic friction between belt and roller μ_k = 0.2.

****

Use the belt-friction equation $\frac{T_2}{T_1} = e^{\mu\theta}$

Since slipping $\mu = \mu_k = 0.2$

Angle of contact $\theta = \pi$ for the given configuration.

With $T_2 = 200$ N, $T_1 = \frac{200}{e^{0.2\times\pi}} = 106.7$ N

Note: Bearing is held stationary.

Apply Newton's 2nd law for the roller:

Σ Torque = J x angular acceleration

$0.1\,T_2 - 0.1\,T_1 = 0.1\,\alpha$

Hence $\alpha = (200 - 106.7)\frac{0.1}{0.1}$ rad/s^2

$\alpha = \underline{93.3}$ rad/s^2

7-35

The bar shown rotates about a fixed axis at C. The bar weighs 128.8 lbs. When the bar is in the position shown, it has an angular velocity of 4 rad/sec and an angular acceleration of 20 rad/sec^2, both counterclockwise.

A) Determine the force P.
B) Determine the pin reaction on the bar at C.

P
5'
3
4
1'
C
200#

****

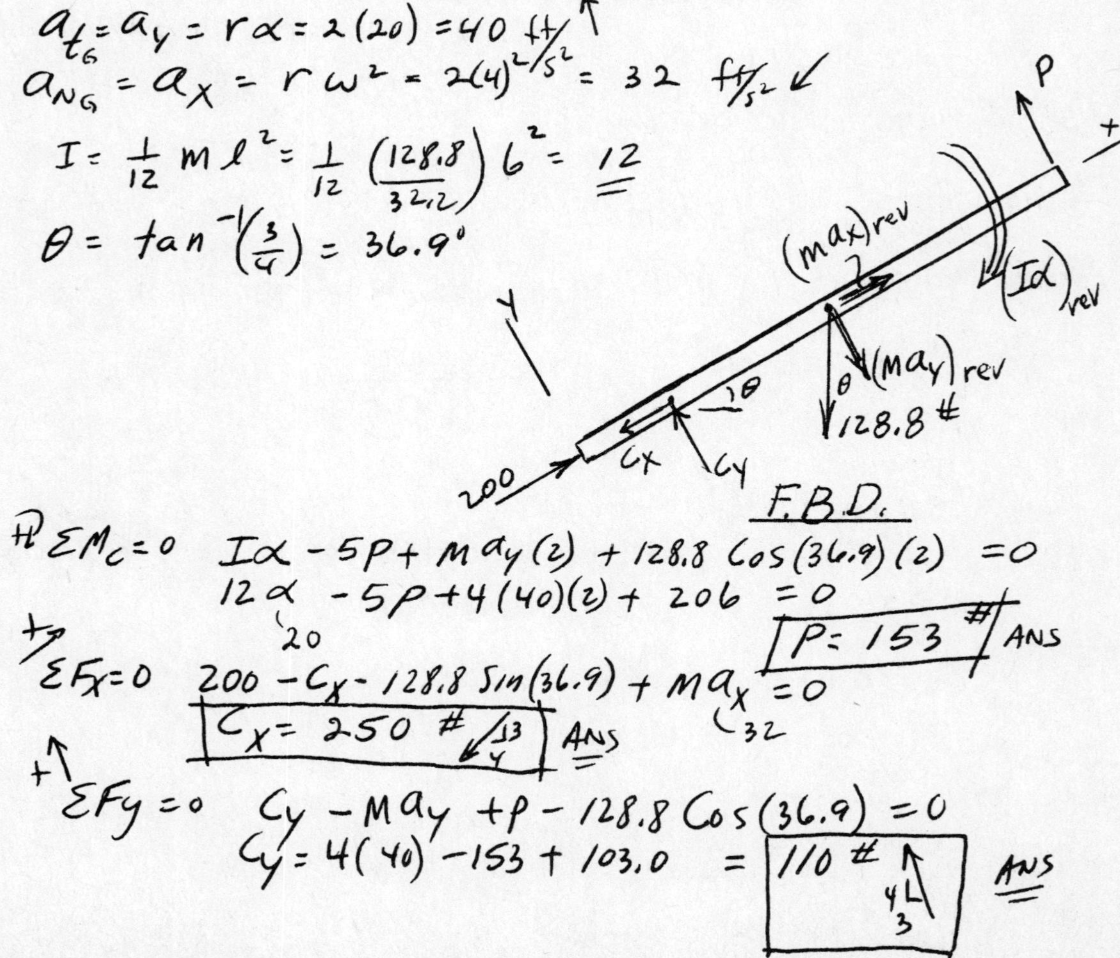

7-36

A warehouse door weighing 300 pounds is suspended by two rollers from a horizontal track as shown below. Due to a malfunction the left wheel slides on the track instead of rolling freely. The coefficient of friction between the track and the left wheel is 0.30. The right wheel rolls freely and may be assumed frictionless. Determine the horizontal force, P, required for the door to be uniformly accelerated at 2 ft/sec^2.

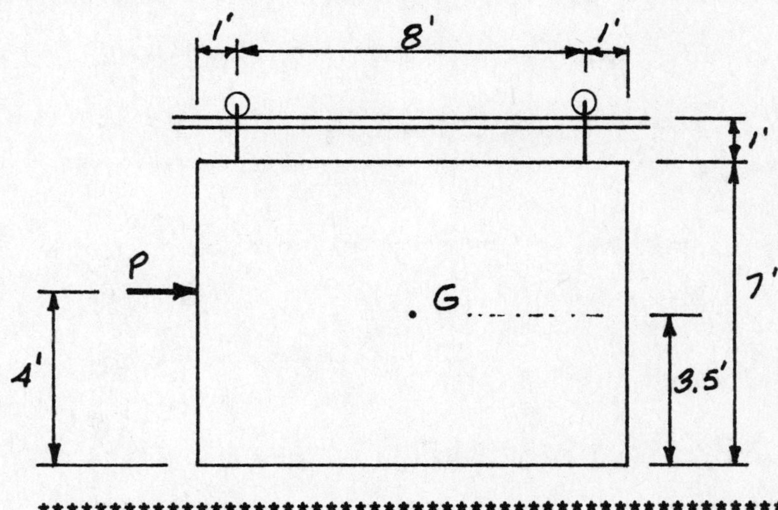

****

FREE BODY DIAGRAM:

$$\xrightarrow{+}\ \Sigma F_x = m(a_G)_x$$

$$-.3N_1 + P = \frac{300}{32.2}(2)$$

$$P = .3N_1 + 18.63 \text{ lb}$$

$$+\uparrow\ \Sigma F_y = m(a_G)_y^{\,0} = 0 \qquad N_1 + N_2 - 300 = 0$$

$$N_2 = 300 - N_1$$

$$\overset{+}{\curvearrowright}\ \Sigma M_G = 0 \qquad P(.5) - .3N_1(4.5) + N_1(4) - N_2(4) = 0$$

$$(.3N_1 + 18.63)(.5) - 1.35N_1 + 4N_1 - (300 - N_1)(4) = 0$$

$$N_1 = 175.1 \text{ lb}$$

$$P = .3N_1 + 18.63 = .3(175.1) + 18.63$$

$$\underline{\underline{P = 71.2 \text{ lb}}}$$

7-37

A 20-lb carriage moves towards an obstruction with a velocity of 40 ft/sec. Calculate the angular velocity and the velocity of the mass center of the carriage after the impact. Determine the impulse during this impact. Assume that the motion is frictionless and the impact is perfectly plastic. Neglect the height of the carriage from the surface.

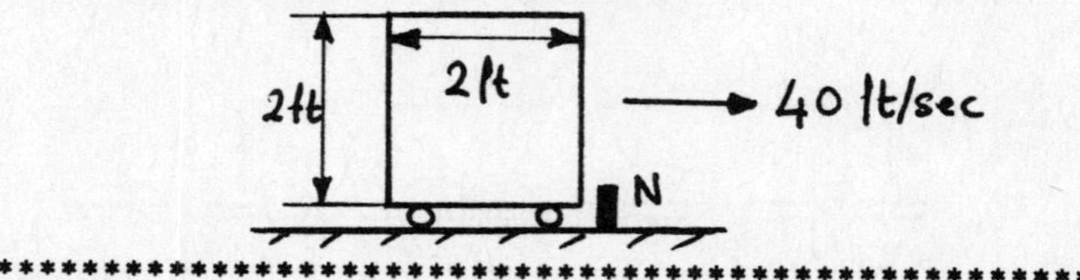

****

The principle of impulse and momentum for a solid body in two dimensional motion :

$$m\vec{V}_{G_1} + \Sigma\vec{F}.\Delta t = m\vec{V}_{G_2} \qquad (1)$$

$$I_G\vec{\omega}_1 + \Sigma\vec{M}_G.\Delta t = I_G\vec{\omega}_2 \qquad (2)$$

$l_1 = \frac{l}{2}$ G mV_{G_1} N + G W N $W.\Delta t \approx 0$ $e = 0$ N_x N_y = mV_{G_2} G α d α N

$\alpha = 45°$
$d = l/\sqrt{2}$

Eqn.(1), when applied, gives (plastic impact, e=0, means the corner N will remain in the obstruction):

$$x : \quad mV_{G_1} - N_x.\Delta t = m.V_{G_2}.\sin\alpha \qquad (1'_x)$$

$$y : \quad 0 + N_y.\Delta t = mV_{G_2}.\cos\alpha \qquad (1'_y)$$

$$\text{Eqn.}(2) : \quad 0 + \left(N_x.\frac{l}{2}.\Delta t - N_y.\frac{l}{2}.\Delta t\right) = I_G.\omega_2 \qquad (2')$$

$$I_G = \frac{1}{12}m(2l^2) = \frac{ml^2}{6} \qquad \therefore\ I_G.\omega_2 = \frac{ml^2}{6}.\omega_2$$

Also, from kinematics : $V_{G_2} = d.\omega_2 = \frac{l}{\sqrt{2}}.\omega_2$ can be written. Introducing these into Eqn. (2'), we can get

$$\omega_2 = \frac{3V_{G_1}}{4l} \quad . \quad \therefore\ V_{G_2} = d.\omega_2 = \frac{l}{\sqrt{2}}.\frac{3V_{G_1}}{4l} = \frac{3V_{G_1}}{4\sqrt{2}}$$

CHECK: Moment equation can be applied about N; then,

$$I_G.\vec{\omega}_1 + \Sigma \vec{M}_N.\Delta t + (m\vec{V}_{G_1}.l_1) = I_G.\vec{\omega}_2 + (m\vec{V}_{G_2}.l_2)$$

where ( ) terms represent moment of linear momentum.

$$\therefore \quad 0 + 0 + mV_{G_1}.\frac{l}{2} = \frac{ml^2}{6}.\omega_2 + mV_{G_2}.d$$

$$d = l/\sqrt{2} \quad , \quad V_{G_2} = l\omega_2/\sqrt{2}$$

$\therefore$ we can obtain $\omega_2 = \dfrac{3V_{G1}}{4l}$

If we substitute the values given:

$$\omega_2 = \frac{3\times 40}{4\times 2} = 15 \text{ rad/sec}$$

$$V_{G_2} = \frac{3\times 40}{4\sqrt{2}} = 21.21 \text{ ft/sec}$$

$$\therefore N_x.\Delta t = mV_{G_1} - mV_{G_2}\sin 45° = \frac{20}{32.2}\left(40 - \frac{21.21}{\sqrt{2}}\right) = 15.52 \text{ lb-sec}$$

$$N_y.\Delta t = mV_{G_2}.\cos 45° = \frac{20}{32.2}\times\frac{21.21}{\sqrt{2}} = 9.31 \text{ lb-sec}$$

$$(N_x.\Delta t)^2 + (N_y.\Delta t)^2 = (N.\Delta t)^2$$

$$15.52^2 + 9.31^2 = N^2\Delta t^2 \rightarrow N.\Delta t = 18.1 \text{ lb-sec}$$

7-38

A cylinder of radius 1 ft and weight 32.2 lbs is rolling without slipping with a constant clockwise angular velocity of 2 rad/sec. when, at t= 0, a force P is applied at the center of the cylinder. The variation of P with time is shown below. It is assumed that the cylinder continues to roll without slipping. Determine the angular velocity of the cylinder for

a. t= 1 sec
b. t= 5 sec
c. t= 10 sec

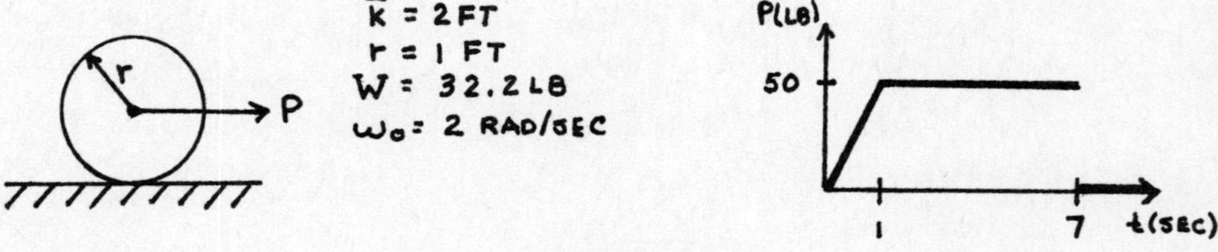

****

THE VARIABLES INVOLVED IN THIS PROBLEM ARE FORCE, VELOCITY, AND TIME. THUS AN IMPULSE MOMENTUM TECHNIQUE IS MOST SUITABLE FOR THE SOLUTION OF THE PROBLEM. SINCE THE CYLINDER IS ASSUMED TO ROLL WITHOUT SLIP

$$\bar{v} = r\omega$$

ALSO

$$\bar{I} = \bar{k}^2 m = (2\text{ FT})^2(1\text{ SLUG}) = 4\text{ SLUG-FT}^2$$

$\bar{I}\omega_o$, $m\bar{v} = mr\omega_o$ + $\int_0^t Wdt$, $\int_0^t Pdt$, $\int_0^t Fdt$, $\int_0^t Ndt$ = $\bar{I}\omega(t)$, $m\bar{v}(t) = mr\omega(t)$

SYSTEM MOMENTA AT t=0

SYSTEM EXTERNAL IMPULSES BETWEEN 0 AND TIME t

SYSTEM MOMENTA AT TIME t

THE FOLLOWING PRINCIPLE OF ANGULAR IMPULSE - ANGULAR MOMENTUM CAN BE WRITTEN FOR THE CONTACT POINT C:

ANGULAR MOMENTUM ABOUT C AT t=0 + SYSTEM EXTERNAL ANGULAR IMPULSES ABOUT C BETWEEN 0 AND TIME t = ANGULAR MOMENTUM ABOUT C AT TIME t

OR

$$\bar{I}\omega_0 + (mr\omega_0)r + r\int_0^t p\,dt = \bar{I}\omega(t) + (mr\omega(t))r$$

OR

$$\omega(t) = \frac{(\bar{I} + mr^2)\omega_0 + r\int_0^t p\,dt}{\bar{I} + mr^2}$$

$\int_0^t p\,dt$ CAN BE EVALUATED EITHER ANALYTICALLY OR GEOMETRICALLY. GEOMETRICALLY IT CAN BE INTERPRETED AS THE AREA UNDER THE P VS t CURVE BETWEEN TIME 0 AND TIME t

$$\int_0^1 P\,dt = \tfrac{1}{2}(50\text{ LB})(1\text{ SEC}) = 25\text{ LB-SEC}$$

$$\int_0^5 P\,dt = 25\text{ LB-SEC} + (50\text{ LB})(4\text{ SEC}) = 225\text{ LB-SEC}$$

$$\int_0^{10} P\,dt = 25\text{ LB-SEC} + (50\text{ LB})(6\text{ SEC}) = 325\text{ LB-SEC}$$

THEN $\bar{I} + mr^2 = 4\text{ SLUG-FT}^2 + (1\text{ SLUG})(1\text{FT}^2) = 5\text{ SLUG-FT}^2$

AND

$$\omega(1) = \frac{1}{5\text{ SLUG-FT}^2}\left[5\text{ SLUG-FT}^2\left(2\,\frac{\text{RAD}}{\text{SEC}}\right) + 1\text{FT}(25\text{ LB-SEC})\right]$$
$$= 7\text{ RAD/SEC}$$

$$\omega(5) = \frac{1}{5\text{ SLUG-FT}^2}\left[5\text{ SLUG-FT}^2\left(2\,\frac{\text{RAD}}{\text{SEC}}\right) + 1\text{ FT}(225\text{ LB-SEC})\right]$$
$$= 47\text{ RAD/SEC}$$

$$\omega(10) = \frac{1}{5\text{ SLUG-FT}^2}\left[5\text{ SLUG-FT}^2\left(2\,\frac{\text{RAD}}{\text{SEC}}\right) + 1\text{FT}(325\text{ LB-SEC})\right]$$
$$= 67\text{ RAD/SEC}$$

7-39

The pulley system indicated has mass 4 kg. and a radius of gyration of 1.5 meters. Find the tensions in the cords and the angular acceleration of the pulleys when the masses are released.

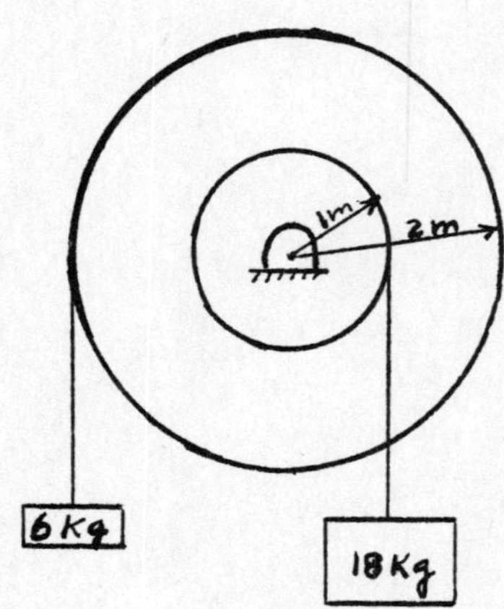

****

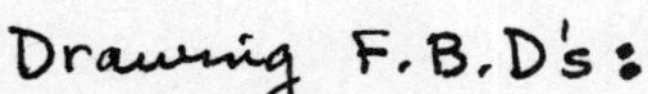

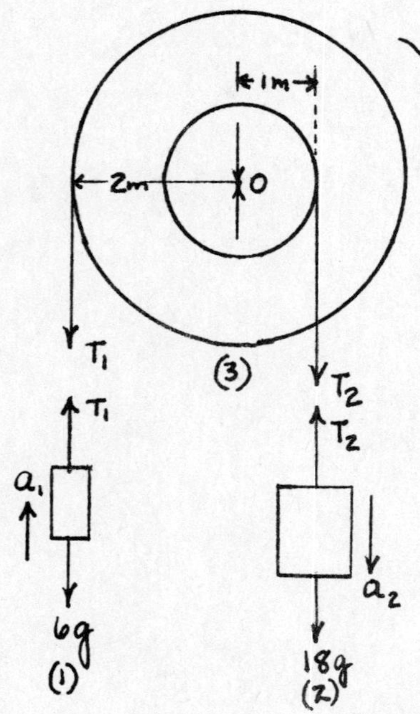

We observe: $a_1 = r_1\alpha = 2\alpha$ $(*_1)$

$a_2 = r_2\alpha = 1\alpha$ $(*_2)$

From (1): $+\uparrow \Sigma F = T_1 - 6g = 6a_1 \overset{(*_1)}{=} 12\alpha$

or $T_1 = 6g + 12\alpha$ $(\not{*}_1)$

From (2): $+\downarrow \Sigma F = 18g - T_2 = 18a_2 \overset{(*_2)}{=} 18\alpha$

or $T_2 = 18g - 18\alpha$ $(\not{*}_2)$

From (3): $\curvearrowright \Sigma M = I\alpha$

$T_2(1) - T_1(2) = 4(1.5)^2\alpha$

$T_2 - 2T_1 = 9\alpha$

using $\not{*}_1, \not{*}_2$: $18g - 18\alpha - 2(6g + 12\alpha) = 9\alpha$

$6g - 42\alpha = 9\alpha$

$\Rightarrow \quad 6g = 51\alpha$

or $\alpha = \frac{6g}{51} = \frac{6(9.8)}{51}$

$\alpha = 1.15 \text{ rad/sec}^2$

$\not{*}_1 \Rightarrow \therefore T_1 = 6(9.8) + 12(1.15) = 72.6 \text{ (N)}$

$\not{*}_2 \Rightarrow \therefore T_2 = 18(9.8) - 18(1.15) = 155.7 \text{ (N)}$

7-40

A 50-kg rod has its mass center at G and rotates around the point A as shown. Determine the angular acceleration of the rod at the instant shown. Assume the radius of gyration of the rod to be 1.16 m.

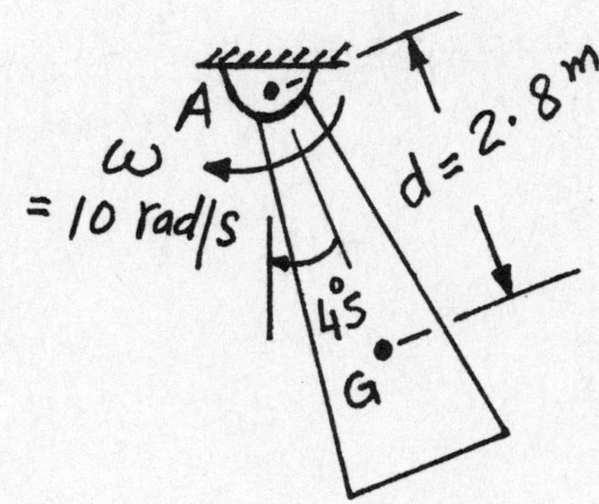

****

Mass moment of rod about $A = \sum M_A = mg\sin 45^\circ \cdot d$

Moment of inertia of rod about $A = I_A = m(k_G^2 + d^2)$

Also, $\sum M_A = I_A \alpha$

or, $mg\sin 45^\circ \cdot d = m(k_G^2 + d^2)\alpha$

or, angular acceleration $= \alpha = \dfrac{g\sin 45^\circ \cdot d}{(k_G^2 + d^2)}$

or, $\alpha = \dfrac{9.81 \times 0.707 \times 2.8}{(1.16^2 + 2.8^2)} = 2.1 \text{ rad/sec}^2$

PLANE MOTION: WORK AND KINETIC ENERGY

7-41

A slender rod of length 2 feet and mass 3 slugs is released from rest in the horizontal position and swings freely. Find:

(a) the angular velocity of the rod when it passes a vertical position

(b) the reaction at A at this instant.

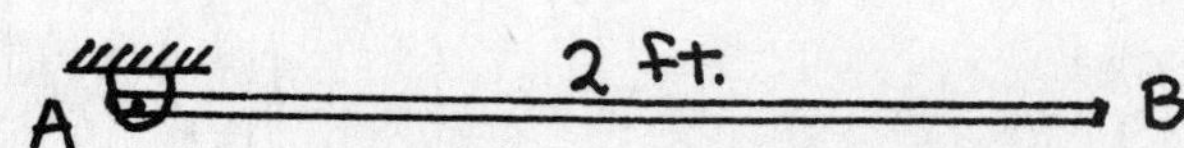

****

(a)

A, G_1, B, G_2, B

BY CONSERVATION OF ENERGY, WITH DATUM IN ORIGINAL POSITION:

$$V_1 + T_1 = V_2 + T_2$$

$$0 + 0 = -3g \cdot 1 + \frac{1}{2} \cdot 3 \cdot \omega^2 + \frac{1}{2}\left[\frac{1}{12} \cdot 3 \cdot 2^2\right]\omega^2$$

$$\omega^2 = 48.3$$

$$\omega = 6.95 \text{ rad/sec}$$

(b) FBD OF ROD

R, 3g

$$(a_G)_y = 1 \cdot \frac{3}{2} g = \frac{3}{2} g \uparrow$$

$$\Sigma F_y = m (a_G)_y$$

$$R - 3g = 3\left(\frac{3}{2} g\right)$$

$$R = \left(\frac{9}{2} + 3\right) g = 241.5 \text{ lb.}$$

7-42

Gears A and B are initially at rest when a moment M is applied to gear A. The magnitude of M is

$$M = \frac{1}{\pi}\left(3\theta_A^{\,2} + 4\pi^2\right) \text{ ft-lb}$$

where θ_A is the amount of rotation of gear A from its initial position measured in radians. Gear A may rotate freely but the support of gear B is not well lubricated and produces a constant frictional moment of $3\pi/2$ ft-lb opposing the rotation of gear B. Determine the angular velocity of gear A after it has completed 1 revolution.

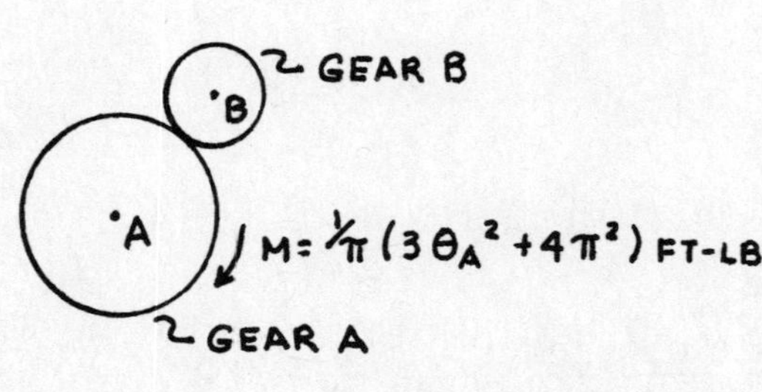

$r_A = 1$ FT
$\bar{k}_A = \sqrt{6}$ FT
$W_A = 64.4$ LB

$r_B = \frac{1}{2}$ FT
$\bar{k}_B = \sqrt{2}$ FT
$W_B = 32.2$ LB

****

KINEMATICS:

$$r_A\theta_A = r_B\theta_B \qquad \theta_B = \frac{r_A}{r_B}\theta_A = \frac{1\text{ FT}}{\frac{1}{2}\text{ FT}}\theta_A = 2\theta_A$$

$$r_A\omega_A = r_B\omega_B \qquad \omega_B = 2\omega_A$$

THE PROBLEM INVOLVES MOMENTS AND ANGULAR DISTANCES THUS IT IS BEST SOLVED BY A WORK-ENERGY ANALYSIS. THE SYSTEM IS NOT CONSERVATIVE DUE TO THE FRICTIONAL MOMENT ON GEAR B.

PRINCIPLE OF WORK ENERGY FOR THE SYSTEM COMPOSED OF BOTH GEARS.

$$T_1 + U_{1\to2} = T_2$$

WHERE T_1 = INITIAL KINETIC ENERGY OF THE SYSTEM = 0

T_2 = KINETIC ENERGY OF THE SYSTEM AFTER GEAR A HAS COMPLETED ONE REVOLUTION

$$= \frac{1}{2}\bar{I}_A\omega_A^2 + \frac{1}{2}\bar{I}_B\omega_B^2$$

$$= \frac{1}{2}(\sqrt{6}\text{ FT})^2(2\text{ SLUGS})\omega_A^2 + \frac{1}{2}(\sqrt{2}\text{ FT})^2(1\text{ SLUG})(2\omega_A)^2$$

$$= 10\,\omega_A^2\text{ SLUGS-FT}^2$$

$U_{1\to2}$ = WORK DONE BY EXTERNAL FORCES AND MOMENTS DURING THE TIME OF INTEREST

$$= \int M\,d\theta_A - \int M_F\,d\theta_B$$

$$= \int_0^{2\pi\text{ RAD}} \frac{1}{\pi}(3\theta_A^2 + 4\pi^2)\text{ FT-LB}\,d\theta_A - \int_0^{4\pi}\frac{3\pi}{2}\text{ FT-LB}\,d\theta_B$$

$$= \left[\frac{1}{\pi}(\theta_A^3 + 4\pi^2\theta_A)\Big|_0^{2\pi} - \frac{3\pi}{2}(4\pi)\right]\text{ FT-LB}$$

$$= 10\pi^2\text{ FT-LB}$$

THUS

$$10\,\omega_A^2\text{ SLUGS-FT}^2 = 10\pi^2\text{ FT-LB}$$

$$\omega_A = \pi\text{ RAD/SEC}$$

7-43

A slender rod of mass m and length L is supported at A by a pin and at B by a wire. The wire at B is suddenly cut and the rod rotates about A. After the rod has rotated through an angle θ determine the reactions at A.

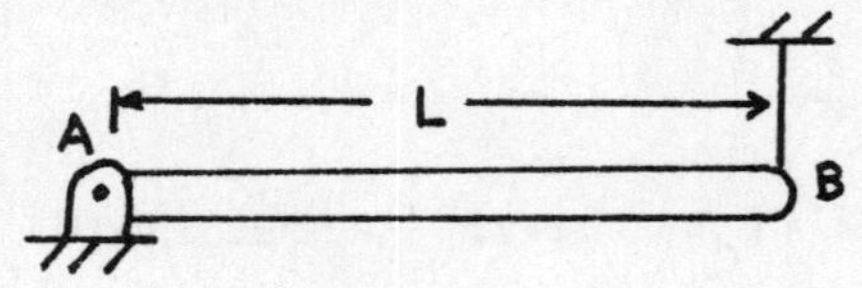

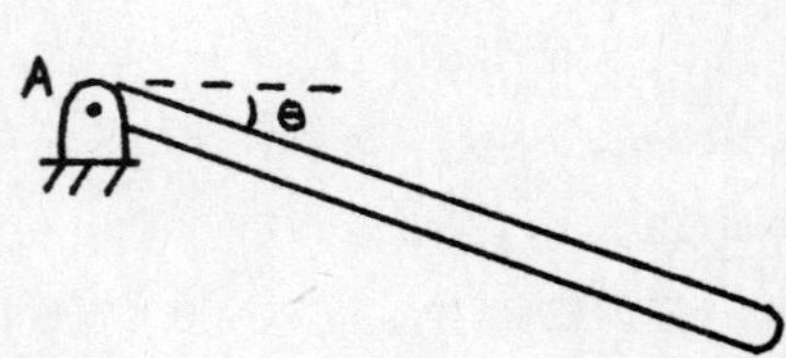

****

FREE BODY DIAGRAM OF ROD

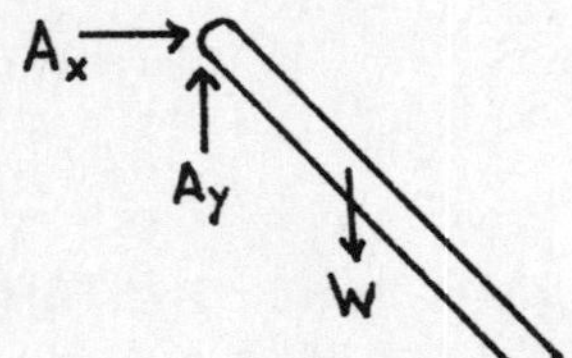

EXTERNAL FORCES

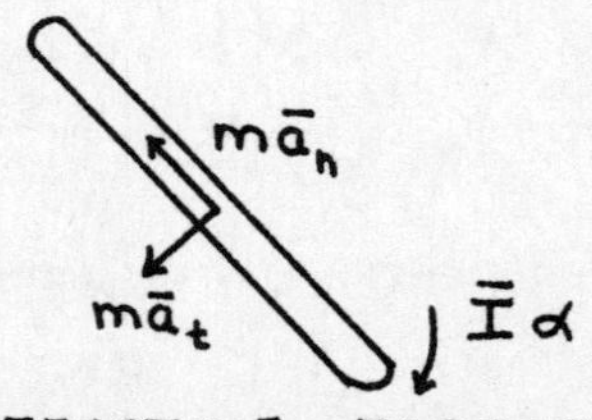

EFFECTIVE FORCES

D'ALEMBERT'S PRINCIPLE:

$$(\Sigma \overset{+}{\curvearrowright} M_A)_{ext} = (\Sigma \overset{+}{\curvearrowright} M_A)_{eff}$$

$$\frac{WL}{2}\cos\theta = m\bar{a}_t \frac{L}{2} + \bar{I}\alpha \qquad (1)$$

$$(\Sigma F_x)_{ext} = (\Sigma F_x)_{eff}$$

$$A_x = -m\bar{a}_n\cos\theta - m\bar{a}_t\sin\theta \qquad (2)$$

$$(\Sigma F_y)_{ext} = (\Sigma F_y)_{eff}$$

$$A_y - W = m\bar{a}_n\sin\theta + m\bar{a}_t\cos\theta \qquad (3)$$

KINEMATICS:

$$\bar{a}_t = \frac{L}{2}\alpha \qquad (4)$$

$$\bar{a}_n = \frac{L}{2}\omega^2 \qquad (5)$$

SUBSTITUTING (4) INTO (1) YIELDS

$$\frac{WL}{2}\cos\theta = \left(\frac{1}{12}mL^2 + m\frac{L^2}{4}\right)\alpha = \frac{mL^2}{3}\alpha$$

$$\alpha = \frac{3g}{2L}\cos\theta \qquad (6)$$

HOWEVER TO FIND A_y AND A_x ONE NEEDS TO CALCULATE ω FROM A WORK-ENERGY ANALYSIS

$$T_1 + U_{1\to 2} = T_2$$

T_1 = KINETIC ENERGY OF SYSTEM IN ITS INITIAL POSITION = 0

T_2 = KINETIC ENERGY OF SYSTEM AFTER IT HAS ROTATED THROUGH ANGLE θ

$$= \frac{1}{2}\bar{I}\omega^2 + \frac{1}{2}m\bar{v}^2 = \frac{1}{2}\left[\frac{1}{12}mL^2\omega^2 + m\frac{L^2}{4}\omega^2\right]$$

$$= \frac{1}{6}mL^2\omega^2$$

$U_{1\to 2}$ = WORK DONE BY ALL EXTERNAL FORCES BETWEEN POSITION 1 AND POSITION 2

$$= W\frac{L}{2}\sin\theta$$

THUS

$$W\frac{L}{2}\sin\theta = \frac{1}{6}mL^2\omega^2$$

$$\omega = \left(\frac{3g}{L}\sin\theta\right)^{1/2} \qquad (7)$$

SUBSTITUTING (4)-(7) INTO (2) AND (3) YIELDS

$$A_x = -m\frac{L}{2}\left(\frac{3g}{L}\sin\theta\right)\cos\theta - m\frac{L}{2}\left(\frac{3g}{2L}\cos\theta\right)\sin\theta = \frac{-9mg}{4}\sin\theta\cos\theta$$

$$A_y = mg + \frac{mL}{2}\left(\frac{3g}{L}\sin\theta\right)\sin\theta + \frac{mL}{2}\left(\frac{3g}{2L}\cos\theta\right)\cos\theta$$

7-44

The 100-lb pulley has a centroidal radius of gyration of $k_o = 0.7$ ft as shown. Compute the total kinetic energy of the system if the pulley is rotating at a constant angular velocity of 30 rad/s in an anticlockwise direction. Assume the load cable does not slip on the pulley.

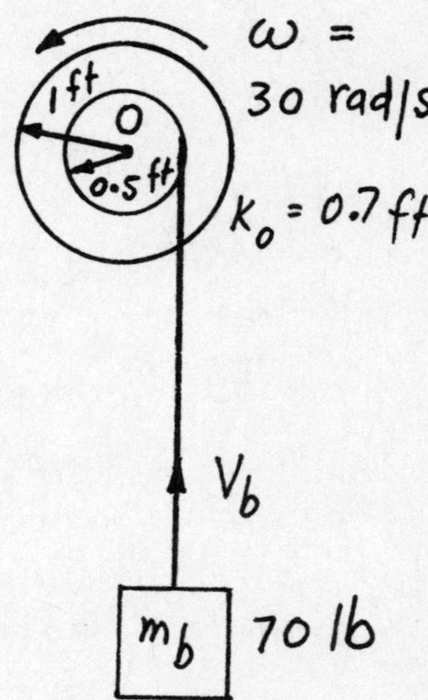

****

Total kinetic energy of the system = Rotational energy of pulley + translational energy of 70 lb mass

or, $$T.K.E. = \frac{1}{2} I_o \omega^2 + \frac{1}{2} m_b V_b^2$$

$$= \frac{1}{2} m_p k_o^2 \omega^2 + \frac{1}{2} m_b V_b^2$$

$$= \frac{1}{2} \frac{100}{32.2} (0.7)^2 (30)^2 + \frac{1}{2} \frac{70}{32.2} (0.5 \times 30)^2$$

$$= 930 \text{ ft-lb}$$

7-45

The brake drum shown in the figure rotates initially with 100 rpm in the clockwise direction. The moment of inertia of the drum is 10 kg.m^2. The drum is stopped by applying a force of 200 N on the brake. How many revolutions will be completed before this drum stops? Note: μ_k=0.38 .

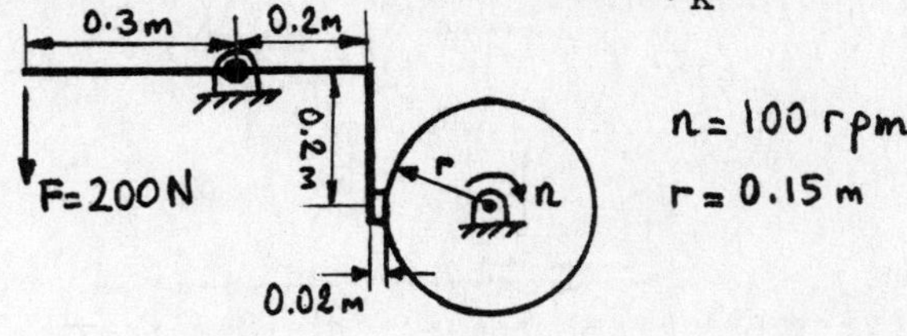

****

Energy method can be applied in this problem : Considering firstly the disk,

$$T_1 + U_{12} = T_2$$

Angular speed $\omega_1 = \frac{2\pi n_1}{60} = \frac{2\pi \times 100}{60} = 10.47$ rad/sec

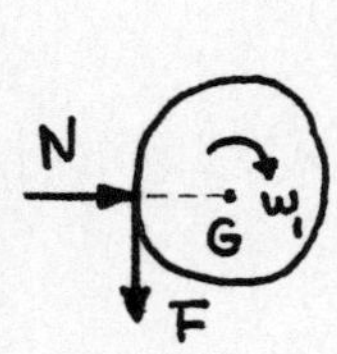

Frictional force $F = \mu_k N$

Initial kinetic energy $T_1 = \frac{1}{2} I_G . \omega_1^2 = \frac{1}{2} \times 10 \times 10.47^2 = 548.3 \ kg \frac{m^2}{sec^2}$

The work done : $U_{12} = -M . \Delta\theta = -F . r . \Delta\theta = -\mu_k . N . r . \Delta\theta = -0.38 \times 0.15 \times N . \Delta\theta$

$\therefore \quad 548.3 - 0.057 \ N . \Delta\theta = 0 \quad \rightarrow N . \Delta\theta = 9619.55$

Now, consider the brake : moment about K

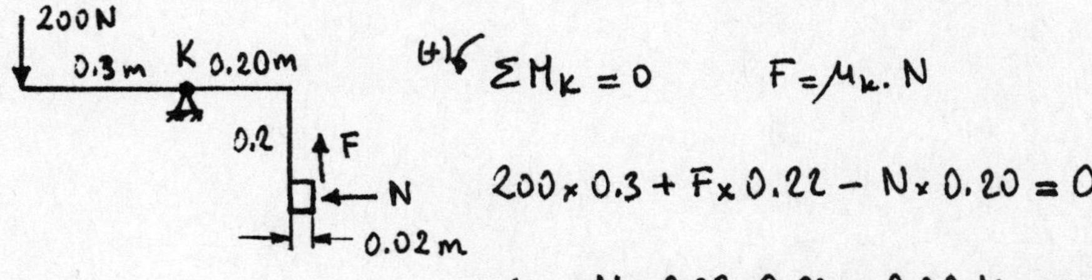

$(+\circlearrowright) \ \Sigma M_K = 0 \qquad F = \mu_k . N$

$$200 \times 0.3 + F \times 0.22 - N \times 0.20 = 0$$

$$60 + N \times 0.38 \times 0.22 - 0.20 \ N = 0 \rightarrow N = 515.46 N$$

$N . \Delta\theta = 9619.55 \quad \rightarrow 515.46 \times \Delta\theta = 9619.55$

$\Delta\theta = 18.66$ rad $= 18.66/2\pi = 2.97$ revolutions

7-46

The surface below the disk moves in the right direction with a constant velocity U. The disk is of constant thickness and is initally at rest. It is dropped on to the surface. After two revolutions, the disk starts rotating with a constant rpm. What is the velocity U of the bottom surface ?

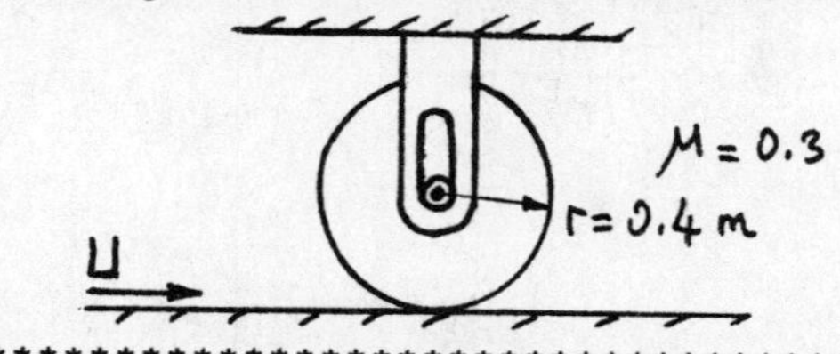

****

$$\Sigma F_y = 0 = W - N \rightarrow N = W$$

$$\Sigma M_G = I_G . \alpha \rightarrow r.F = I_G . \alpha \qquad F = \mu N = \mu . mg$$

$$r . \mu mg = \left(\tfrac{1}{2} m r^2\right) . \alpha$$

$$\alpha = \frac{2 \mu g}{r} = \frac{2 \times 0.3 \times 32.2}{0.4} = 48.3 \text{ rad/sec}^2$$

At point P (the point of contact), velocities will be same.

$\therefore \omega = U/r$ for the disk.

$$\omega^2 = \omega_0^2 + 2 . \alpha . \Delta\theta = (U/r)^2 \qquad \omega_0 = 0 : \text{initially disk is not rotating.}$$

$$0 + 2 \times 48.3 \times (2 \times 2\pi) = U^2 / 0.4^2 \rightarrow U = 13.94 \text{ m/s}$$

CHECK : This problem can be solved in a different way by using the energy method :

$$T_1 + U_{12} = T_2 \ , \quad T_1 = 0 \quad \text{initially disk is at rest.}$$

$$M . \Delta\theta = \frac{1}{2} I_G . \omega^2 \qquad M : \text{moment}$$

$$F . r . \Delta\theta = \frac{1}{2} \left(\frac{1}{2} m r^2\right) . \omega^2 = \frac{1}{2} \left(\frac{m r^2}{2}\right) \frac{U^2}{r^2} \qquad \text{since } \omega = U/r .$$

$$F = \mu . N = \mu . mg$$

$$(\mu . m . g) . r . \Delta\theta = \frac{m}{4} U^2 \rightarrow U^2 = 4 \mu r g . \Delta\theta$$

$$U^2 = 4 \times 0.3 \times 0.4 \times 32.2 \times (2 \times 2\pi)$$

$$U = 13.94 \text{ m/sec} .$$

7-47

The homogeneous bar AB weighs 96 pounds and has an angular velocity of 2.0 radians per second clockwise in the position shown. As the bar rotates it strikes the spring, which is initially uncompressed. It then comes to rest in the vertical position. Determine the spring modulus.

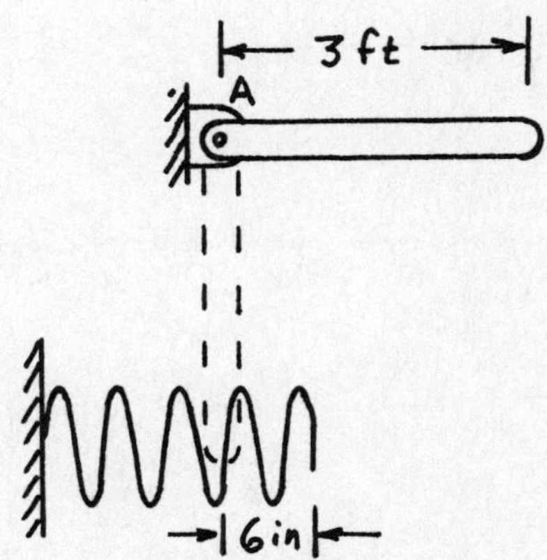

****

Work of weight $= mg\Delta h = 96\text{ lb}(1.5\text{ ft}) = +144\text{ ft-lb}$

Work of spring $= -\frac{1}{2}kS^2\Big|_{S=0}^{1/2} = -\left(\frac{1}{2}\right)K\left(\frac{1}{4}\right) = -\frac{K}{8}\text{ ft-lb}$

Kinetic energy $= \frac{1}{2}mV^2 + \frac{1}{2}I_G\omega^2$ or $\frac{1}{2}I_A\omega^2$

where $I_G = \frac{mL^2}{12}$ and $I_A = \frac{mL^2}{3} = 9$

$K.E._i = \frac{1}{2}(9)(2)^2 = 18$, $K.E._f = \frac{1}{2}(9)(0)^2$

Work $= \Delta$ K.E.

$144 - \frac{K}{8} = 0 - 18$

$$K = 1296\text{ lb/ft}$$

7-48

The spring has an initial tension of 20 lb. in the at rest position shown. The modulus of the spring is 20 lb/ft. The homogeneous bar AB is welded to the 100 lb homogeneous bar BC at B.

Determine the angular velocity of bar AB after the system rotates 90° clockwise.

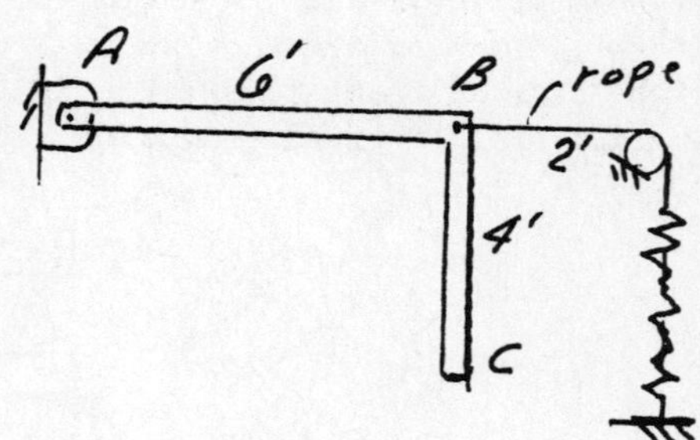

****

Use Work and Energy - Draw initial and final positions

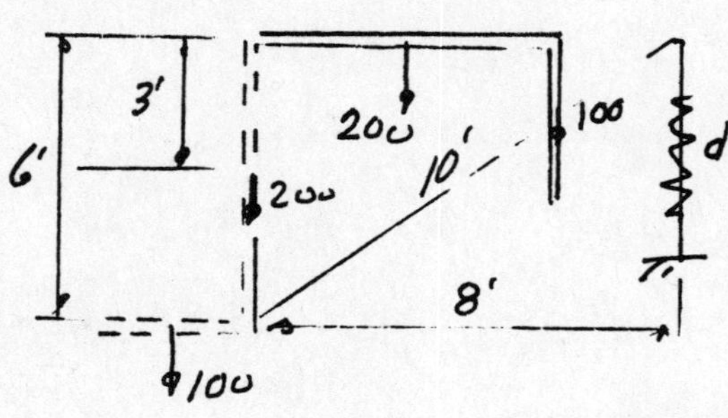

To determine the work done by the spring draw a Force distance diagram

Initial stretch = 20/20 or 1'

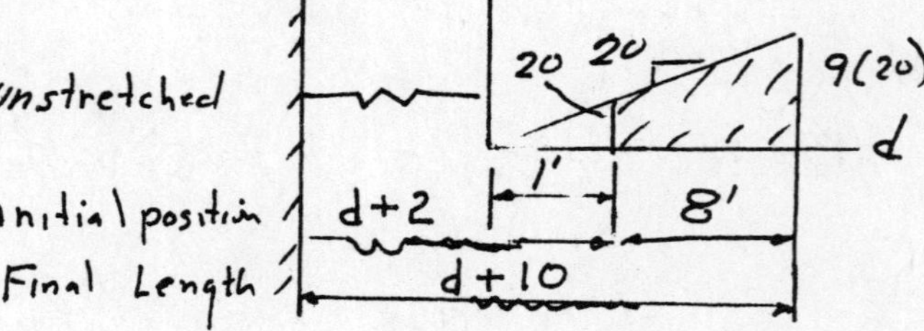

$$U_{1-2} + T_1 = T_2$$

$$200(3) + 100(4) - 800$$

Area = Work of Spring

$$\frac{(20+180)}{2}(8) = 800$$

Since spring holds system back it does negative work

$$+0 = \frac{1}{2}\frac{200}{g}(3\omega)^2 + \frac{1}{2}\frac{1}{12}\frac{200}{g}6^2\omega^2 + \frac{1}{2}\frac{100}{g}\left(\sqrt{6^2+2^2}\right)^2\omega^2 + \frac{1}{2}\frac{100}{g}\frac{1}{12}4^2\omega^2$$

$$200g = 100(9\omega^2) + 300\omega^2 + 2000\omega^2 + 66.7\omega^2 = 3267\omega^2$$

$$\omega = \sqrt{\frac{200}{3267}g} \text{ or } \underline{1.4 \text{ rad/sec}}$$

7-49

A steel plate 24 inches in diameter and three inches thick has four 2-inch diameter holes equally spaced on an 18-inch diameter circle as shown below. The net weight of the plate is 375 pounds. The plate is rotating about an axis through its center at 400 rpm. Determine the kinetic energy stored in the plate.

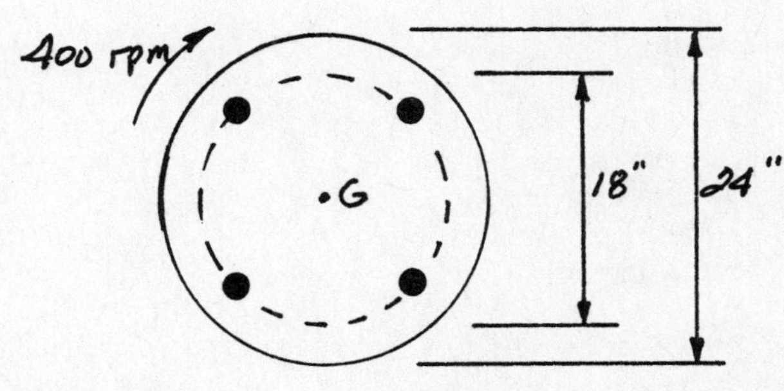

****

$$K.E. = \frac{1}{2} I_G \omega^2 \qquad \omega = 400 \frac{rev}{min} \times \frac{1\ min}{60\ sec} \times \frac{2\pi\ rad}{rev} = 41.89 \frac{rad}{sec}$$

$$I_G = I_G\ (\text{large circle}) - I_G\ (4 \text{ small circles})$$

$$= \frac{1}{2} m_1 R_1^2 - 4\left[\frac{1}{2} m_2 R_2^2 + m_2 d^2\right] = \frac{1}{2} m_1 R_1^2 - 4 m_2 \left[\frac{1}{2} R_2^2 + d^2\right]$$

$$\gamma_{plate} = \frac{\text{weight}}{\text{volume}} = \frac{375\ lb}{3\ in\left[\frac{\pi}{4}(24\ in)^2 - 4 \cdot \frac{\pi}{4}(2\ in)^2\right]} = .2842 \frac{lb}{in^3}$$

$$m_1 = \frac{\frac{\pi}{4}(24\ in)^2 (3\ in)(.2842 \frac{lb}{in^3})}{32.2\ ft/sec^2} = 11.98 \frac{lb\text{-}sec^2}{ft}$$

$$m_2 = \frac{\frac{\pi}{4}(2\ in)^2 (3\ in)(.2842 \frac{lb}{in^3})}{32.2\ ft/sec^2} = .0832 \frac{lb\text{-}sec^2}{ft}$$

$$I_G = \frac{1}{2}\left(11.98 \frac{lb\text{-}sec^2}{ft}\right)(1 ft)^2 - 4\left(.0832 \frac{lb\text{-}sec^2}{ft}\right)\left[\frac{1}{2}\left(\frac{1}{12} ft\right)^2 + \left(\frac{9}{12} ft\right)^2\right]$$

$$= (5.99 - .19)\ lb \cdot sec^2 \cdot ft = 5.80\ lb \cdot sec^2 \cdot ft$$

$$K.E. = \frac{1}{2}(5.80\ lb \cdot sec^2 \cdot ft)\left(41.89 \frac{rad}{sec}\right)^2$$

$$\underline{\underline{K.E. = 5090\ ft \cdot lb}}$$

7-50

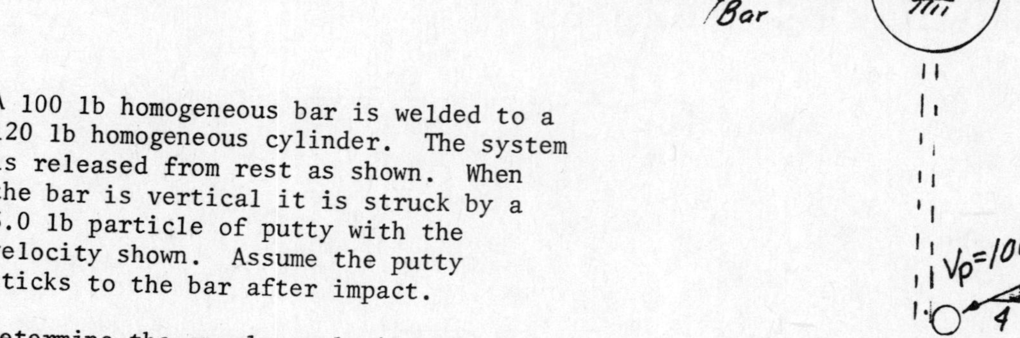

A 100 lb homogeneous bar is welded to a 120 lb homogeneous cylinder. The system is released from rest as shown. When the bar is vertical it is struck by a 3.0 lb particle of putty with the velocity shown. Assume the putty sticks to the bar after impact.

Determine the angular velocity of the bar immediately after impact.

****

Use Work and Energy to determine velocity at impact.

$$V_{1-2} + T_1 = T_2$$

$$100(5) + 0 = \tfrac{1}{2}\tfrac{100}{g}(5\omega)^2 + \tfrac{1}{2}\tfrac{1}{12}\tfrac{100}{g}6^2\omega^2 + \tfrac{1}{2}\tfrac{1}{2}\tfrac{120}{g}2^2\omega^2$$

$$500(2)g = (2500 + 300 + 240)\omega^2$$

$$\omega = \sqrt{\frac{100g}{304}} \text{ or } 3.25$$

Use Impulse and Momentum at impact

O_y, O_x, 100, 3, time $\quad + \quad$ $\tfrac{1}{2}\tfrac{120}{g}2^2(3.25)$, $\tfrac{100}{g}(5)(3.25)$, $\tfrac{1}{12}\tfrac{100}{g}6^2\,3.25$, $\tfrac{3}{g}100$ $\quad = \quad$ $\tfrac{1}{2}\tfrac{120}{g}2^2\omega$, $\tfrac{100}{g}5(\omega)$, $\tfrac{1}{12}\tfrac{100}{g}6^2\omega$, $\tfrac{3}{g}8\omega$

$\circlearrowright + \sum M_O(\text{time})$

$$0 + \tfrac{1}{12}\tfrac{100}{g}6^2(3.25) + \tfrac{100}{g}(5)(3.25)(5) + \tfrac{1}{2}\tfrac{120}{g}(2)^2(3.25)$$

$$-\tfrac{4}{5}\tfrac{3}{g}(100)(8) = \tfrac{300}{g}(\omega) + \tfrac{2500}{g}(\omega) + \tfrac{240}{g}\omega + \tfrac{3}{g}8(8)\omega$$

$$7973 = 3232\omega$$

$$\omega = \underline{\underline{2.47}} \text{ rad/sec} \circlearrowright$$

7-51

Determine the number of revolutions the brake drum will turn while coming to rest from an angular velocity of 800 revolutions per minute. The coefficient of friction between the drum and each of the brake arms is 0.30. The drum weighs 150 pounds and has a centroidal radius of gyration of 0.8 feet. Neglect the thickness of the brake arms.

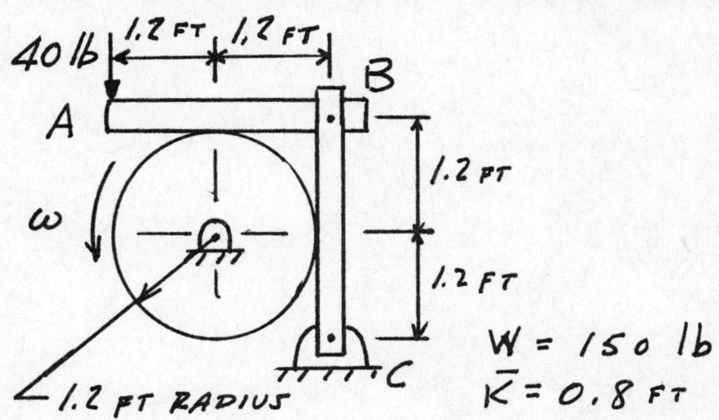

****

ARM A-B: (40; 1.2′, 1.2′; B_y, B_x, F_1, N_1)

$+\circlearrowleft \Sigma M_B = 0 \qquad 40(2.4) - N_1(1.2) = 0$

$N_1 = 80 \text{ lb} \uparrow$

$F_1 = \mu N_1 = 0.3(80) = 24 \text{ lb} \leftarrow$

$\xrightarrow{+} \Sigma F_x = 0 \qquad B_x - 24 = 0 \qquad B_x = 24 \text{ lb} \rightarrow$

ARM B-C: (B_y, $B_x = 24$, 1.2′, 1.2′, N_2, F_2, C_x, C_y)

$+\circlearrowleft \Sigma M_C = 0 \qquad 24(2.4) - N_2(1.2) = 0$

$N_2 = 48 \text{ lb} \rightarrow$

$F_2 = \mu N = 0.3(48) = 14.4 \text{ lb} \uparrow$

DRUM ($F_1 = 24$, $F_2 = 14.4$)

$\bar{I} = m\bar{k}^2 = \frac{W}{g}k^2$

$\bar{I} = \frac{150}{32.2}(0.8)^2 = 2.981 \text{ lb·ft·s}^2$

$800 \text{ RPM} = 83.8 \text{ rad/s}$

WORK-ENERGY

$T_1 + U_{1\to 2} = T_2$

$T_2 = 0$

$T_1 = \frac{1}{2}\bar{I}\omega^2$

$U_{1\to 2} = \Sigma(Fr\theta)$

$\frac{1}{2}\bar{I}\omega^2 - \Sigma(Fr\theta) = 0$

$\frac{1}{2}(2.981)(83.8)^2 - (24 + 14.4)1.2\theta = 0$

$\theta = 227 \text{ rads}$

$\theta = 227 \text{ rads} \times \frac{1 \text{ rev}}{2\pi \text{ rad}} = 36.2 \text{ revs}$

7-52

A circular grinding wheel weighing 200 pounds and 3 feet in diameter is rotating on frictionless bearings at 125 rpm. The wheel is brought to rest by means of a braking lever mechanism as shown below. The radius of gyration of the wheel is 14 inches and the coefficient of friction between the wheel and the braking lever is 0.25. Determine the force, P, applied to the brake lever which will bring the wheel to rest in 10 sec.

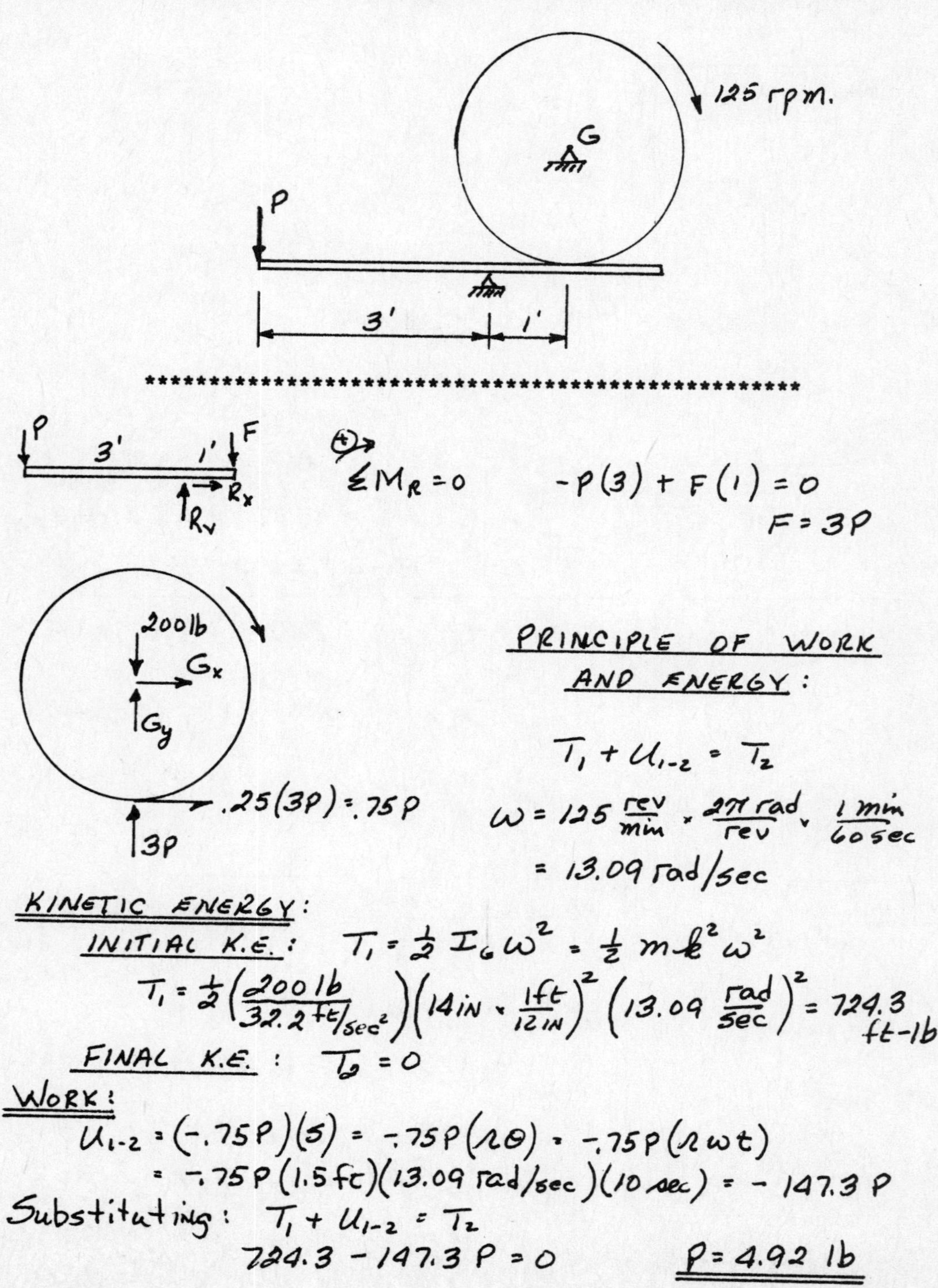

7-53

AB and BC are two identical slender, homogeneous bars of constant cross-section. Each bar weighs W lb. and is 2a ft. long. The end C of bar BC is attached to a slider which slides along a track, and is thus constrained to move vertically downward. If the system is released from rest when A and C are along the horizontal, determine the velocity of points B and C when bar AB is vertical and BC is horizontal. ($I_G = ma^2/3$ for a bar of mass m and length 2a.)

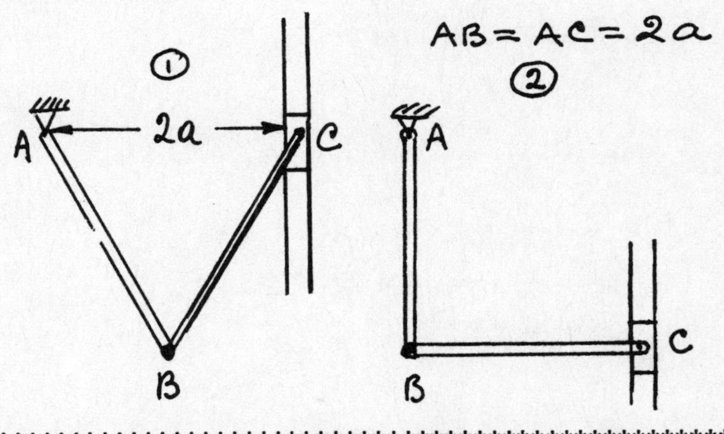

****

Kinematics: Taking datum at AC ($V = 0$) the potential energy of both bars is $V_1 = -\left(mga\frac{\sqrt{3}}{2}\right)2$, the center of gravity of each bar being at the midpoint. $T_1 = 0$ (released from rest). For position 2 the instant center of bar BC is located at B. Hence, $V_B = 0$, $\omega_{AB} = 0$ (extreme point), $T_{AB} = 0$.

$$T_2 = T_{AB} + T_{BC} = T_{BC} = \frac{1}{2}mV_G^2 + \frac{1}{2}I_G\omega^2 \quad \text{for BC}$$

$$T_2 = \frac{1}{2}(\omega a)^2 + \frac{1}{2}\left(\frac{1}{3}ma^2\right)\omega^2 = \frac{2}{3}ma^2\omega^2$$

$$V_2 = V_{AB} + V_{BC} = -mga - mg(2a) = -3mga$$

Conservation of energy gives $\omega^2 = \frac{3}{2}\frac{g}{a}(3-\sqrt{3})$

Finally,

$$\underline{\underline{V_B = 0}} \qquad \underline{V_c} = 2\omega a = \underline{\underline{2.76\sqrt{ga}}}$$

7-54

A circular plate of radius R is retained in the vertical position by pins A and B. Pin A is suddenly removed. Determine the angular velocity of the plate as its diameter BA passes through the vertical position.

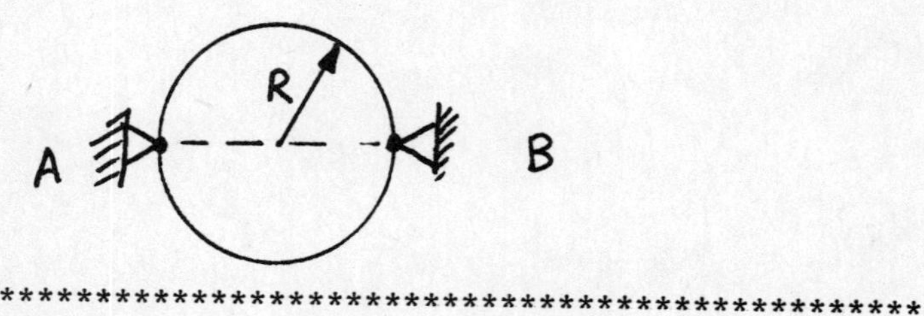

****

$$T_1 + V_1 = T_2 + V_2$$

$$0 + 0 = \frac{1}{2} I_B \omega^2 - W.R$$

$$I_B = \bar{I} + mR^2 = \frac{1}{2}mR^2 + mR^2 = \frac{3}{2}mR^2$$

$$W = mg$$

$$0 = \frac{1}{2}\frac{3}{2}mR^2\omega^2 - mgR$$

or $$\omega = 2\sqrt{g/3R}$$

7-55

A uniform disc of weight W and radius r is pinned at A and lies in a vertical plane. Determine the minimum impulse that must be applied at B in order to make the diameter AB rotate through a right angle and go from the vertical position AB to the horizontal position AB' ($I_G = mr^2/2$ for a circular plate of mass m and radius r.)

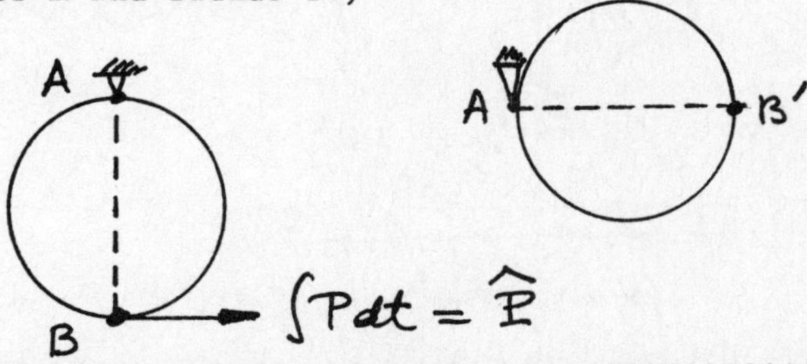

****

Immediately before and after the impulse, we have the diagram

A, B + A, B $\hat{P}$ = A, $I_G\omega$, mV_G, B

Taking moments of momenta about A gives

$$\hat{P}(2r) = I_G\omega + mV_G r \quad (1)$$

Substituting $V_G = \omega r$ and I_G in (1) yields

$$\hat{P} = \frac{3}{4} m\omega r \quad (\omega_{min} \text{ Corresponds to } \hat{P}_{min})$$

Conservation of energy between after impact and when AB' becomes horizontal gives, taking datum through A

$$T_1 = \frac{1}{2} m V_G^2 + \frac{1}{2} I_G \omega^2 = \frac{1}{2} m(\omega r)^2 + \frac{1}{2}\left(\frac{mr^2}{2}\right)\omega^2 = \frac{3}{4} m\omega^2 r^2$$

$V_1 = -mgr$. $T_2 = 0$ (ω_{min} to produce $V_2 = 0$ was calculated). $V_2 = 0$ (datum)

Hence, $\frac{3}{4} m\omega^2 r^2 - mgr = 0$, $\omega^2 = \frac{4}{3}\frac{g}{r}$

Substituting in $\hat{P}$ gives $\underline{\underline{\hat{P} = m\sqrt{\frac{3}{4} gr}}}$

7-56

A solid homogeneous cylinder is released from rest in the position indicated. The cylinder has mass 12 kg. The spring constant is 2 N/m and its unstretched length is 1 meter. What is the angular velocity of the cylinder after it has moved 1 meter? (Assume the cylinder does not slip.)

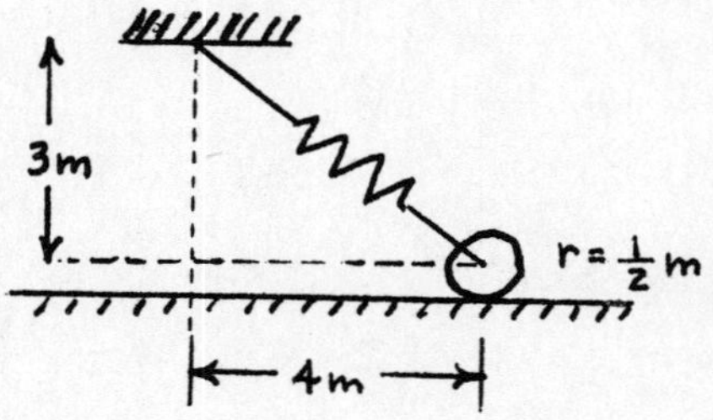

****

Note: $T_1 = 0$, $T_2 = \frac{1}{2} m \bar{v}_2^2 + \frac{1}{2} \bar{I} \omega_2^2$ but $\bar{v}_2 = r\omega_2 = \frac{1}{2}\omega_2$

$\bar{I} = \frac{1}{2} m r^2 = \frac{1}{2}(12)(\frac{1}{2})^2 = \frac{3}{2}$

$$\therefore T_2 = \frac{1}{2}(12)(\frac{1}{2}\omega_2)^2 + \frac{1}{2}(\frac{3}{2})\omega_2^2 = \frac{3}{2}\omega_2^2 + \frac{3}{4}\omega_2^2 = \frac{9}{4}\omega_2^2$$

Since there is no slipping, the point of application of the frictional force does no work. Hence, we may use conservation of energy:

$$T_1 + V_1 = T_2 + V_2$$

$$0 + 16 = \frac{9}{4}\omega_2^2 + 10.51$$

$$\therefore 5.49 = \frac{9}{4}\omega_2^2$$

$$\therefore \omega_2^2 = \frac{4}{9} 5.49 = 2.44$$

$$\therefore \omega_2 = 1.56 \left(\frac{\text{rad}}{\text{sec}}\right)$$

The system possesses potential energy by virtue of the springs configuration

$V_1 = \frac{1}{2} k (d_1 - 1)^2$; $d_1 = 5$

$\underline{V_1} = \frac{1}{2}(2)4^2 = \underline{16}$

$V_2 = \frac{1}{2} k (3\sqrt{2} - 1)^2$, $d_2 = 3\sqrt{2}$

$= \frac{1}{2}(2)(3.2426)^2$

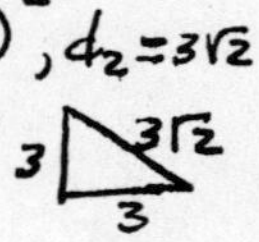

$\underline{V_2} = \underline{10.51}$

7-57

Bar AB has a mass of 75 kg and body C is 20 kg. The velocity of body C in the position shown is 2.4 m/s downward. What is the angular velocity of bar AB when it has moved to a horizontal position?

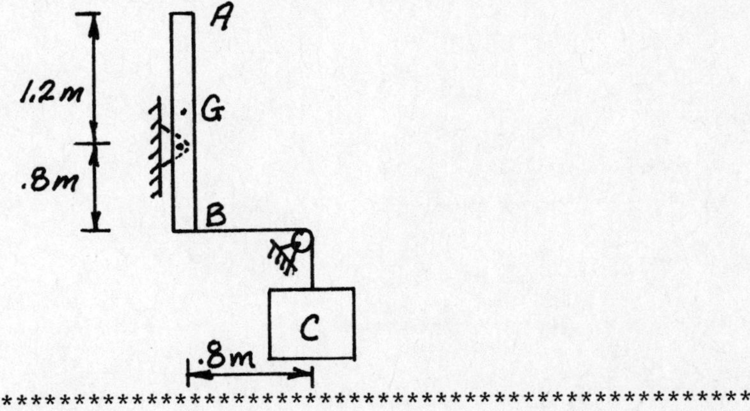

****

$$U_{1-2} = 75(9.81)(.2) = 147.2 \text{ N·m}$$

$V_C = V_B$ THEREFORE $\omega_{AB,1} = \dfrac{2.4}{.8} = 3 \text{ rad/s}$ ↻

$$T_1 = \frac{1}{2} I_o \omega^2 + \frac{1}{2}(20)(2.4)^2$$

$$= \frac{1}{2}\left(\underset{(28)}{\frac{1}{12} \times 75 \times 2^2 + 75(.2)^2}\right) 3^2 + 10(2.4)^2 = 183.6 \text{ N·m}$$

$$T_2 = \frac{1}{2}(28)\,\omega_2^2 + \frac{1}{2}(20) V_C^2$$

$$V_{C_2} = .8\,\omega_2$$

$$T_2 = 14\,\omega_2^2 + \frac{1}{2}(20)(.8\omega_2)^2 = 20.4\,\omega_2^2$$

$U_{1-2} = T_2 - T_1 \qquad 147.2 = 20.4\,\omega_2^2 - 183.6$

$$\omega^2 = 16.2$$

$$\omega = 4.03 \text{ rad/s} \circlearrowright$$

7-58

A 2 m long bar weighing 4 N is used to connect two wheels at their peripheries. If the wheels are 1 m and 0.5 m in diameter respectively, calculate the kinetic energy of the bar when the smaller wheel is rotating at 2500 r.p.m. at the position shown in the accompanying illustration.

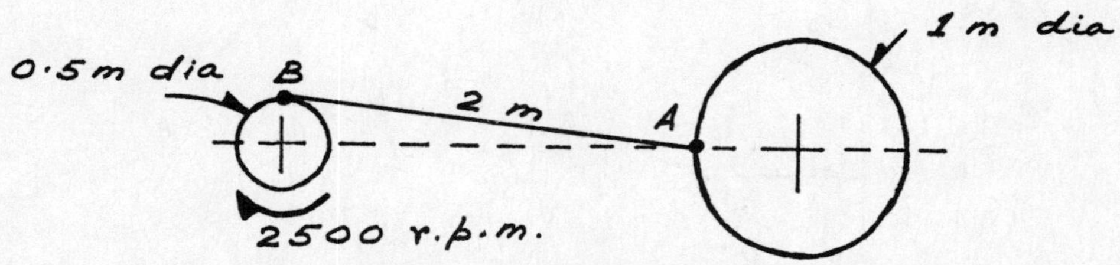

****

Solution

0.25 m, B, θ, C, A, O_B, I.C., r_c, O_A

$$m = \frac{W}{g} = \frac{4\ N}{9.81\ m/s^2} = 0.407\ kg$$

$$\omega = \frac{(2500\ r.p.m)(2\pi\ rad)}{(60\ sec/min)} = 261.87\ rad/s$$

$$\theta = \cos^{-1}\frac{(0.25m)}{(2\ m)} = 82.819°$$

By law of cosines, we have

$$(O_B C)^2 = (O_B B)^2 + (BC)^2 - 2(O_B B)(BC)\cos\theta$$

$$= (0.25m)^2 + \left(\frac{2\ m}{2}\right)^2 - 2(0.25m)\left(\frac{2m}{2}\right)(0.125)$$

$$= 1\ m$$

$$\therefore r_c = O_B C = \sqrt{1 m} = 1\ m$$

$$I_{I.C.} = \frac{1}{12} mL^2 + md^2$$

$$= \frac{1}{12}(0.407 kg)(2\ m)^2 + (0.407\ kg)(1\ m)^2$$

$$= 0.543\ kg.m^2$$

$$\text{Now, } T = \frac{1}{2} I_{I.C.}\ \omega^2$$

$$= \frac{1}{2}(0.543\ kg.m^2)(261.87\ rad/s)^2$$

$$= 18618.356\ kg \cdot m^2/s^2$$

$$= 18618.356\ N.m.$$

Ans.

7-59

The uniform rectangular plate shown is at rest with edge AB horizontal when the supporting string at "B" is cut. Corner "A" is supported by a smooth pin.

(a) Where is the center of mass of the plate when the plate attains its maximum angular speed?

(b) Calculate that maximum angular speed.

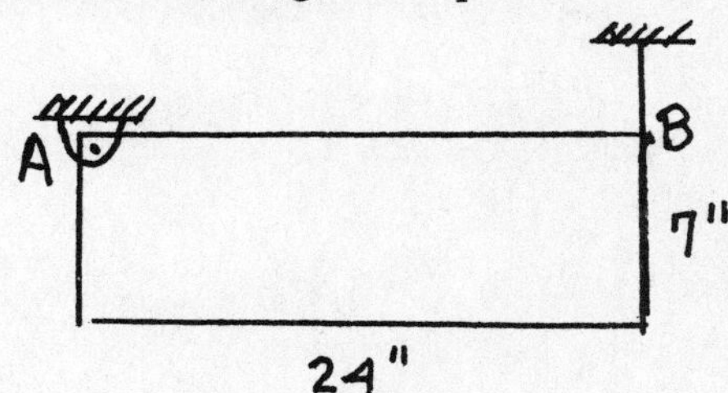

****

(a) MAX ANGULAR SPEED OCCURS WHEN G IS BELOW A.

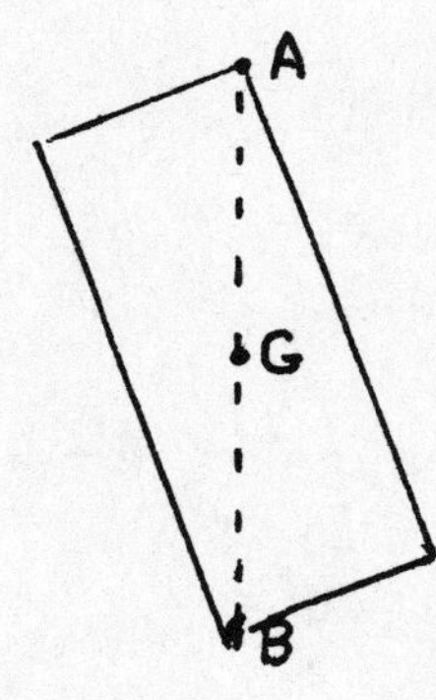

$$AG = \frac{\sqrt{24^2+7^2}}{2} = 12.5''$$

$$\text{and } V_G = r\omega = \frac{25}{24}\omega$$

BY CONSERVATION OF ENERGY: (DATUM ALONG AB)

$$-mg\left(\frac{7}{24}\right) = -mg\left(\frac{25}{24}\right) + \frac{1}{2}\frac{m}{12}\left(\frac{25}{12}\right)^2\cdot\omega^2 + \frac{1}{2}m\left(\frac{25}{24}\right)^2\cdot\omega^2$$

SOLVING FOR ω: $\omega = 5.78 \frac{rad}{sec}$.

7-60

A 10-lb force pulls a carriage of 15 lb which is initially at rest. Each of the four wheels weighs 1 lb and their radius is 2 inches. Assume that there is no friction during the motion. After it has moved 6 ft, find the velocity of the carriage a) considering the system as a particle, b) considering the wheels as solid bodies.

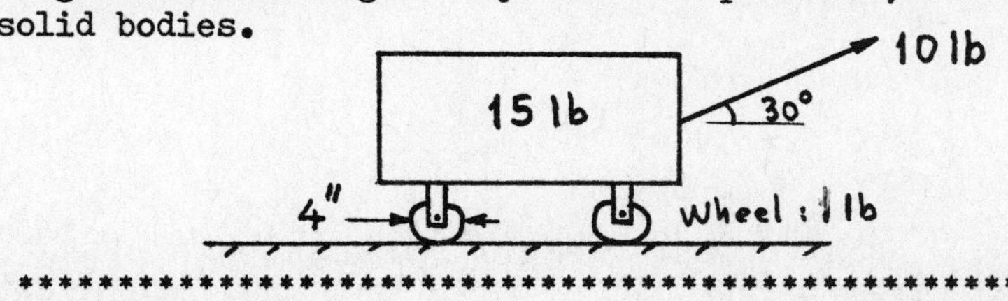

****

a) In this case, we neglect the rotation of the wheels.

Energy method : $T_1 + U_{12} = T_2 \rightarrow 0 + F.\Delta x = \frac{1}{2} m V_2^2$

$m = (15+4)/32.2 = 0.59$

$0 + 10 \times \cos 30° \times 6 = \frac{1}{2} \times 0.59 \times V_2^2$

$V_2 = 13.27$ ft/sec.

b) In this case, the wheels are rotating :

Energy method : $T_1 + U_{12} = T_2$

$$0 + F.\Delta x = \frac{1}{2} m_c V_2^2 + 4\left(\frac{1}{2} m_w V_2^2 + \frac{1}{2} I_G.\omega^2\right)$$

where $\omega = V_2 / r$: angular speed of the wheels ; r : radius.

$$\frac{1}{2} I_G.\omega^2 = \frac{1}{2} \cdot \left(\frac{1}{2} m_w r^2\right) . \frac{V_2^2}{r^2} = \frac{m_w}{4} . V_2^2$$

$$10 \times \cos 30° \times 6 = \frac{1}{2} \times \frac{15}{32.2} V_2^2 + 4\left(\frac{1}{2} \times \frac{1}{32.2} V_2^2 + \frac{1}{4} \times \frac{1}{32.2} V_2^2\right)$$

$V_2 = 12.62$ ft/sec < 13.27 ft/sec

since some of the work given is used to rotate the wheels.

GENERAL (3-D) MOTION: INERTIA TENSOR

7-61

The gimbal frame β rotates at the rate $\omega_p = \omega_{po} + \alpha_p t$ while the flywheel spins relative to the gimbal frame at the rate $\omega_s = \omega_{so} + \alpha_s t$. ($\omega_{po}$, α_p, ω_{so}, and α_s are all constant.) The moments of inertia of the flywheel about the x, y, and z axes are I, I, and J, respectively.

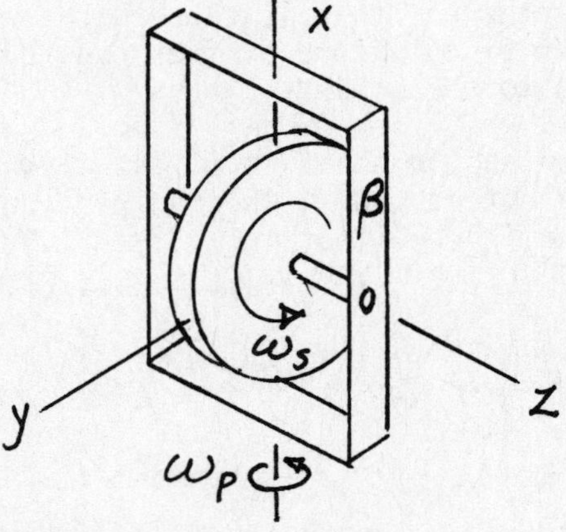

In terms of these parameters, what is the moment applied to the flywheel?

****

The angular momentum is

$$\mathbf{H} = I\omega_p \mathbf{i} + J\omega_s \mathbf{k}$$

The gimbal-observed derivative of this is

$$\frac{d_\beta \mathbf{H}}{dt} = I\alpha_p \mathbf{i} + J\alpha_s \mathbf{k}$$

and the ground-observed derivative of $\mathbf{H}$ is

$$\frac{d_g \mathbf{H}}{dt} = \frac{d_\beta \mathbf{H}}{dt} + \boldsymbol{\omega}_{\beta/g} \times \mathbf{H}$$

$$= I\alpha_p \mathbf{i} + J\alpha_s \mathbf{k} + \omega_p \mathbf{i} \times (I\omega_p \mathbf{i} + J\omega_s \mathbf{k})$$

$$= I\alpha_p \mathbf{i} - J\omega_p\omega_s \mathbf{j} + J\alpha_s \mathbf{k}$$

This must equal the moment applied to the flywheel.

7-62

A spinning flywheel is to be used for energy storage in an automobile. The bearing supports are to be oriented so that the gyroscopic moment will aid in cornering; that is, we want the side of the car that is closer to the center of curvature of the road to tend to move downward.

How should the axis of the flywheel be oriented, and in which direction should the flywheel rotate?

****

Referring to the orientation in the above sketch, the moment that the flywheel transmits to the bearing supports should be in the ↶ direction.

The moment applied to the flywheel by the bearing supports must be opposite to this, i.e., in the ↷ direction. The direction of $d\mathbf{H}/dt$ must therefore be <u>forward</u> for this turn.

This will be induced by the changing direction of the car if the angular momentum $\mathbf{H}$ is directed toward the <u>right</u>

Thus, the axis should be horizontal, right-left, and the flywheel spinning such that its angular velocity is directed to the right.

This will also aid cornering the other way.

7-63

If the weight of the cone shown in the accompanying illustration is 8 N, its height h = 0.25 m and its radius r = 0.15 m, calculate the moment of inertia of the cone about the z' axis.

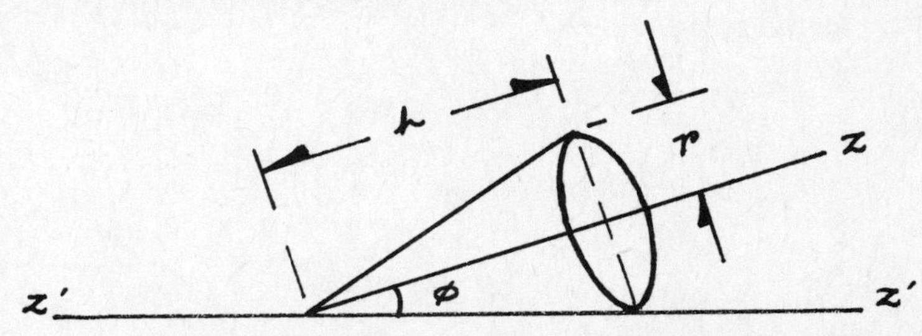

****

Solution

$$\tan\phi = \frac{r}{h} = \frac{0.15\,m}{0.25\,m} = 0.6$$

$$\sin\phi = \frac{r}{\sqrt{r^2+h^2}} = \frac{0.15\,m}{\sqrt{(0.15\,m)^2+(0.25\,m)^2}} = 0.514$$

$$\cos\phi = \frac{h}{\sqrt{r^2+h^2}} = \frac{0.25\,m}{\sqrt{(0.15\,m)^2+(0.25\,m)^2}} = 0.857$$

$$I_{xx} = I_{yy} = \left[\frac{3}{80}M(4r^2+h^2)\right] + \left[M\left(h-\frac{h}{4}\right)^2\right]$$

where M (kg) is the mass of the cone

$$= \left[\frac{3}{80}M\{4(0.15\,m)^2+(0.25\,m)^2\}\right] + \left[M\left(0.25\,m-\frac{0.25\,m}{4}\right)^2\right]$$

$$= 0.041\,M \quad m^2$$

$$I_{zz} = \frac{3}{10}Mr^2 = \frac{3}{10}M(0.15\,m)^2 = 0.007\,M\ m^2$$

$$I_{xy} = I_{yz} = I_{zx} = 0$$

Hence, $I_{z'z'} = u_{z'x}^2 I_{xx} + u_{z'y}^2 I_{yy} + u_{z'z}^2 I_{zz}$

where $u_{z'x} = \frac{r}{\sqrt{r^2+h^2}}$; $u_{z'y} = 0$; $u_{z'z} = \frac{h}{\sqrt{r^2+h^2}}$

∴ By substitution, we get

$$I_{z'z'} = (0.514)^2(0.041\,m^2)\left(\frac{8\,N}{9.81\,m/s^2}\right) + 0 + (0.857)^2(0.007\,m^2)\left(\frac{8\,N}{9.81\,m/s^2}\right)$$

$$= 0.013 \quad kg \cdot m^2$$ Ans.

7-64

Determine the mass moment of inertia as well as the radius of gyration about the x-axis for the paraboloid volume as shown. Assume the density of the material to be ρ_m slug/ft^3.

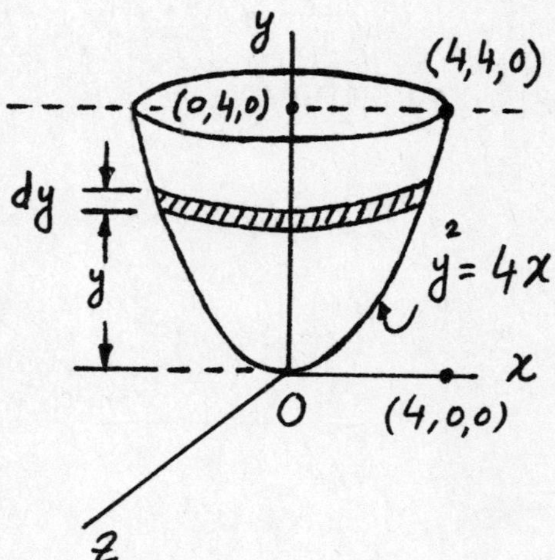

****

Mass of elemental volume $= dm = 2\pi x \, dy \, \rho_m$

Mass of paraboloid $= m = \int_0^4 2\pi \left(\frac{y^2}{4}\right) dy \, \rho_m = 33.5 \rho_m$ slug

Mass moment of inertia of paraboloid about x-axis

$$I_{xx} = \int dm \, y^2 = \int_0^4 \frac{\pi}{2} \rho_m y^4 \, dy = 321.7 \rho_m \text{ slug-ft}^2$$

Radius of gyration $k_x = \sqrt{\frac{I_{xx}}{m}} = \sqrt{\frac{321.7 \rho_m}{33.5 \rho_m}} = 3.1$ ft

GENERAL (3-D) MOTION: MOMENT AND RATE OF CHANGE OF ANGULAR MOMENTUM

7-65

A thin rectangular metal plate, 0.2 m x 0.3 m, is held in a vertical position in a frame that is rotating about a vertical axis at 3 rad/s, as shown in figure 1. A latching mechanism on the frame (not shown) is released and the plate assumes a horizontal position as shown in figure 2. Find the angular velocity of the frame and plate in the second position. Assume the mass of the frame is small compared to the mass of the plate.

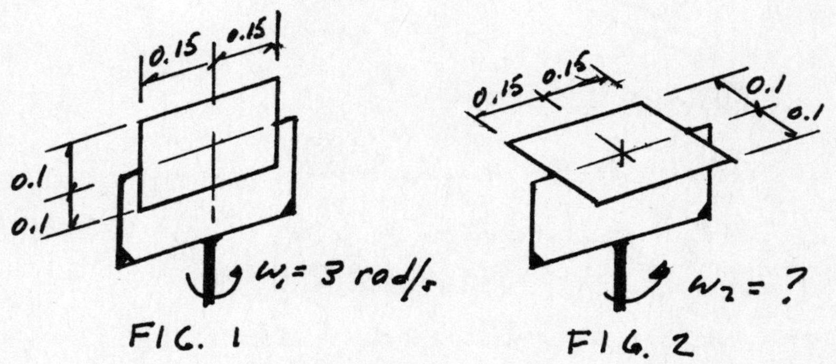

****

CONSERVATION OF ANGULAR MOMENTUM:

$$I_1 \omega_1 = I_2 \omega_2 \Rightarrow \frac{1}{12} m b^2 \omega_1 = \frac{1}{12} m (a^2 + b^2) \omega_2$$

$$\frac{1}{12} m (0.3)^2 \omega_1 = \frac{1}{12} m (0.2^2 + 0.3^2) \omega_2$$

$$0.09\ \omega_1 = 0.13\ \omega_2$$

$$\omega_2 = \frac{0.09}{0.13} \omega_1 = \frac{0.09}{0.13}(3) = 2.08 \text{ rad/s}$$

Note: For a thin rectangular plate:

$I_y = mb^2/12$

$I_z = m(a^2 + b^2)/12$

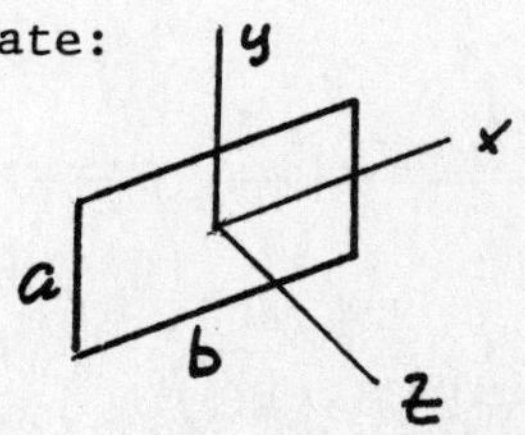

7-66

A crushing mill of weight W and radius r spins freely about its axle to which it is attached by means of a ball bearing. The axle is attached by means of a ball and socket joint at A, and the assembly is made to rotate at a constant angular velocity ω_o about the vertical axis passing through A. The axle has negligible weight. Determine a) the angular velocity of the mill with respect to the axle, b) the velocity of the center of the mill, c) the acceleration of the center of the mill, d) the normal reaction exerted by the track on the mill, and e) the reaction of the pivot on the axle. ($I_G = mr^2/4$, and $mr^2/2$ are the moments of inertia with respect to a diameter, and perpendicular to any two diameters respectively, for a disc of mass m and radius r.)

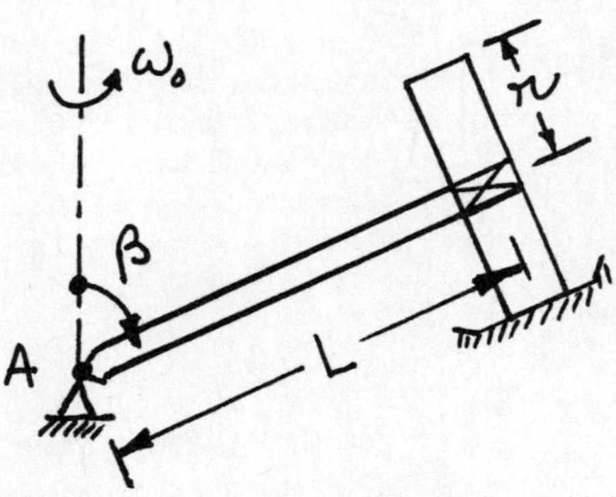

****

a) Taking the coordinate axes (not fixed to the mill, the velocity of the point of contact can be expressed as

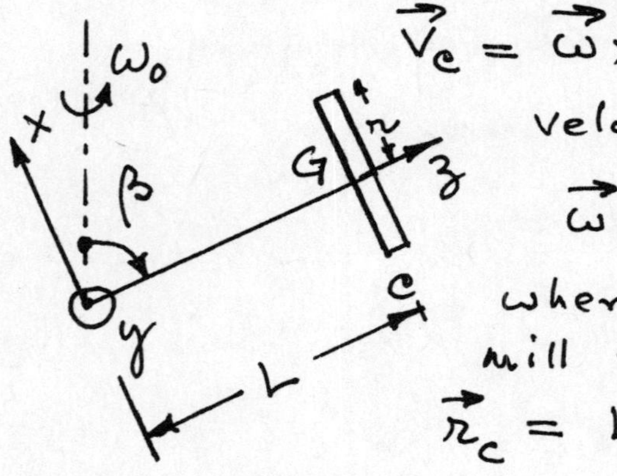

$$\vec{V}_c = \vec{\omega} \times \vec{r}_c = 0$$. The angular velocity is given by

$$\vec{\omega} = (\omega_o \sin\beta)\vec{i} + (\omega_o \cos\beta + s)\vec{k}$$

where s is the spin of the mill relative to the axle.

$$\vec{r}_c = L\vec{k} - r\vec{i}$$

Substitution yields

$$s = -\frac{\omega_o}{r}(L\sin\beta + r\cos\beta)$$

b) The velocity of G is given by

$$\vec{V}_G = \frac{d\vec{r}_G}{dt} = \frac{\delta \vec{r}_G}{\delta t} + \vec{\Omega} \times \vec{r}_G$$

where $\vec{r}_G = L\vec{k}$, $\vec{\Omega} = (\omega_o \sin\beta)\vec{i} + (\omega_o \cos\beta)\vec{k}$

Note that $\vec{\Omega}$ (frame) $\neq \vec{\omega}$ (body), since Axyz is not rigidly attached to the mill.

$\frac{\delta \vec{r}_G}{\delta t} = 0$ (relative to moving axes in which x, y, z are fixed.

$$\vec{\Omega} \times \vec{r}_G = -(\omega_o L \sin\beta)\vec{j}$$

$$\underline{\underline{\vec{V}_G = -(\omega_o L \sin\beta)\vec{j}}}$$

c) To find $\vec{a}_G$, apply the basic formula

$$\vec{a}_G = \frac{d\vec{V}_G}{dt} = \frac{\delta \vec{V}_G}{\delta t} + \vec{\Omega} \times \vec{V}_G$$

But $\frac{\delta \vec{V}_G}{\delta t} = 0$, $\vec{\Omega} \times \vec{V}_G = \omega_o^2 L \sin\beta[(\cos\beta)\vec{i} - (\sin\beta)\vec{k}]$

So that $\underline{\underline{\vec{a}_G = \omega_o^2 L \sin\beta\,[(\cos\beta)\vec{i} - (\sin\beta)\vec{j}\,]}}$

d) The components of angular momentum about G are $H_x = I_x \omega_x = \frac{1}{4} m r^2 \omega_o \sin\beta$

$H_y = I_y \omega_y = 0$, $H_z = \frac{1}{2} m r^2 (\omega_o \cos\beta + S) = I_z \omega_z$

$H_z = -\frac{1}{2} m r \omega_o L \sin\beta$. $\vec{H}_G = H_x \vec{i} + H_z \vec{k}$

$\dot{\vec{H}} = \frac{\delta \vec{H}_G}{\delta t} + \vec{\Omega} \times \vec{H}_G$ $\qquad$ $\frac{\delta \vec{H}_G}{\delta t} = 0$ ($\vec{i}, \vec{k}$ fixed in frame)

$\vec{\Omega} \times \vec{H}_G = \omega_o [H_x \cos\beta - H_z \sin\beta] \vec{j}$, collecting terms

$$\dot{\vec{H}}_G = \frac{m\omega_o^2 r \sin\beta}{2} \left[L \sin\beta + \frac{r}{2} \cos\beta \right] \vec{j}$$

A free-body diagram yields

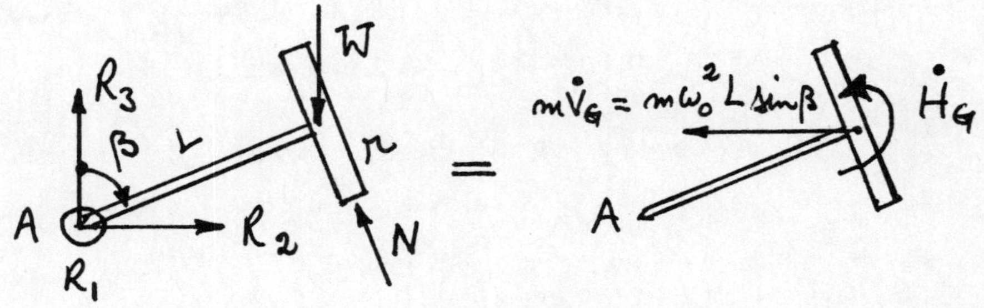

Summing the moments of all the vectors on the LHS = Same on the RHS, yields

$$NL - WL\sin\beta = (m\omega_o^2 L \sin\beta)(L\cos\beta) + \dot{H}_G$$

Substituting for $\dot{H}_G$ and solving for N gives

$$N = W\sin\beta + m\omega_o^2 L \sin\beta \left[\frac{r}{2L} \sin\beta + \left(1 + \frac{r^2}{4L^2}\right) \cos\beta \right]$$

e) Summation of all the vectors on the LHS = Same on the RHS gives

$R_1 = 0 \qquad R_3 + N\sin\beta - W = 0$

$R_2 - N\cos\beta = -m\omega_o^2 L \sin\beta$, Therefore

$R_1 = 0$, $R_2 = N\cos\beta - m\omega_o^2 L\sin\beta$

$R_3 = W - N\sin\beta$, where N is as given in part (d).

7-67

A vertical axle AB rotating at a constant angular velocity ω = 19.7 rad/s carries a 6 ft long arm CD weighing 37.5 lb pinned to the axle at C at an angle of 75° to the horizontal, as shown in the accompanying illustration. If a horizontal cable DE maintains the rod in a steady state during rotation, calculate (a) the tension in the cable, and (b) the reaction at C.

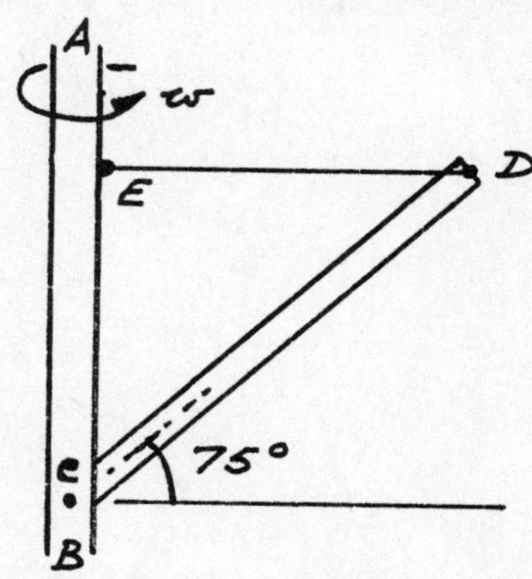

****

Solution

From geometry, if F is the mid point of the arm (a slender rod), we have

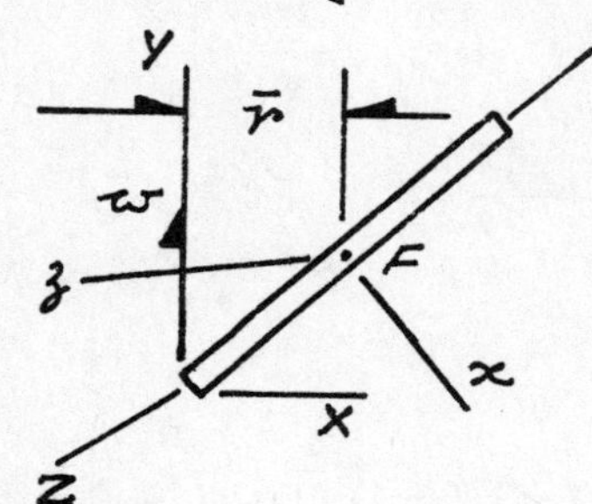

$$\bar{r} = \frac{1}{2}\ CD\ \cos 75°$$
$$= \frac{1}{2}(6\ ft)(0.258) = 0.776\ ft$$

Hence, vector $m\bar{a}$ attached at F is given by

$$m\bar{a} = m(-\bar{r}\omega^2 I)$$
$$= \left(\frac{37.5\ lb}{32.2\ ft/s^2}\right)(-0.776\ ft)(19.7\ rad/s)^2 I$$
$$= -(350.727\ lb)I$$

Angular momentum H_F:

$$H_F = \bar{I}_x \omega_x i + \bar{I}_y \omega_y j + \bar{I}_z \omega_z k$$

where $\bar{I}_x = \bar{I}_z = \frac{1}{12} mL^2$; $\bar{I}_y = 0$

also, $\omega_x = -\omega \cos 75°$; $\omega_y = \omega \sin 75°$; $\omega_z = 0$

Hence,
$$H_F = \left(\frac{1}{12} mL^2\right)(-\omega \cos 75°)i + (0)(\omega \sin 75°)j + \left(\frac{1}{12} mL^2\right)(0)k$$
$$= -\frac{1}{12} mL^2 \omega \cos 75°\ i$$
$$= -\frac{1}{12}\left(\frac{37.5\ lb}{32.2\ ft/s^2}\right)(6\ ft)^2(19.7\ rad/s)(0.258)\ i$$
$$= -17.757\ i$$

Couple $\dot{H}_F$

The rate of change $\dot{H}_F$ of H_F with respect to the fixed axes XYZ is given by

$$\dot{H}_F = (\dot{H}_F)_{Fxyz} + \omega \times H_F$$

$$\dot{H}_F = 0 + (-\omega \cos 75° i + \omega \sin 75° j) \times (-17.757\, i)$$
$$= 17.757\ \omega\ \sin 75°\ k$$
$$= 17.757\,(19.7 \text{ rad/s})(0.965)k \quad = (337.893 \text{ lb.ft})\ k$$

Equations of motion:

$T = -TI$, D, F, E, $W = -37.5$ lb J, $C_x I$, $C_z K$, C, $75°$, $C_y J$, X, Y, Z

$=$

$m\bar{a} = -350.727\ I$, $\dot{H}_F = 337.893\ K$, F, X, Y, 0.776 ft

From geometry, we have

$$EC = CD \sin 75° = (6 \text{ ft})(0.965) = 5.795 \text{ ft}$$
$$GC = \frac{1}{2} EC = \frac{1}{2}(5.795 \text{ ft}) = 2.897 \text{ ft}$$

(a) The system of external forces is equal to the system of effective forces; hence,

$$\Sigma M_C = \Sigma (M_C)_{eff}$$

OR, $(0.776\ I) \times (-37.5\ J) + (5.795\ J) \times (-T\ I)$

$$= (337.893\ K) + (2.897\ J) \times (-350.727\ I)$$

OR, $(-29.1 + 5.795T)\ K = (337.893 + 1016.056)\ K$

$\therefore \quad T = 238.662$ lb. <u>Ans.</u>

(b) $\quad \Sigma F = \Sigma F_{eff}$

OR, $C_x I + C_y J + C_z K - 238.662\ I - 37.5 J = -350.727\ I$

OR, $\quad C = (-112.065)\ I + 37.5$ lb$)\ J$ <u>Ans.</u>

7-68

In the illustration shown, a rotating bar OA is connected to a thin rod OB at point O, which also carries a torsional spring whose modulus is given by the expression C_m N.m/rad. If the bar OA is considered to be without mass and the rod OB has a mass m, determine by using Euler's equations of motion an expression for the rod's motion.

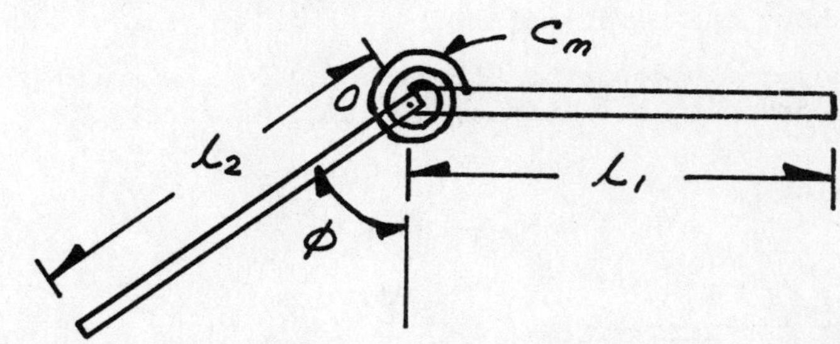

****

<u>Solution</u>

$$\omega_x = \dot{\Psi} \cos \phi$$
$$\omega_y = \dot{\Psi} \sin \phi$$
$$\omega_z = -\dot{\phi}$$

Also, $I_x = 0$

$$I_y = I_z = \frac{1}{12} m \ell_2^2 = I$$

By applying moments about the center of mass, we get

$$M_x = 0$$
$$M_y = 0$$
$$M_z = C_m \phi + O_y \frac{\ell_2}{2}$$

By substituting the above into Euler's equation, we get

$$I \frac{d}{dt} (\dot{\Psi} \sin \phi) + I \dot{\Psi} \dot{\phi} \cos \phi = 0$$

OR, $M_z = -I\ddot{\phi} + I \dot{\Psi}^2 \cos \phi \sin \phi$

$$= C_m \phi + O_y \frac{\ell_2}{2}$$

From which we can rewrite

$$\frac{d}{dt} (\dot{\Psi} \sin \phi) + \dot{\Psi} \dot{\phi} \cos \phi = 0 \qquad \text{\underline{Ans}.}$$

and $\ddot{\phi} - \dot{\Psi}^2 \cos \phi \sin \phi + \frac{C_m \phi}{I} + O_y \frac{\ell_2}{2I} = 0$ <u>Ans</u>.

7-69

A car wheel is mounted on an axle for testing purposes and constrained to rotate on a horizontal plane, as shown in the accompanying illustration. The wheel (a homogeneous disk) has a weight of 20 lb and a diameter of 3 ft, while the 5 ft long axle is of negligible mass. The wheel rotates about the axle at an anticlockwise angular velocity ω_d = 12 rad/s, while the axle rotates clockwise about the pivot point at an angular velocity ω_a = 15 rad/s. Calculate (a) the angular velocity of the wheel, (b) the angular momentum of the wheel about the pivot point, and (c) its kinetic energy.

O, C, ω_a, ω_d

****

Solution

y, L, r, C, x, $\omega_d i$, r_D, z, $-\omega_a j$, D

(a) Angular velocity:

Total angular velocity of the wheel:

$$\omega = \omega_d i - \omega_a j \quad \dots (1)$$

The velocity of point D = 0

Hence, $V_D = \omega \times r_D = 0$

$$V_D = (\omega_d i - \omega_a j) \times (L i - r j) = (L\omega_a - r\omega_d)\,k$$

OR, $\omega_a = \dfrac{r\,\omega_d}{L}$

Substituting ω_a in Eq. (1), we get

$$\omega = \omega_d i - \left(\frac{r\omega_d}{L}\right) j$$

$$= 12 i - \left(\frac{3\,ft}{2}\right)\left(\frac{12}{5\,ft}\right) j = (12 i - 3{\cdot}6 j) \quad \underline{\text{Ans}}.$$

(b) Angular momentum:

The axle is massless and can be assumed to be a part of the wheel, with a fixed point at O. Considering x, y, z as principal axes of inertia for the wheel, we have

$$H_x = I_x \omega_x = \left(\tfrac{1}{2} m r^2\right)\omega_d$$

$$H_y = I_y \omega_y = \left(mL^2 + \tfrac{1}{4} m r^2\right)\left(-\frac{r\omega_d}{L}\right)$$

$$H_z = I_z \omega_z = \left(mL^2 + \tfrac{1}{4} m r^2\right)(0) = 0$$

Hence,

$$H_O = \left[\left(\tfrac{1}{2} m r^2\right)\omega_d\right] i - \left[m\left(L^2 + \frac{r^2}{4}\right)\left(\frac{r\,\omega_d}{L}\right)\right] j$$

$$= \left[\tfrac{1}{2}\left(\frac{50}{32{\cdot}2}\right)\left(\frac{3'}{2}\right)^2 (12)\right] i - \left[\frac{50}{32{\cdot}2}\left\{5^2 + \left(\frac{3}{2}\right)^2\left(\frac{1}{4}\right)\right\}\left(\frac{3}{2}\right)\left(\frac{12}{5}\right)\right] j$$

$$= \{20{\cdot}962\, i - 142{\cdot}895\, j\} \quad \underline{\text{Ans}}.$$

(c) Kinetic Energy:

$$T = \frac{1}{2}\left(I_x \omega_x^2 + I_y \omega_y^2 + I_z \omega_z^2\right)$$

$$= \frac{1}{2}\left[\left(\frac{1}{2} m r^2 \omega_d^2\right) + \left(m L^2 + \frac{1}{4} m r^2\right)\left(-\frac{r\,\omega_d}{L}\right) + 0\right]$$

$$= \frac{1}{8} m r^2 \left(6 + \frac{r^2}{L^2}\right) \omega_d^2$$

$$= \frac{1}{8}\left(\frac{50}{32.2}\right)\left(\frac{3}{2}\right)^2 \left\{6 + \left(\frac{3}{2}\right)^2\left(\frac{1}{5}\right)^2\right\} 12^2$$

$$= 382.989 \text{ ft. lb.} \qquad \underline{\text{Ans}}.$$

GENERAL (3-D) MOTION: WORK AND KINETIC ENERGY

7-70

A uniform disk D made of a special alloy is under test in a laboratory for energy transfer purposes. The mass of the very thin disk is 12.5 kg and its radius is 0.25 m. It is mounted on an axle at a point P such that its coordinate is 0.45 m j and the axle is freely pivoted in a ball and socket joint at point Q. If the disk is being rotated with an angular velocity ω_y = 13.5 rad/s j and ω_z = 3.75 rad/s k, calculate (a) the angular and the linear momenta of the disk about point Q, and (b) the kinetic energy of the disk.

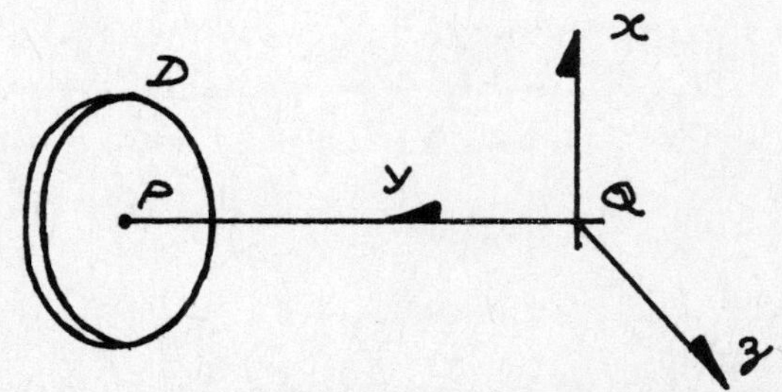

****

Solution

(a) Angular momentum

$H_Q = I_x \omega_x + I_y \omega_y + I_z \omega_z$; where $\omega_x = 0$;

$I_y = \frac{1}{2} m r^2$; $I_z = \left(\frac{m r^2}{4}\right) + m (PQ)^2$

Hence, $H_Q = 0\, i + \left[\frac{1}{2}(12.5\ kg)(0.25\ m)^2 (13.5\ rad/s)\, j\right] + \left[\left\{\frac{(12.5\ kg)(0.25\ m)^2}{4} + (12.5\ kg)(0.45\ m)^2\right\} 3.75\ rad/s\right] k$

$= (5.273\, j + 10.223\, k)\ kg \cdot m^2/s$ Ans.

Linear momentum

$V_P = \omega \times r_{P/Q}$

$= (13.5\ rad/s\ j + 3.75\ rad/s\ k) \times (0.45\ m\ j)$

$= -1.687\ m/s\ i$

Hence, $L = m V_P$

$= (12.5\ kg)(-1.687\ m/s\ i)$

$= -21.087\ kg \cdot m/s\ i$ Ans.

(b) Kinetic energy

Considering x, y, z as the reference axes and attaching the principal and parallel body axes x′, y′, z′ at P, we have

$T = \frac{1}{2} m v_Q^2 + \frac{1}{2}\left[I_{x'} \omega_{x'}^2 + I_{y'} \omega_{y'}^2 + I_{z'} \omega_{z'}^2\right]$

$= \frac{1}{2}(12.5\ kg)(-1.687\ m/s)^2 + \frac{1}{2}\left[0 + \frac{1}{2}\{(12.5\ kg)(0.25\ m)^2 (13.5\ rad/s)^2\} + \frac{1}{4}(12.5\ kg)(0.25\ m)^2 (3.75\ rad/s)^2\right]$

$= 54.755\ kg \cdot m^2/s^2 = 54.755\ N.m.$ Ans.

CHECK : Using x, y, z as the reference axes we can also write

$T = \frac{1}{2}\, \omega\, H_Q$

$= \frac{1}{2}(13.5\, j + 3.75\, k)\ rad/s\ (5.273\, j + 10.223\, k)\ kg\ m^2/s$

$= 54.760\ kg \cdot m^2/s^2 = 54.760\ N \cdot m$ Ans.

The small difference between the values of T by two different methods is in rounding off the computation.

8
MECHANCIAL VIBRATIONS

8-1

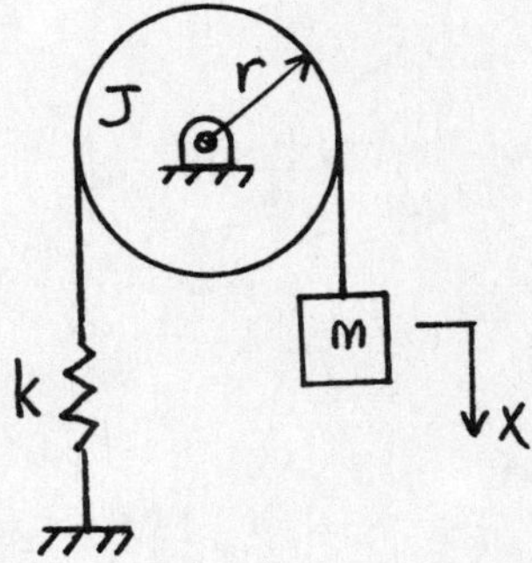

A light inextensible cord passes over a pulley of radius r and polar moment of inertia J, that is supported on frictionless bearings. One end of the cord is anchored through a vertical spring of stiffness k and the other end carries a mass m. If the mass is pulled down and released, what is the frequency of oscillation of ensuing motion?

****

Consider a displacement x from the unstretched position of the spring.

Velocity of the mass $= \frac{dx}{dt} = \dot{x}$

Angular velocity of the pulley $= \frac{1}{r}\frac{dx}{dt} = \frac{\dot{x}}{r}$

Total kinetic energy of the system $= \frac{1}{2}m\dot{x}^2 + \frac{1}{2}J\left(\frac{\dot{x}}{r}\right)^2$

Elastic potential energy $= \frac{1}{2}kx^2$

Gravitational potential energy $= -mgx$ $\left\{ x=0 \text{ is the datum} \right.$

Since there is no frictional dissipation, energy is conserved.

$$\frac{1}{2}m\dot{x}^2 + \frac{1}{2}J\left(\frac{\dot{x}}{r}\right)^2 + \frac{1}{2}kx^2 - mgx = \text{Constant}$$

Differentiate with respect to time;

$$m\dot{x}\ddot{x} + \frac{J}{r^2}\dot{x}\ddot{x} + kx\dot{x} - mg\dot{x} = 0$$

Note: $\ddot{x} = \frac{d^2x}{dt^2}$

Since $\dot{x} \neq 0$ in general,

$$\left(m + \frac{J}{r^2}\right)\ddot{x} + kx = mg$$

This is of the form $\ddot{x} + \omega^2 x = u(t)$ where ω is the natural frequency of undamped oscillations. Thus,

$$\omega = \sqrt{\frac{k}{m + \frac{J}{r^2}}}$$

8-2

Find the period of small oscillations of the composite rod shown in the figure. OC and AB are identical slender homogeneous bars of uniform cross section welded at right angles at C. The mass of each bar is m, and the assembly oscillates in the vertical plane. ($I_G = ma^2/3$ for a bar of mass m and length 2a.)

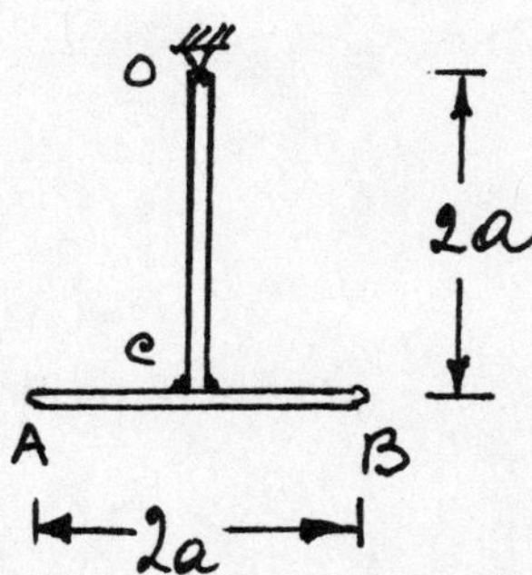

****

Consider the system when OC makes an angle θ with the vertical. The kinetic energy of each bar is $T = \frac{1}{2}mV_G^2 + \frac{1}{2}I_G\omega^2$.

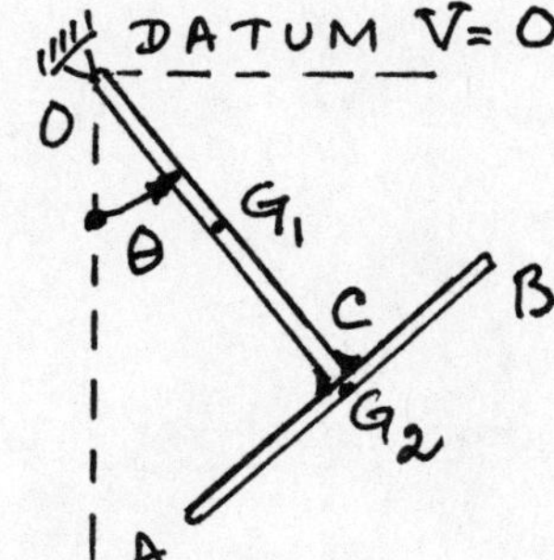

Hence, $T_{OC} = \frac{m}{2}(a\dot{\theta})^2 + \frac{1}{2}\left(\frac{ma^2}{3}\right)\dot{\theta}^2$

$T_{AB} = \frac{m}{2}(2a\dot{\theta})^2 + \frac{1}{2}\left(\frac{ma^2}{3}\right)\dot{\theta}^2$

$T = T_{OC} + T_{AB} = \frac{2}{3}ma^2\dot{\theta}^2 + \frac{13}{6}ma^2\dot{\theta}^2$

$T = \frac{17}{6}ma^2\dot{\theta}^2$; $V = V_{OC} + V_{AB}$

$$V = -mga\cos\theta - 2mga\cos\theta = -3mga\cos\theta$$

Conservation of energy gives $T + V = \text{Const.}$

or

$$\frac{17}{6}ma^2\dot{\theta}^2 - 3mga\cos\theta = \text{const.}$$

Differentiating w.r.t. time gives

with $\dot{\theta} \neq 0$, $\dot{\theta}\left[\ddot{\theta} + \frac{9}{17}\frac{g}{a}\sin\theta\right] = 0$, for small θ

$$\ddot{\theta} + \frac{9}{17}\frac{g}{a}\theta = 0, \quad \omega^2 = \frac{9}{17}\frac{g}{a}; \quad T = 2\pi\left(\frac{17a}{9g}\right)^{1/2}$$

8-3

A 5 kg mass is attached to a pulley of radius 500 mm and centroidal mass moment of inertia 1 kg-m^2. The pulley is pinned at its center. The mass is also attached to a spring of stiffness 200 N/m. The pulley is rotated 10° clockwise from the equilibrium position of the system and the system is released from this position. Determine

a. The deflection of the spring when the system is in equilibrium.
b. The period of free vibration for the system.
c. The maximum acceleration of the mass for the situation described.

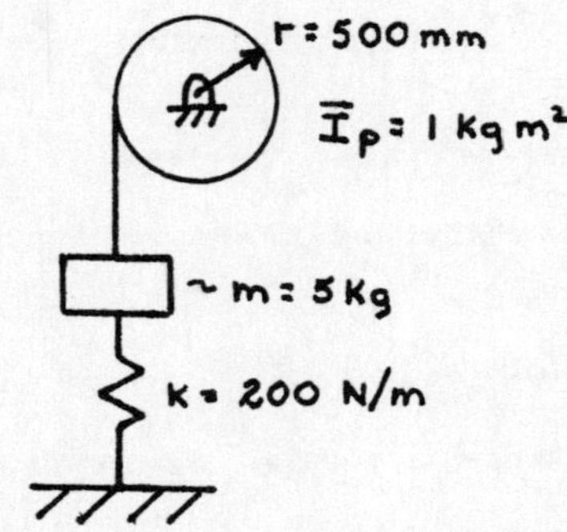

****

a) FREE BODY DIAGRAM OF SYSTEM WHEN IT IS IN EQUILIBRIUM

m_p, A_x, A_y, mg, $K\delta_{ST}$

$$\Sigma M_A = 0 = -K\delta_{ST}\, r + mgr$$

$$\delta_{ST} = \frac{mg}{K} = \frac{(5\ \text{Kg})(9.81\ \text{m/sec}^2)}{200\ \text{N/m}} = 0.245\ \text{m}$$

b) LET $x(t)$ BE THE DOWNWARD DISPLACEMENT OF THE MASS FROM THE EQUILIBRIUM POSITION OF THE SYSTEM. THE FREE BODY DIAGRAMS FOR THE SYSTEM FOR THE DYNAMIC ANALYSIS ARE

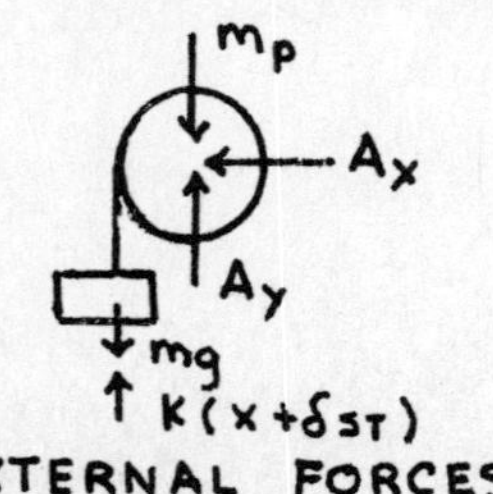

EXTERNAL FORCES

$\bar{I}_P\ddot{\theta}$

$m\ddot{x}$

EFFECTIVE FORCES

USING D'ALEMBERT'S PRINCIPLE

$$(\Sigma M_A)_{ext} = (\Sigma M_A)_{eff}$$

$$mgr - K(x+\delta_{ST})r = m\ddot{x}r + \bar{I}_P\ddot{\theta}$$

BUT $K\delta_{ST} = mg$, $\ddot{\theta} = \ddot{x}/r$

HENCE

$$(\bar{I}_{P/r} + mr)\ddot{x} + Kxr = 0$$

OR

$$\ddot{x} + \left(\frac{Kr^2}{\bar{I}_P + mr^2}\right)x = 0$$

DIFFERENTIAL EQUATIONS FOR FREE VIBRATIONS CAN BE WRITTEN IN THE FORM

$$\ddot{x} + \omega_n^2 x = 0$$

WHERE ω_n IS THE NATURAL FREQUENCY. FOR THIS PROBLEM

$$\omega_n = \sqrt{\frac{Kr^2}{\bar{I}_P + mr^2}} = \sqrt{\frac{(200\ N/m)(.5\ m)^2}{1\ Kg\,m^2 + 5\ Kg\,(.5m)^2}} = 4.71\ RAD/SEC$$

THE PERIOD OF VIBRATION T IS GIVEN BY

$$T = \frac{2\pi}{\omega_n} = \frac{2\pi}{4.71\ RAD/SEC} = 1.33\ SEC$$

c) THE SOLUTION OF THE DIFFERENTIAL EQUATION IS

$$x(t) = A\cos\omega_n t + B\sin\omega_n t$$

WHERE A AND B ARE CONSTANTS OF INTEGRATION TO BE DETERMINED FROM THE INITIAL CONDITIONS. THE PULLEY IS ROTATED 10° CLOCKWISE FROM EQUILIBRIUM CAUSING THE MASS TO MOVE UPWARD $(.5\ m)(.349\ RAD) = .175\ m$

THUS $x(0) = -0.175\ m$

THE SYSTEM IS RELEASED FROM REST. THUS

$$\dot{x}(0) = 0$$

$$x(0) = -0.175\ m = A$$

$$\dot{x}(0) = 0 = B\omega_n$$

THUS $x(t) = -0.175\cos 4.71t$

THE ACCELERATION OF THE MASS IS

$$\ddot{x}(t) = -A\omega_n^2\cos\omega_n t = 3.88\cos\omega_n t$$

THE MAXIMUM ACCELERATION IS 3.88 m/sec^2

8-4

A 30-kg mass rests in equilibrium position on a frictionless surface restrained on the two sides by the springs as shown in the figure. The block is moved 0.1 meter to the right and left to oscillate. Determine the period and frequency of the motion. Three springs are always in tension.

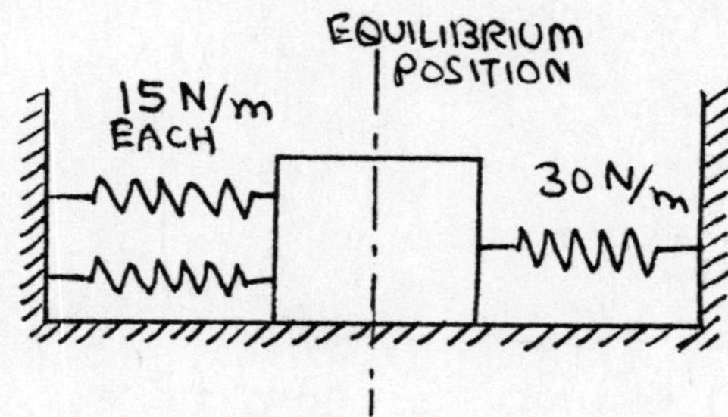

****

To write differential equation of motion, we draw the free body diagram of the block.

$\frac{T}{2}+k'x$ ← [block] → $T-kx$

$\frac{T}{2}+k'x$ ←

T is the tension in the right spring at equilib. T/2 must be the tension in each left spring

$$\Sigma F_x = m a_x$$

$$(T-kx) - 2\left(T/2 + k'x\right) = m a_x$$

$$T - kx - T - 2k'x = m\ddot{x}$$

$$-30x - 30x = 30\ddot{x}$$

$$30\ddot{x} + 60x = 0 \quad \text{or} \quad \ddot{x} + 2x = 0$$

$$\omega_n = \sqrt{2} = 1.41 \text{ rad/sec}$$

$$\text{Period } \tau = \frac{2\pi}{\omega_n} = 4.46 \text{ Sec.}$$

$$f = \frac{1}{\tau} = \frac{1}{4.46} = 0.224 \text{ cps}$$

8-5

A ball B of weight W_B is attached to a rod of weight W. Find the natural frequency of vibration in terms of k, a, l, W, W_B, and g.

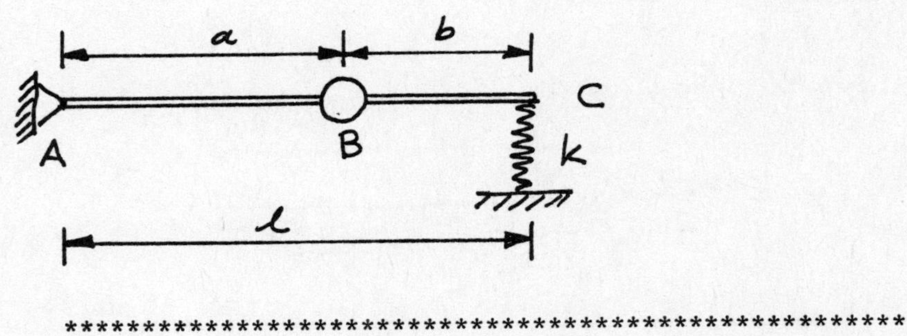

****

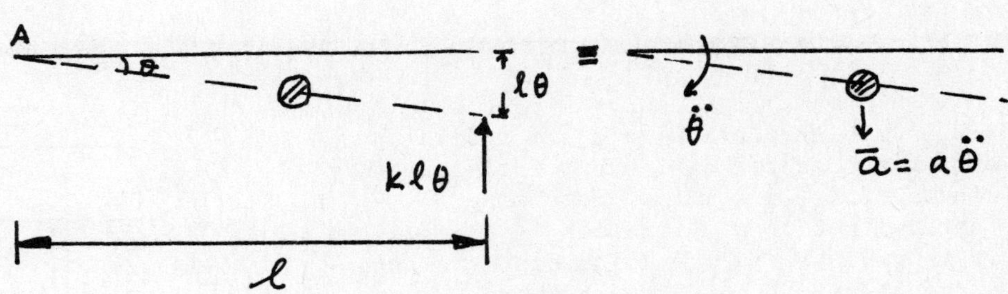

$$+\circlearrowright \Sigma M_A = I_A \ddot{\theta} + \frac{W_B}{g}(a\ddot{\theta})(a)$$

$$-k\ell\theta(\ell) = I_A \ddot{\theta} + \frac{W_B}{g} a^2 \ddot{\theta}$$

$$I_A = \frac{1}{12}\frac{W}{g}\ell^2 + \frac{W}{g}\left(\frac{\ell}{2}\right)^2 = \frac{1}{3}\frac{W}{g}\ell^2$$

Substituting :

$$\ddot{\theta}\left(\frac{1}{3}\frac{W}{g}\ell^2 + \frac{W_B}{g}a^2\right) + k\ell^2\theta = 0$$

or

$$\ddot{\theta} + \frac{k\ell^2}{\frac{1}{3}\frac{W}{g}\ell^2 + \frac{W_B}{g}a^2}\theta = 0$$

$$p^2 = \frac{k\ell^2}{\frac{1}{3}\frac{W}{g}\ell^2 + \frac{W_B}{g}a^2}$$

The natural frequency:

$$f = \frac{p}{2\pi} = \frac{1}{2\pi}\sqrt{\frac{k\ell^2}{\frac{1}{3}\frac{W}{g}\ell^2 + \frac{W_B}{g}a^2}}$$

8-6

A 50-kg circular plate with a 3 m radius is attached to a rod having a rotational stiffness of k = 6 N m/rad. Determine the natural period of vibration of the plate if it is given a small angular displacement θ in the plane of the plate as shown.

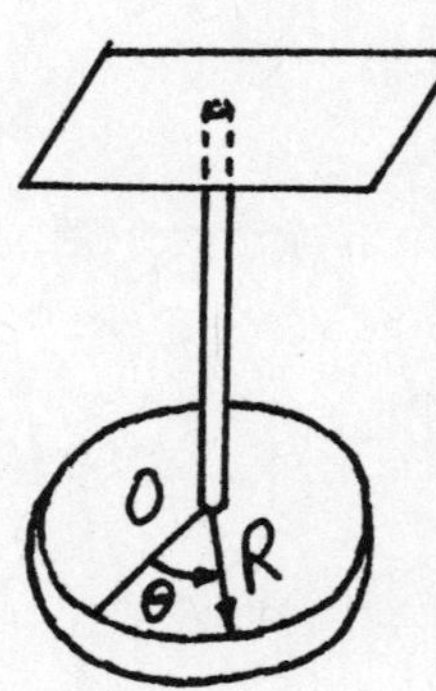

****

The small circular displacement θ of the plate is described by the differential equation

$$\frac{d^2\theta}{dt^2} + \frac{K}{I_0}\theta = 0$$

Where the circular frequency $p = \sqrt{\frac{K}{I_0}}$

Now mass moment of inertia of plate $= I_0 = \frac{1}{2}mR^2$

or, $I_0 = \frac{1}{2}(50)(3)^2 = 225$ kg-m^2

Natural period of vibration $= \tau = \dfrac{2\pi}{p}$

or, $\tau = 2\pi\sqrt{\dfrac{I_0}{k}} = 2\pi\sqrt{\dfrac{225}{6}} = 38.5$ sec

8-7

The circular cylinder of mass m rolls back and forth without slipping along the horizontal surface under the influence of the spring of stiffness k. The radius of the cylinder is a and the radius of gyration about the center is r_C.

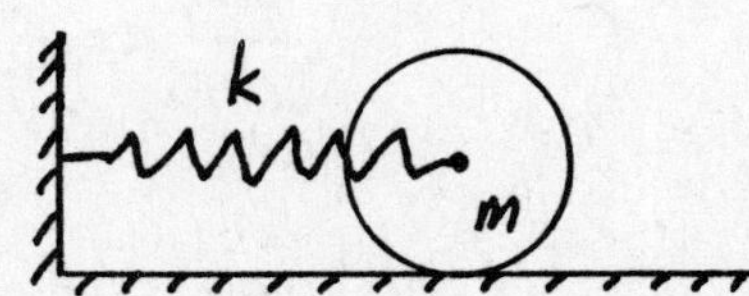

What is the natural frequency of vibration?

****

Let x = displacement of the center C, to the right of equilibrium.

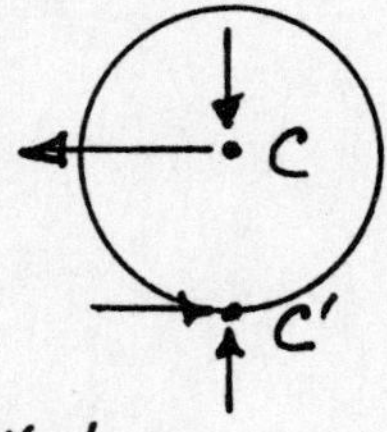

Summing moments about C':

$$-(kx)(a) = I_{C'}\frac{d\omega}{dt} = (m r_C^2 + m a^2)\frac{d}{dt}\left(\frac{\dot{x}}{a}\right)$$

Rearranging,

$$\frac{d^2x}{dt^2} + \frac{ka^2}{m(a^2+r_C^2)}\,x = 0$$

The natural frequency is thus

$$\omega_0 = \sqrt{\frac{ka^2}{m(a^2+r_C^2)}}$$

The force of gravity extends the spring 5/8 of an inch. What will be the natural frequency of vertical vibration?

If the system has damping with c = 0.27 lbf·s/in, what is the natural frequency?

3 lb

****

The static deflection is $x_g = \frac{mg}{k}$. The natural frequency may be determined from

$$\omega_o^2 = \frac{k}{m} = \frac{k}{k x_g/g} = \frac{g}{x_g}$$

$$f_o = \frac{\omega_o}{2\pi} = \frac{1}{2\pi}\sqrt{\frac{g}{x_g}} = \frac{1}{2\pi}\sqrt{\frac{386\ in/s^2}{5/8\ in}} = 3.955\ Hz$$

With damping, the damping ratio is

$$\zeta = \frac{c}{2\sqrt{mk}} = \frac{c}{2m\omega_o} = \frac{c}{2 f_g/g}\sqrt{\frac{x_g}{g}} = \frac{c\sqrt{g x_g}}{2 f_g}$$

$$= \frac{0.27\ lbf\cdot s/in}{2\ (3\ lbf)}\sqrt{(386\ in/s^2)(\tfrac{5}{8}\ in)}$$

$$= 0.699$$

The damped natural frequency is then

$$f = \sqrt{1-\zeta^2}\, f_o$$

$$= \sqrt{1-(0.699)^2}\ (3.955\ Hz)$$

$$= 2.838\ Hz$$

9 OTHER PROBLEMS

9-1

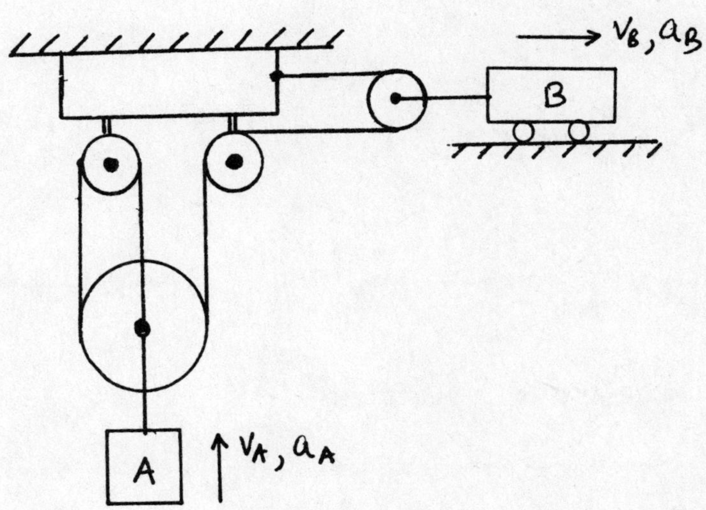

Determine the velocity v_B and acceleration a_B of mass B in terms of the velocity v_A and acceleration a_A of mass A.

****

If we raise A through a distance x, we would release a length 3x of the cord. If we move Mass B through y to the right, we would take up length 2y of the cord. For compatibility (i.e. cord remains taut), we must have

$$2y = 3x \qquad \text{or} \qquad y = 1.5x$$

$$V_B = \frac{dy}{dt} = 1.5\frac{dx}{dt} = \underline{1.5\,V_A}$$

$$a_B = \frac{dV_B}{dt} = 1.5\frac{dV_A}{dt} = \underline{1.5\,a_A}$$

9-2

Particle A is moving in a horizontal frictionless plane with a velocity of 5 m/sec $\vec{i}$ when it collides with particle B at the origin of the coordinate system shown in Figure 1. After the collision particle B is observed to move with a velocity of 2m/sec $\vec{i}$. The mass of particle A is 1 kg and the mass of particle B is 2 kg. What is the coefficient of restitution between the particles?

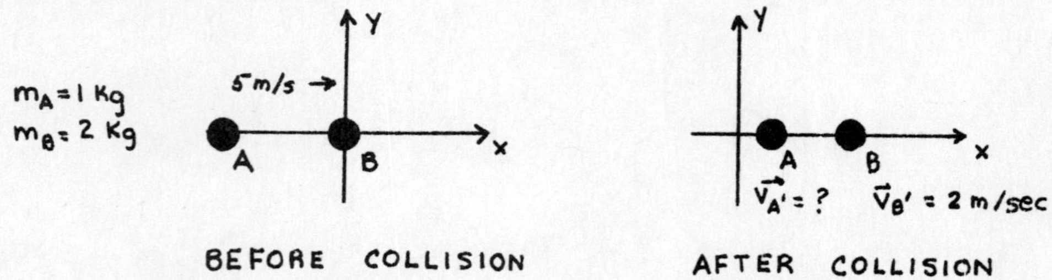

****

CONSERVATION OF MOMENTUM

$$m_A\vec{V}_A + m_B\vec{V}_B^{\,\nearrow 0} = m_A\vec{V}_A' + m_B\vec{V}_B'$$

$$(1\,\text{Kg})(5\vec{i}) = (1\,\text{Kg})\,\vec{V}_A' + 2\,\text{Kg}\,(2\,\text{m/sec})\,\vec{i}$$

$$\vec{V}_A' = 1\,\vec{i}\ \text{m/sec}$$

RELATIVE VELOCITY EQUATION ALONG LINE OF IMPACT

$$V_B' - V_A' = e(V_A - V_B)$$

WHERE e IS THE COEFFICIENT OF RESTITUTION

$$(2\ m/sec) - (1\ m/sec) = e\,(5\ m/sec - 0)$$

$$e = 0.2$$

9-3

The spherical particle of mass m bounces on a smooth horizontal surface. The speeds before and after impact and their directions are as indicated on the figure. Determine the angle θ and the value of the coefficient of restitution e between the particle and the surface.

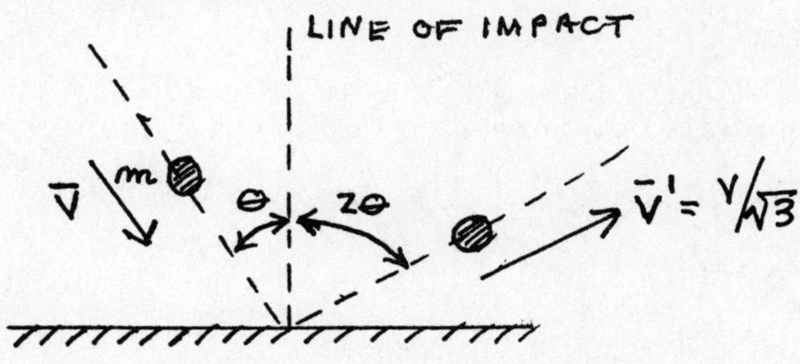

****

Let (y up, x right)

$$V_y = V_y' = V \sin\theta = \frac{V}{\sqrt{3}} \sin 2\theta$$

$$= \frac{2V}{\sqrt{3}} \sin\theta \cos\theta$$

SOLVING $\cos\theta = \frac{\sqrt{3}}{2}$ OR $\theta = 30°$

$$e = -\frac{V_x'}{V_x} = \frac{-\frac{V}{\sqrt{3}}\cos 2\theta}{-V\cos\theta} = \frac{1}{\sqrt{3}}\,\frac{\cos 60°}{\cos 30°}$$

$$= \frac{1}{\sqrt{3}} \cdot \frac{1/2}{\sqrt{3}/2} = \frac{1}{3}$$

9-4

A uniform rod of length 2 m and mass 5 kg is hanging vertically from a hinge at one end. An impulsive force of magnitude 2000 N is applied horizontally at the centroid of the rod in a time interval of 0.01 s, by striking with a club. Determine the initial angular velocity ω of ensuing motion of the rod, and the impulsive reaction at the hinge.

****

Neglect nonimpulsive forces such as weight and friction in comparison to the impulsive forces.

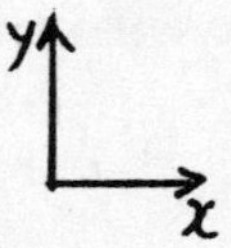

Applied impulse at G $= F.\Delta t = 2000 \times .01$ N.S
$= 200$ N.S

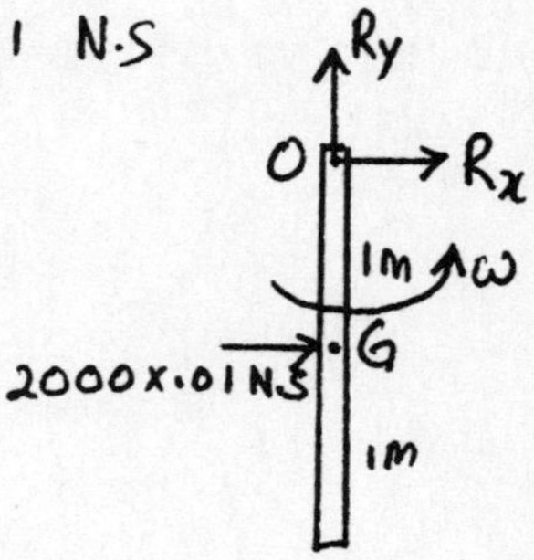

Angular impulse about O $= 1 \times 20$ m.N.S

$=$ change in angular momentum

$= I_o\,\omega - 0$

Note: O is a fixed point

Substitute $I_o = \frac{1}{12}5.2^2 + 5.1^2$ {parallel axis theorem

or,

Moment of inertia about O $\quad I_o = \frac{20}{3}$ kg.m²

in the previous equation

$$1 \times 20 = \frac{20}{3} \times \omega$$

or,

$$\omega = \underline{3 \text{ rad/s}}$$

9-5

Knowing that block B moves downward with a constant velocity of 180 mm/s, determine a) the velocity of Block A and, b) the tension in the cable attached to the 50 kg block B.

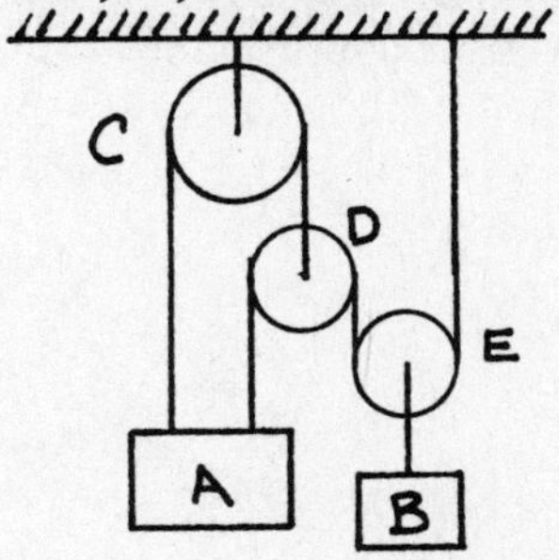

****

THE LENGTH OF THE CABLE OUER D AND E IS A CONSTANT USEING THE SUPPORT SURFACE AS A REFERENCE

$x_B + (x_B - x_D) + (x_A - x_D) = \text{CONSTANT}$

$2x_B - 2x_D + x_A = \text{CONSTANT}$, TAKEING DERIUATIUE

$\frac{dx}{dt} = \text{UELOCITY}$ $\quad 2V_B - 2V_D + V_A = 0$

FOR OTHER CABLE $\quad x_A + x_D = \text{CONSTANT}$

AND $V_A + V_D = 0$, $V_A = -V_D$

FROM ABOUE $-V_D = -\frac{V_A}{2} - V_B$, $V_A = -\frac{V_A}{2} - V_B$ $\quad V_A = \frac{2}{3} V_B$

SINCE $V_B = 180 \frac{mm}{s} \downarrow$ $\quad V_A = 120 \frac{mm}{s} \uparrow$

SINCE THE UELOCITIES ARE CONSTANT, ACCELERATION IS ZERO SO

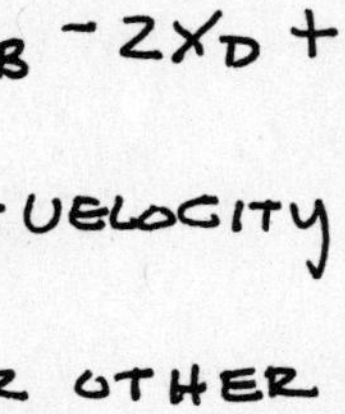

$\Sigma F_y = m a_y \overset{0}{} = T - 50(9.81)$

$T = 491\ N$

9-6

Block A weighing 25 lbs collides with block B with a velocity of 30 ft/sec. Block B weighs 10 lbs. Assuming perfectly elastic collision find the velocity of block B after impact.

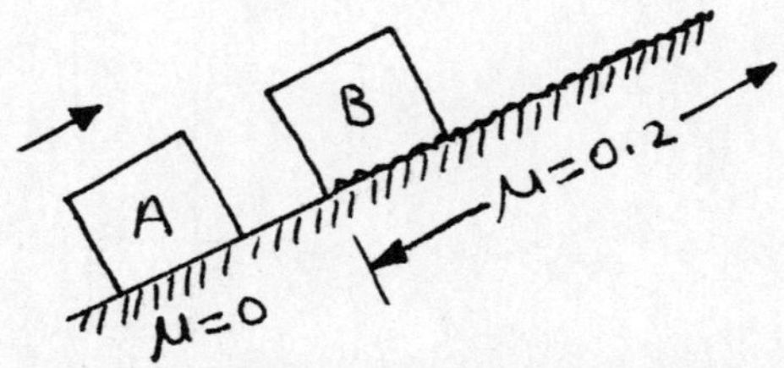

****

Impact -

From conservation of linear momentum

$$m_A v_A + m_B v_B = m_A v_A' + m_B v_B'$$

v_A' and v_B' are velocities after impact.

$v_B = 0$

$$\frac{25}{g}(30) = \frac{25}{g} v_A' + \frac{10}{g} v_B'$$

$$750 = 25 v_A' + 10 v_B' \quad \text{—— (1)}$$

From the definition of coefficient of restitution.

$$e = -\frac{v_B' - v_A'}{v_B - v_A} \qquad \text{For elastic collision } e = 1$$

$$1 = -\frac{v_B' - v_A'}{-30}$$

$$30 = v_B' - v_A' \quad \text{—— (2)}$$

Solve equations (1) & (2) simultaneously

$$v_B' = 42.85 \text{ ft/sec}$$

9-7

An 8 lb. ball is moving horizontally at a speed of 80 ft/s., when it strikes a slender homogeneous rod AB suspended from pin A in the vertical position. After impact, it is observed that the rod acquires an angular velocity of 10 rad/s. Determine the coefficient of restitution e between the ball and the rod. ($I_G = mL^2/12$ for a rod of mass m and length L.)

A

5 ft

80 ft/s

8 lb.

B

$L_{AB} = 6$ ft

$W_{AB} = 30$ lb.

$\omega_0 = 0$

$\omega_{AB} = 10$ rad/s (after impact)

****

Taking moments of momenta before and after impact about the mass center yields

$$\frac{8}{g}(80)(5) = \frac{8}{g}V_1'(5) + \frac{90}{g}\omega + \frac{30}{g}V_G(3)$$

A

$\frac{90}{g}\omega$

$\frac{30}{g}V_G$

$\frac{8}{g}V_1'$

B

But $V_G = 3\omega$ (Kinematics)

The restitution equation gives

$$e(80) + V_1' = eV_2 + 5\omega \quad (V_2 = 0)$$

Solving these equations simultaneously and eliminating V_1' between them gives $\underline{\underline{e = 0.75}}$

Note that the angular momentum $I_G\omega$ and the linear momentum mV_G associated with the mass center, as well as the momentum of the ball after impact are shown on the diagram. The initial momenta are on the original figure.

9-8

A slender rod is pinned to a cart which is free to roll on a frictionless surface. The rod-cart system is at rest when it is struck by a ball at A. The coefficient of restitution between the ball and the rod is 0.5. Calculate the angular velocity of the rod and the velocity of the cart immediately after the impact.

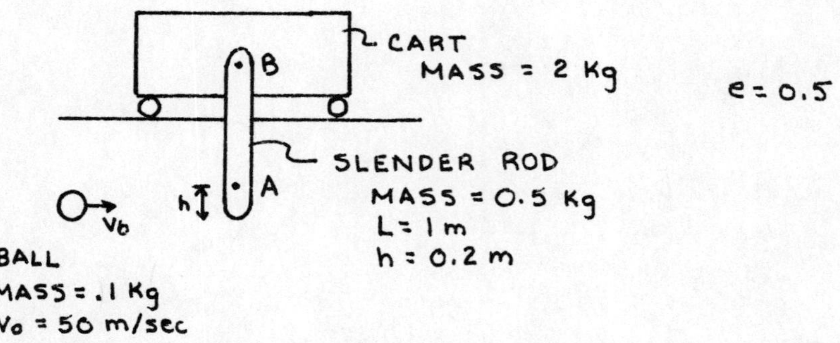

****

DIAGRAMS FOR PRINCIPLE OF IMPULSE - MOMENTUM

$m_c v_c = 0$, $m_b v_b$, $m_r \bar{v}_r = 0$, $\bar{I}\omega_r = 0$ + = $m_c v_c'$, $m_b v_b'$, $m_r \bar{v}_r'$, $\bar{I}\omega_r'$

SYSTEM MOMENTA AT $t = 0$

SYSTEM EXTERNAL IMPULSE

PRINCIPLE OF LINEAR IMPULSE AND MOMENTUM

$$m_b v_b = m_c v_c' + m_r \bar{v}_r' + m_b v_b'$$

$$(.1\text{ Kg})(50\text{ m/sec}) = (2\text{ Kg})\, v_c' + (0.5\text{ Kg})\, \bar{v}_r' + (0.1\text{ Kg})\, v_b' \qquad (1)$$

PRINCIPLE OF ANGULAR - IMPULSE MOMENTUM ABOUT B

$$m_b v_b (L-h) = m_b v_b' (L-h) + m_r \bar{v}_r' (L/2) + \bar{I}\omega_r'$$

$$(.1\text{ Kg})(50\text{ m/sec})(.8\text{ m}) = (.1\text{ Kg})\, v_b' (.8\text{ m}) + (.5\text{ Kg})(\bar{v}_r')(.5\text{ m}) + \tfrac{1}{12}(.5\text{ m})(1\text{ m})^2 \omega_r' \qquad (2)$$

RELATIVE VELOCITY EQUATION

$$v_A' - v_b' = e(v_b - v_A)$$

$$v_A' - v_b' = .5(50\text{ m/sec} - 0) \qquad (3)$$

KINEMATICS:

$$\bar{V}_{r'} = \tfrac{L}{2}\,\omega_r' = .5\,\omega_r' \qquad (4)$$

$$V_A' = V_c' + (L-h)\,\omega_r' = 0.8\,\omega_r' + V_c' \qquad (5)$$

SUBSTITUTING (4) AND (5) INTO (1)-(3) YIELDS

$$5 = 2V_c' + 0.25\,\omega_r' + 0.1\,V_b' \qquad (6)$$

$$4 = 0.167\,\omega_r' + 0.08\,V_b' \qquad (7)$$

$$25 = V_c' + 0.8\,\omega_r' - V_b' \qquad (8)$$

EQUATIONS (6)-(8) CAN BE SOLVED SIMULTANEOUSLY BY CRAMER'S RULE:

$$V_c' = \frac{\begin{vmatrix} 5 & 0.25 & 0.1 \\ 4 & 0.167 & 0.08 \\ 25 & 0.8 & -1 \end{vmatrix}}{\begin{vmatrix} 2 & 0.25 & 0.1 \\ 0 & 0.167 & 0.08 \\ 1 & 0.8 & -1 \end{vmatrix}} \qquad \omega_r' = \frac{\begin{vmatrix} 2 & 5 & 0.1 \\ 0 & 4 & 0.08 \\ 1 & 25 & -1 \end{vmatrix}}{\begin{vmatrix} 2 & 0.25 & 0.1 \\ 0 & 0.167 & 0.08 \\ 1 & 0.8 & -1 \end{vmatrix}}$$

THE DETERMINANTS CAN BE EXPANDED BY MINORS:

$$\begin{vmatrix} 5 & 0.25 & 0.1 \\ 4 & 0.167 & 0.08 \\ 25 & 0.8 & -1 \end{vmatrix} = 5\begin{vmatrix} 0.167 & 0.08 \\ 0.8 & -1 \end{vmatrix} - 0.25\begin{vmatrix} 4 & 0.08 \\ 25 & -1 \end{vmatrix} + .1\begin{vmatrix} 4 & 0.167 \\ 25 & 0.8 \end{vmatrix}$$

$$= 5\left[0.167(-1) - 0.08(.8)\right] - 0.25\left[4(-1) - 0.08(25)\right] + 0.1\left[4(.8) - 25(0.167)\right]$$

$$= 0.2475$$

$$\begin{vmatrix} 2 & 0.25 & 0.1 \\ 0 & 0.167 & 0.08 \\ 1 & 0.8 & -1 \end{vmatrix} = 2\begin{vmatrix} 0.167 & 0.08 \\ 0.8 & -1 \end{vmatrix} - 0.25\begin{vmatrix} 0 & 0.08 \\ 1 & -1 \end{vmatrix} + 0.1\begin{vmatrix} 0 & 0.167 \\ 1 & 0.8 \end{vmatrix}$$

$$= 2\left[0.167(-1) - 0.08(.8)\right] - 0.25\left[0(-1) - 0.08(1)\right] + 0.1\left[0(0.8) - 1(0.167)\right]$$

$$= -0.4253$$

$$\begin{vmatrix} 2 & 5 & 0.1 \\ 0 & 4 & 0.08 \\ 1 & 25 & -1 \end{vmatrix} = 2\begin{vmatrix} 4 & 0.08 \\ 25 & -1 \end{vmatrix} - 5\begin{vmatrix} 0 & 0.08 \\ 1 & -1 \end{vmatrix} + 0.1\begin{vmatrix} 0 & 4 \\ 1 & 25 \end{vmatrix}$$

$$= 2[4(-1)-0.08(25)] - 5[0(-1)-0.08(1)] + 0.1[0(25)-4]$$

$$= -8$$

$$V_c' = \frac{0.2475}{-0.4253} = -0.582 \text{ m/sec}$$

$$\omega_r' = \frac{-8}{-0.4253} = 18.8 \text{ m/sec}$$

9-9

A bar of length l and weight W is released from the position shown. It is observed that the rod rebounds to a horizontal position after striking the inclined plane. Determine **the coe**fficient of restitution between the knob and the inclined surface. $\bar{I} = \frac{1}{12} m l^2$

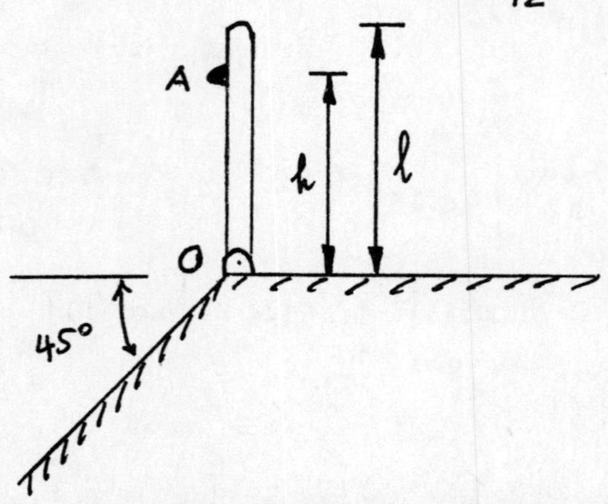

****

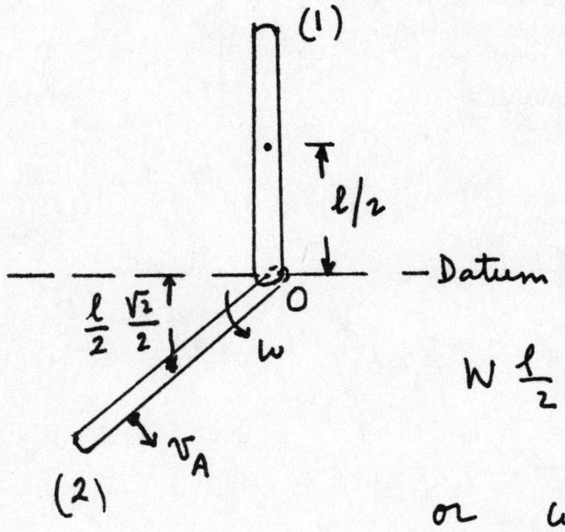

Conservation of energy:

$$T_1 + V_1 = T_2 + V_2$$

$$0 + W \cdot \frac{l}{2} = \frac{1}{2} I_0 \omega^2 - W \frac{l}{2} \frac{\sqrt{2}}{2}$$

$$I_0 = \frac{1}{12} m l^2 + m \left(\frac{l}{2}\right)^2 = \frac{1}{3} m l^2$$

$$W \frac{l}{2} = \frac{1}{2} \frac{1}{3} \frac{W}{g} l^2 \omega^2 - W \frac{l}{4} \sqrt{2}$$

$$\text{or} \quad \omega = 2.263 \sqrt{\frac{g}{l}} \circlearrowright$$

$$v_A = \omega h = 2.263\, h \sqrt{\frac{g}{l}}$$

After rebound:

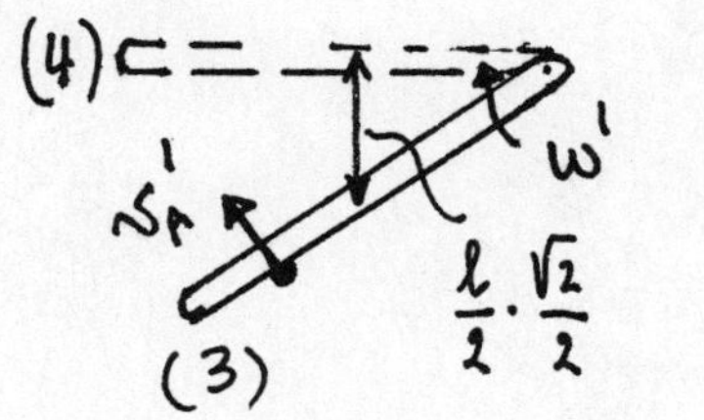

$$v_A' = e\, v_A = 2.263\, h\, e \sqrt{\frac{g}{l}}$$

$$\omega' = \frac{v_A'}{h} = 2.263\, e \sqrt{\frac{g}{l}} \circlearrowleft$$

Conservation of energy between positions 3 and 4:

$$T_3 + V_3 = T_4 + V_4$$

$$\frac{1}{2} I_0 \omega'^2 - W \frac{l}{2} \frac{\sqrt{2}}{2} = 0 + 0$$

$$\frac{1}{2} \frac{1}{3} m l^2 (2.263)^2 e^2 \frac{g}{l} - mg \frac{l}{4} \sqrt{2} = 0$$

$$0.854\, e^2 - 0.354 = 0 \quad \text{or} \quad e = 0.644$$

9-10

A 2-Mg boat is traveling upstream at a speed of 5 m/s relative to the water, which is moving at 2 m/s. The engine suddenly quits, leaving the boat under the influence of a drag force that varies with the boat's relative speed according to the law shown at the right.

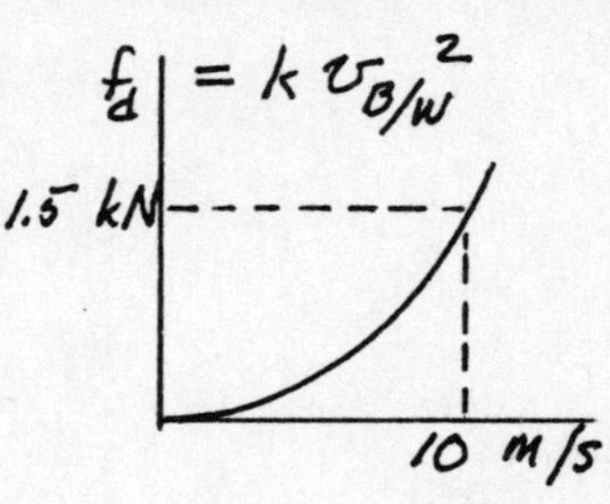

What is the farthest distance the boat travels upstream beyond the point of engine cutoff?

****

Let v = ground-observed velocity of the boat

$$m\frac{dv}{dt} = m v \frac{dv}{dx} = -k(v+v_w)$$

$$\frac{v\,dv}{(v+v_w)} = -\frac{k}{m}dx$$

$$\log(v+v_w) + \frac{v_w}{v+v_w} + C = -\frac{kx}{m}$$

When $x = 0$, $v = v_0 = 3$ m/s

$$C = -\log(v_0+v_w) - \frac{v_w}{v_0+v_w}$$

$$x = \frac{m}{k}\left[\log\left(\frac{v_0+v_w}{v+v_w}\right) + \frac{v_w}{v_0+v_w} - \frac{v_w}{v+v_w}\right]$$

To determine maximum distance, set $v = 0$:

$$x_{max} = \frac{m}{k}\left[\log\left(\frac{v_0+v_w}{v_w}\right) + \frac{v_w}{v_0+v_w} - 1\right]$$

From the graph,

$$1500\text{ N} = k(10\text{ m/s})^2 \rightarrow k = 1.5\text{ kg/m}$$

$$x_{max} = \frac{2000\text{ kg}}{1.5\text{ kg/m}}\left[\log\left(\frac{5}{2}\right) + \frac{2}{5} - 1\right] = \underline{421.7\text{ m}}$$